OIL AND POLITICS IN THE GULF OF GUINEA

To my parents, José António and Adelina

Ricardo Soares de Oliveira

Oil and Politics in
the Gulf of Guinea

HURST AND COMPANY, LONDON

First published in the United Kingdom by
C. Hurst & Co. (Publishers) Ltd
41 Great Russell Street, London WC1B 3PL
© Ricardo Soares de Oliveira, 2007
All rights reserved.
Printed in India

A catalogue record for this volume is available at
the British Library.

ISBNs
978-1-85065-857-3 *casebound*
978-1-85065-858-0 *paperback*

www.hurstpublishers.com

CONTENTS

ACKNOWLEDGEMENTS

Research for this book started more than six years ago. In the time that has elapsed, I have discussed the politics of oil in the Gulf of Guinea with hundreds of people in Angola, Nigeria, São Tomé and Príncipe, Chad, Cameroon, Congo-Brazzaville, and also Belgium, Portugal, France, the US and Great-Britain, most of whom have asked me not to acknowledge them individually. It is only fitting that I start by thanking them all for having shared their time, thoughts and experience with me. As this book only scratches the surface of what is a tremendously important, large-scale process of political and economic change, I expect to see more of them soon.

Thanks are due to the Ministry for Science and Technology, Portugal, which, through its research program PRAXIS XXI, funded the first three years of my research. In Cambridge, the Centre of International Studies and Sidney Sussex College were the ideal setting for doctoral research. In 2004, I had the privilege of being welcomed into the Sidney Sussex Fellowship by the Master and Fellows, whom I thank for the support and congenial atmosphere provided ever since. I owe a debt of gratitude to Charles Larkum, the late College Bursar, who was especially supportive over the past two years. The Center for International and Area Studies at Yale University, where my presence was made possible through a Joseph C. Fox Fellowship, and the Centre d'études et de recherches internationales (CERI-Sciences Po) in Paris, where I thank Christophe Jaffrelot for the hospitality, were two intellectually exciting institutions in which to carry out additional research. Colleagues and friends at the Global Public Policy Institute (GPPi) in Berlin offered a stimulating environment for developing and discussing the broader policy implications of my research. The work conducted for a number of institutions also

viii

proved invaluable. Charles McPherson and Robert Bacon at the World Bank and Johannes Maters and Pedro de Sampaio Nunes at the Directorate for Transports and Energies of the European Commission afforded me with hands-on contexts in which to pursue my research interests. The CRS report project[1] allowed me to make sense of previous professional and fieldwork experience through close work with Caritas International and its grassroots partners in the Gulf of Guinea. I learned greatly from this and more recent fieldwork for the National Democratic Institute of International Affairs.

The list of other individuals I can thank is almost too long to do justice to, but here it goes. Reverend Kevin O'Hara and his Port Harcourt team, particularly Emmanuel Emmanuel and Abba Ayemi, shared their knowledge of Nigeria, as did Richard Synge, Esohe Iyamu and Antony Goldman. Nicole Poirier and her family were excellent hosts in N'Djamena in late September and October 2002, and so were Alejandro and Almodena Diz in São Tomé and Príncipe in March 2002. In Luanda, I thank the kind help of Jerónimo Belo. On matters Angolan, I am thankful to the late Christine Messiant, one of the finest analysts of Angolan politics, Manuel Ennes Ferreira, and Alex Vines for sharing with me their wealth of experience. Paris-based Gabon experts François Gaulme and Douglas Yates gave me some of their time to discuss the intricacies of the Bongo regime. A conversation with Rémy Bazenguissa-Ganga cleared a number of doubts about oil politics in Congo-Brazzaville. Over an (ongoing) series of lunches in Lisbon, Gerhard Seibert has provided valuable insights and material on São Tomé and Príncipe's oil imbroglio and the complex politics of the archipelago. Chad's pipeline was the subject of interesting conversations with Korinna Horta and Jane Guyer as well as CRS report team Terry Lynn Karl, Philippe Copinschi and Ian Gary. Bishop (now Archbishop) Ngartheri of Mondou invited me for an exquisite dinner at his home where I learned greatly about oil politics in Logone and Chad more generally. Oliver Mokomo of CRS Cameroon talked about his country's national oil company with a fine sense of humor. Andrew

---

1 The result of which is Gary and Karl (2003), for which the author was a contributing writer.

Apter and the late Jean-François Médard sent me their fine, and then still unpublished, work. Roland Marchal gave me good PhD and life advice. I gained important insights into Gulf of Guinea economies from Ulrich Bartsch at the IMF. Chester Crocker kindly accepted to discuss US-Angolan relations in the 1980s. Sefakor Ashiagbor, Alison Paul and particularly Dileepan Sivapathasundaram were good travel partners while on assignment for NDI in Nigeria and Congo-Brazzaville. George Frynas was always ready to discuss the region and especially Nigeria, and we had a great time researching a joint article in São Tomé. Nick Shaxson, while involved in writing his own excellent book on African oil, was not only a well-informed sparring partner but also an example of openness and generosity. Stephen Ellis read and commented extensively on the manuscript. Ian Gary was a pleasure to work with and helped me in understanding the politics of oil in the Gulf of Guinea more than anyone else.

At Cambridge, I have incurred countless debts. James Mayall was a thought-provoking and responsive supervisor throughout, as well as a friend. Christopher Clapham and Christopher Alden, who examined the doctoral dissertation that resulted in this book, were highly encouraging and very meticulous in their criticism. Yezid Sayigh was enthusiastic about this project since we first discussed it years ago and kindly read several chapters. Marta Sofia Magalhães also read part of the dissertation and offered valuable comments. Helen Thompson invited me to teach in her course on International Political Economy which proved essential in working out some of the broader implications of my research. Zaheer Kazmi listened endlessly as I grappled with my subject and coaxed me forward in the last months of dissertation write-up. Thorsten Benner and Daniel Large went through the book's several incarnations with unmatched patience. They have also taken time off from their busy lives to lend me a hand at a moment of great distress. Words cannot express how indebted I am to both.

My family—especially José António, Adelina, Tomás, Cristina and António Miguel Soares de Oliveira—were tireless in their support throughout and remained strong in the face of the great tragedy that fell upon us. I also thank Anne de Henning, whom I owe unflinching assistance. My greatest debt, however, is to Devika Singh, who has provided me with the love and inspiration to get the job done despite events that could have held me back. She was the redeeming presence in a terrible, bad-news year, and to her goes my boundless love.

ACRONYMS

AEF	Afrique Equatoriale Française
AGIP	Azienda Generale Italiana Petroli
AGOA	Africa Growth and Opportunity Act
ASODEGUE	Associacíon para la Solidaridad Democrática con Guinea Ecuatorial
BCCI	Bank of Credit and Commerce International
BP	British Petroleum
CNOOC	China National Offshore Oil Company
CNPC	China National Petroleum Corporation
CRS	Catholic Relief Services
CSR	Corporate Social Responsibility
DRC	Democratic Republic of Congo
EFCC	Economic and Financial Crimes Commission
ECMG	External Compliance Monitoring Group
EIR	Extractive Industries Review
EITI	Extractive Industries Transparency Initiative
ENI	Ente Nazionale Idrocarburi
EO	Executive Outcomes
ERCH	Holding Corporation Environmental Remediation Holding Corporation
EIU	Economist Intelligence Unit
FAA	Forças Armadas de Angola
FESA	Fundação Eduardo dos Santos
FLEC	Frente de Libertação do Estado de Cabinda
ICC	International Criminal Court
IMF	International Monetary Fund
JDA	Joint Development Authority
JDZ-	Joint Development Zone

MAIB	Movimento de Autodeterminación de la Isla de Bioko
MEND	Movement for the Emancipation of the Niger Delta
MOSOP	Movement for the Survival of Ogoni People
MoU	Memorandum of Understanding
MPLA	Movimento Popular de Libertação de Angola
NDV	Niger Delta Vigilantes
NDPVF	Niger Delta People's Volunteer Force
NOC	National Oil Company
NPN	National Party of Nigeria
NNPC	Nigerian National Petroleum Corporation
PDP	People's Democratic Party
PEMEX	Petróleos Mexicanos
PGS	Petroleum Geo-Services
PMF	Private Military Firm
PWYP	Publish What You Pay
OECD	Organisation for Economic Co-operation and Development
OAU	Organisation of African Unity
ONGC	Oil and Natural Gas Corporation Limited
OPEC	Organization of Petroleum Exporting Countries
PRGF	Poverty Reduction and Growth Facility
RBDAS	River Basin Development Authorities
SAP	Structural Adjustment Program
SNH	Société Nationale des Hydrocarbures
SNPC	Societé Nationale des Pétroles du Congo
SONANGOL	Sociedade Nacional de Combustíveis de Angola
SONATRACH	Societé Nationale pour la Recherche, la Production, le Transport, la Transformation et la Commercialisation des Hydrocarbures
UNITA	União Nacional para a Independência Total de Angola
UPC	Union des Peuples Camerounais
UNDP	United Nations Development Program
WBG	World Bank Group

LIST OF TABLES

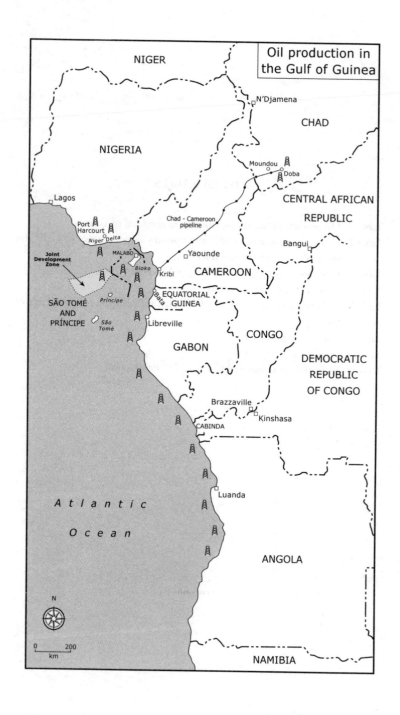

OIL AND POLITICS IN THE GULF OF
GUINEA: AN INTRODUCTION

The poor, associating by an obscure instinct of consolation the ideas of evil
and wealth, will tell you that it is deadly because of its forbidden treasures.

Joseph Conrad, *Nostromo*

Petroleum: an inclusive term to the complex mixture of hydrocarbons that occurs in the earth in liquid, gaseous, or solid forms. As generally used [...] petroleum is a liquid (crude oil) that is recovered from some depth within the Earth
through boreholes. [...] Its greatest use today is in the manufacture of energy
upon which mankind's present and future economy depends.

Encyclopaedia Britannica

Central Africa's Atlantic coastline has in recent times become the subject
of interest for a myriad of constituencies whose attention is normally
elsewhere. Western and Asian governments, international financial institutions, news media and the private sector's many incarnations are
all flocking to the region, courting it, discussing it, shedding light on
its usually "dark" recesses.[1] The reason for this is the presence of untold
wealth, in the form of oil deposits, underground and under the sea. The

---

1 See, amongst many others, James Dao, "Oil puts Africa in spotlight", *New York
Times*, 20 September 2002; Jean-Christophe Servant, "Offensive sur l'or noir
africain", *Le Monde diplomatique*, January 2003; "Oil empires", *Africa Confidential*,
21 March 2002; Ken Silverstein, "Africa drowns in a pool of oil", *Los Angeles
Times*, 1 July 2003.

1

purpose of this book is to describe and analyse this momentous process, its principal actors and its consequences.

With a century-long history of exploration for petroleum and about forty years of sizeable production, the region is not new to the oil business. But only in the last decade has its true worth been accurately gauged. For the first time, the Gulf of Guinea is being approached as a vast oil province that is ignorant of borders and conventions. As if it were dots being connected, the blind spots that existed in between oil emirates (in the form of pauperised non-producers) are being brought into international petroleum markets at astonishing speed. All major multinational companies and scores of smaller firms are making bets and investing their billions across the region's promising acreage. Cutting edge technology now allows the exploitation of oil wells in the ocean's ultra deep waters and enhanced seismic research has reduced the margin of prospecting error to previously unrealistic standards. With current production an estimated 4.5 million barrels of good-quality crude per day and expected to reach the 7 million mark by 2010, and holding still under-assessed natural gas reserves,[2] the Gulf of Guinea is undoubtedly one of the "world's fastest-growing sources of energy".[3] The region is also perceived as a politically pliant alternative to radicalised oil exporters elsewhere, despite the prevalent brutality and unrest. While Nigeria is a long-standing OPEC member and Angola joined the cartel in 2007, neither is keen on overly restrictive production quotas or the curbing of foreign investment. The politics of oil regional producers are heterogeneous and unfettered by any sort of shared antagonistic rhetoric vis-à-vis the West, so that participation in a joint price hike, 1973-style, is not a credible scenario.[4]

The importance of the up-and-coming West-Central African rim of oil-producing states is heightened by the world's growing demand for oil, including that of fast growing economies such as China and India,

---

2   On the mounting importance of natural gas for the energy economy see Yergin and Stoppard (2003).

3   Author interview with senior Transports and Energy Directorate official, European Commission, Brussels, 21 June 2001.

4   David White, James Harding and John Reed, "Oilmen flock to an alternative source", *Financial Times*, 7 July 2003.

and the post-11 September 2001 US imperative to diversify energy supplies away from over-reliance on the Persian Gulf.[5] While the Gulf of Guinea presently holds a modest 5 to 7 per cent of proven petroleum world reserves, there is a general understanding that this is an underestimate. Furthermore, when put in the context of a broader diversification of supply policy involving Russia, Central Asia and Latin America, the contribution of the Gulf of Guinea for consumer states seems weighty indeed. It is already supplying some 15 per cent of US oil imports and the figure will likely be raised to 25 per cent by 2015. The Gulf of Guinea is proportionally as important to China, with Angola having bypassed Saudi Arabia as its main source of crude oil in 2006. This new relevance has led the US to forcefully reevaluate its regional engagement, counter long-standing French influence and rising Chinese stakes there, reopen deserted diplomatic outposts, establish a Africa Command and plan a regional military base, and restate its view of the Gulf of Guinea as an area of "national strategic interest".[6] The "rise" of the Gulf of Guinea has provided a set of *Realpolitik* arguments that overrides lingering concerns with human rights standards across the region and prioritises alliances with reliable partners instead.

Oil multinationals, in their drive to minimise the geological risk of research and increase profits,[7] and as beneficiaries of a densely networked rapport with local regimes, are both precursors and keen emulators of

---

5   See, amongst many, Mike Crawley, "With Mideast uncertainty, US turns to Africa for oil", *Christian Science Monitor,* 23 May 2002; Paul Maidment, "The Other Gulf", *Forbes,* 10 March 2003; Rosa Townsend, "EE UU aspira a controlar el petróleo de África en un giro estratégico de su política energética", *El País,* 12 July 2003; Brigitte Breuillac and Alain Faujas, "Pourquoi le pétrole africain intéresse les États-Unis", *Le Monde,* 23 April 2003; and James Dao, "In Quietly Courting Africa, White House Likes Dowry", *New York Times,* 19 September 2002. Approving references to the Gulf of Guinea appeared in the so-called "Cheney Report" of May 2001. See White House (2001).

6   Assistant Secretary of State for Africa Walter Kansteiner, quoted in "Black Gold", *The Economist,* 24 October 2002.

7   As US Vice-President Dick Cheney put it in his more outspoken days as CEO of oil services giant Halliburton: "you've got to go where the oil is. I don't think about it [political volatility] very much" (quoted in Christian Aid (2003:3).

this novel focus on the region. Its importance for companies far surpasses that of the region's share of world reserves: most alternative sources of oil are situated in locations to which, for political reasons, they have only limited access, whereas the Gulf of Guinea is wide open to investors. Oil companies are thus undeterred by the ubiquitous corruption, political violence and administrative chaos—all countries in the region are high in sensible risk assessment charts—and may occasionally perceive these circumstances as conducive to business-friendly tax regimes by weak, investment-hungry states. The Bretton Woods organisations are equally enthusiastic, on account of the size and legality of the ongoing invest- ment, the profitability of lending to the oil sector and the fear of being side-stepped as the continent's key development institutions were they not to support the rush for oil. This political backing has created the context for private sector access to very favourable financing and export insurance guarantees, the viability of projects that would otherwise not have been pursued, such as the Chad-Cameroon pipeline, and a compro- mised and permissive attitude towards the macroeconomic conduct of oil exporters.

As a major recipient of (exclusively oil-related) foreign direct invest- ment, the Gulf of Guinea is, with South Africa, the strategic exception to the otherwise almost forgotten economic landscape of a continent swiftly disappearing from the formal international economy to occupy niches in unrecorded and illegal sectors. In this context, the Gulf of Guinea is ever more significant to the world's key business, and with collapsing states, no industrial production and agricultural commodities of dubious value, "useful" Africa is increasingly confined to her petroleum reserves. For those possessing them, this means that extraordinary amounts of money are suddenly conjured out of the ground. A 2004 study of Africa's oil boom conservatively estimated that host-states would receive $349 bil- lion from the sale of their oil in the next decade alone.[8] This far exceeds any previous or intended aid receipts, and is in fact Africa's largest inflow of money in history.[9]

---

8   PFC Energy (2004). This was before the all-time high oil prices reached in 2005 and 2006. The real amount is probably closer to $500 billion or more.
9   Gary and Karl (2003).

The new Gulf of Guinea

The catapulting of the Gulf of Guinea from strategic neglect to geopolitical stardom in the last few years is illustrative of how space is easily re-
conceptualised by capital and politics. The image now conveys wealth,
security and opportunity, unconvincingly superimposed on local poverty
and the authoritarianism of host governments. It brings to mind, for
companies, phenomenal rewards; for ruling elites, undreamed riches;
for diplomatic pundits, diversification of energy supply and freedom
from Middle East bondage. But this set of hopeful visions and terms of
reference, which will likely frame discussions of the region in the next
decades, is also very recent. In fact, to someone outside the oil loop, "the
Gulf of Guinea" may be no more than a vague geographical reference.

Only ten years ago, it did not convey its present meaning—denomina-
tions such as West or Equatorial Africa usually sufficed. Historically, the
term "Guinea" was used in reference to most of the western coast of Africa from Senegal to Gabon.[10] More narrowly, "Gulf of Guinea" appears
in contemporary cartography to delineate the maritime space between
Benin and Equatorial Guinea's continental enclave of Rio Muni. The Gulf
of Guinea where oil gushes out endlessly from the ultra deep, ultra safe
offshore deposits did not yet exist: one could find, alternatively stated,
a selection of modest or up-and-coming producers bordering big player
Nigeria along the South Atlantic coast. Those few with regional experience might see it in a more adjoined manner; and some companies were
busily carving for themselves a cross-boundaries stake in oil and politics.
And yet, no group of countries perceived itself as "Gulf of Guinea", no
international organisation addressed it as a unit and, more importantly,
no one spoke of it. While "the Horn of Africa" and the "Frontline States"
conveyed clear geographical or political images, the Gulf of Guinea rang
no bells. It can be argued that it was the oil business (with characteristic
disregard for conventional geography) that proceeded to deploy it as an
all-encompassing tag for the coastal region that runs across Central Africa
from Nigeria to southern Angola. In this book I will keep to this inac-

---

10   Dictionnaire Encyclopédique Quillet, "Guinée", volume 3.

curate but broadly shared usage as it now conveys, in Achille Mbembe's
words, a "major territorial figure".[11]

	Angola	Cameroon	Chad	Congo Brazzaville	Equatorial Guinea	Gabon	Nigeria
Crude Oil Production 2006 ('000 bp|	1400	90	170	245	350	237	2600

Table 1. Oil producers and oil production in the Gulf of Guinea[12]

Oil-rich countries of the Gulf of Guinea are as divergent as any random
assortment of sub-Saharan African countries could be. They come from
different colonial backgrounds (Lusophone, Anglophone, Francophone
and Hispanophone) and are of all shapes and sizes, from large entities to
islets, from demographic powerhouses to scantily populated states.[13] Nigeria, in particular, is seen to be on a class of its own and rarely discussed
in a regional context. But something is shared by all.[14] What they have in
common is, firstly, reliance on the legal extraction of petroleum and the
rents that accrue from its sale. Petroleum is the lifeline of these states to
an unprecedented extent: on average, 68 per cent of government receipts

---

11   Mbembe (2000).

12   Source: Energy Information Administration, US Department of Energy,
www.eia.doe.gov (accessed 11 February 2007)

13   The oil-producing states in the region include Angola, Gabon, Congo-Brazzaville, Nigeria, Cameroon, Equatorial Guinea and Chad. São Tomé and Príncipe
is expected to become one in 2006-7. The Democratic Republic of Congo is a
minor producer as well but oil plays a negligible role in the country's economy,
so I will not discuss it here.

14   Two caveats. The first relates to Nigeria. Any regional analysis that brings
it in needs to contend with the attendant "imbalance", as Ralph Austen noted
(1987: 234). While I consciously sought not to walk excessively into the well-
threaded ground of Nigeria's political economy of oil, the Nigerian experience
shares many traits with others in the Gulf of Guinea and is essential for getting
the regional picture right. The second caveat is that Cameroon is in many ways
not a typical oil state owing to the comparatively small size of the sector and the
country's moderately diversified economy. However, there are aspects (e.g. the
politics of the national oil company, the oil-related presidential slush funds, relations with foreign oil companies) that merit a comparative approach.

emanate from the oil sector.[15] Secondly, they share a relationship struck with a small number of oil majors holding a *de facto* monopoly over the technology and finance needed to exploit the increasingly remote and, more often than not, offshore petroleum deposits.[16] The capital-intensive nature of oil investment, its unusual long-term prospects and the need to legally protect it mean that multinational corporations require the collaboration of internationally recognised sovereign partners. In turn, the latter benefit from webs of international complicity deployed by their business allies and respective home governments to guarantee their credibility on the world stage.

State experiences

The postcolonial trajectories of the states studied in this book are varied and better understood if placed in a continuum. At one end, there is the deceitful soundness of Gabon or Cameroon, where civil conflict is absent; at the other, the brutality of Equatorial Guinea, the recently discontinued war economy of Angola, or the militia politics of Congo-Brazzaville. Despite these nuances, the oil states of the Gulf of Guinea jointly present some of the most pronounced instances of state weakness or failure in the developing world, in a noxious encounter of two of the worst experiences of postcolonial statehood. The first is that of the African postcolonial state, its legacies and a more recent pattern of institutional dereliction. It is noteworthy that the pre-oil context of Gulf of Guinea states was already characterised by authoritarian politics, poor leadership, weak economies and fragile institutions, scarcely good harbingers for the oil era. In addition, over the past two decades, states of the Gulf of Guinea have suffered from deficiencies common to other

---

15 Katz *et al.,* (2004: 4).

16 Although I discuss multinational oil firms in general, particular attention is paid to five companies that are together responsible for more than 90 per cent of regional oil extraction, have a presence in at least three of the region's states, and have been present in the Gulf of Guinea for decades. These are Total, Chevron, ExxonMobil, Royal Dutch/Shell, and ENI. In addition to these and other western companies, the mostly state-owned companies from the developing world are fast becoming key actors across the region.

African states such as crumbling bureaucracies, decaying infrastructure, irregular administration, and the virtual abandonment of swathes of their national territory.

The other experience is that of the petro-state. With few exceptions, the worldwide impact of oil revenues on oil states has been problematic. Oil rents have not only shaped the way producers engage with the international economy, but sent them down remarkably "common development paths, similar trajectories, and [...] generally perverse outcomes".[17] Invariably, petroleum has led to the entertaining of ideas of endless wealth and power and to a belief in the magic of oil windfalls to spontaneously generate sustainable development. But careless contraction of debt, "Dutch Disease"—whereby an appreciated exchange rate favours imports and undermines domestic productive sectors—corruption and sheer wastage have dealt severe blows to oil-rich states. By providing the government with an easily obtainable, if unstable, stream of hard currency, oil has precluded the need for domestic taxation and, therefore, for any type of engagement, however exploitative, with the population. The environmental impact of oil production unmitigated by effective state regulation has everywhere been harmful, and in seasoned oil provinces such as the Niger Delta it amounts to grave damage for the ecosystem and local health and livelihood. As recent research shows, oil states are also more prone than their oil-free counterparts to suffer from civil conflict, and their populations are poorer. While there seems to be nothing intrinsically wrong with petroleum or any other commodity, in actuality very few states have been successful in tapping it for conventionally productive ends. The structures of the petro-state, when added to the weaknesses and pathologies of the African postcolonial state—from which most structural prerequisites for sound use of oil revenues were missing from the start—have exacerbated previous shortcomings and created new ones.

With these legacies and their attendant results, how do Gulf of Guinea oil states endure and why is the oil partnership a rewarding one for its key players? The outcomes of this process—for both people and institutions—seem so dispiriting that it is hard to conceive the future in terms

---

17 Karl (1997: xv).

other than those of the implosion and extinction of such brittle polities. This explains why few studies of the subject seek to go much beyond the appearance of chaos and crisis. But one should not be kept back by the analytical limitation that sees "crisis" in exclusively pathological ways. Alternatively, if one perceives it as "the normal manifestation of political and economic change"[18] (to use a medical analogy, as serious illness without the attendant demise), one will be better placed to understand both the destabilising trends and the sources of viability that keep the oil state from coming apart.[19] It is this paradoxical sustainability of the Gulf of Guinea state—a sovereign entity that may be empirically feeble but whose existence and reputation is nonetheless credited and nourished—and the "permanent crisis"[20] in which it thrives that my analysis will be focusing on.

The Gulf of Guinea may well be the new "El Dorado" for oil companies and their home states:[21] for the majority of its inhabitants it is not. For them, oil has meant a fall into oblivion, especially away from urban areas; further decay in living conditions; and passer-by victimhood when caught in the feverish struggle for control of the state that different claimants rightly see as the ultimate prize. For petro-elites congregating around the "national cake", in contrast, the oil economy is an unparalleled device for internal domination and external recognition; the wellspring of unexpected, unearned wealth; the instrument to fulfil dreams and the source of the coercive means to keep those dreams real. The deportment and choices of the new entrants to the petroleum club seem premised on imitating the intemperance of older members such as Nigeria. And with oil company plans for the region currently thought out against a five-decades backdrop,[22] it seems right to say that the experience of oil

---

18   Hont (1994: a169).

19   See Soares de Oliveira (2001).

20   Hont (1994).

21   Charlotte Denny, "Corruption and Chaos - true cost of new oil rush", *The Guardian*, 17 June 2003.

22   During this time the oil deposits of other states such as Niger and the Central African Republic will certainly be brought on stream, while current producers such as Gabon and Cameroon may see their petroleum era come to an abrupt end. The thriving of the oil economy at the regional level will thus en-

will be a formative, crucial one for the political trajectory of host states well into the twenty-first century.

Aims and Plan of the Book

The purpose of this book is to explore the centrality of oil for the political economy of Gulf of Guinea states that depend on it for their prosperity or survival as well as the paradoxical nature and consequences of the large-scale process of change this unleashes. This is the first study on the subject to take on the regional, as opposed to the country-specific, dimension. It has four aims. The first is to bring out the extent to which oil has forged the interaction of the region with the rest of the world and how the ongoing expansion of the sector will deepen this pivotal role. The second is to investigate how this international relevance of petroleum has shaped the domestic realm, how it has transpired into internal political calculations, and how it is constitutive of the postcolonial moulding of these polities. I am particularly interested here in studying the institutional impact of the firms-state oil partnership. Thirdly, the study examines the interests of different sets of *empowered* actors in the oil economy, their interplay, and the manner and the contexts in which their goals diverge or converge. Finally, it analyses the paradoxical sources of long-term sustainability of the political economy of oil in the Gulf of Guinea amidst seemingly unmanageable chaos.

The book is structured around theme-based chapters that address the Gulf of Guinea as a whole, rather than the more conventional country case-study format. The latter is inappropriate to my purposes as it fails to grasp the broader implications of the processes studied here while not being a substitute, owing to their necessarily modest length, for country monographs. The regional and thematic approach allows the presentation of the big picture, making it possible to zoom in on individual experiences where this is useful and necessary but allowing it to zoom out and address wider themes more effectively. Because of the amount of

---

compass very different experiences for individual states. While few if any public statements exist that mirror such a timeline, most interviews I have conducted (especially, but not exclusively, in regard to Angola and Nigeria) point to very forward-looking planning horizons.

material and ambitious scope, the narrative is episodic and a function of the issues I have prioritised. While an understanding of the region's postcolonial history is welcome, the absence of country-specific contextualization should not be a hindrance. This approach is, of course, not without risks, but its benefits will hopefully become readily apparent to the reader.

The book's structure is as follows. Chapter 1 provides the framework for analysing the paradox of the successful failed state in the Gulf of Guinea amidst seemingly inauspicious circumstances: while "failed" because sharing the worst elements of the petro-state and the African postcolonial state, it is also "successful" because it is cash-rich, preserved from actual demise and surprisingly aloof from domestic and external pressures. Turning to key domestic arenas, Chapter 2 seeks to understand the role of oil in structuring the prevalent forms of governance in the Gulf of Guinea, the type of state goals that are pursued under the shadow of the oil economy and its implications for life in the oil state. The political behaviour of the elites of the oil state is investigated in Chapter 3. Chapter 4 is a historical overview of oil company involvement in the Gulf of Guinea from the colonial era up to the early 1990s, when momentous changes in offshore technology allowed a qualitative jump in the oil industry's interest in the region. The subject of Chapter 5 is the current "scramble" for the Gulf of Guinea's oil, the way oil firms are thinking about their long-term engagement there, and the manner in which the devastation wrought upon the region by decades of mismanagement and oil-related decline casts a shadow over upbeat expectations. Chapter 6 studies the way in which the political economy of oil of the Gulf of Guinea is embedded in the global economy, the *Realpolitik* agendas that for long brought together exporter states, oil companies, and importer states, and the progressive agendas that have come to challenge them in the past decade. The chapter then inquires about the political possibilities open to the oil states within the constraints and opportunities created by petroleum, suggesting that political strategies pursued on the basis of oil wealth will collapse when it is exhausted.

I am not centrally concerned with making critical statements on local rulers, oil firms and the many legal and illegal entities that make up the vast networks of complicity in and around the oil-producing state. But the implications of oil dependence I uncover—cronyism, lack of freedom and utmost material deprivation—are indicative of forms of resource exploitation sadly prevalent in the more distant corners of the global economy. Therefore an ethical concern certainly underlies my inquiry. Its primary intent is to discern what is going on in the Gulf of Guinea, and this is no small task in view of what we currently do not know: the key actors, their lines of reasoning and, whenever possible, the history that partly gives meaning to contingency. I seek to do so within the framework of "universal political theory and not theories of third world political development" and other exceptionalist arguments on Africa's politics.[23] This said there is an equal commitment to refraining from the excesses of "common problematisation" whereby methodological universalism is equated with a universalism of experience. That all politics is amenable to a comparative framework is a truism. But in an effort to render hitherto exotic goings-on digestible, scholars of different persuasions have fallen into the trap of over-familiarity, whereby Africa's previously "unique" politics are labelled "banal".[24]

This study, on the contrary, is ambivalent towards the idea that what it describes is "ordinary", mere "business and politics by other means"— even while views of African politics as sunk in chaos and irrationality are given short shrift. There is of course lucrative sense to a few in the commerce of Gulf of Guinea oil, as there once was in the transatlantic business of slavery: much of my work is precisely to delineate such rational strategies. However much sense there may be, alas, one must not let its politics be explained away in those terms alone. Profit and unbearable suffering may not be antithetical, but neither can one dispose of the linkage in a functional way. In seeking to make sense of politics in the frontier and tie all loose ends, analysts have often failed to see that something extraordinary is afoot. The statistician's sensibility, or that of the overly rational social scientist, may be a prerequisite for a good

---

23 This follows the methodological assumption of the studies in Chabal (1986).

24 The expression is Jean-François Bayart's. A similar view from a then-leftist perspective is that of the "anti-mystificatory intent" in Halliday (2002). In a different context, Geoffrey Hawthorn (that of a history of social theory) rejects a "forced intimacy" with his subject and affirms the goal of "making it strange". See Hawthorn (1986: x).

study. But by itself, it ultimately fails to address the human dimension of oil in the Gulf of Guinea, because the mark of decades-long exploitation is to be found not in easily graspable items but in the more sedimentary, elusive imagination of those who live with its consequences.[25] The aim is thus to provide as accurate a study as currently possible of petroleum's circulation in the region's body politic,[26] but in the knowledge that it falls short of conveying reality in its full unwholesomeness.

The conceptual lens: four key principles

I will now briefly explore four key ideas that underpin this study, both in the manner in which it was conducted and in its ultimate shape as a text. It is appropriate to discuss these ideas at this stage as they form the bedrock of my argumentation.

Dependency without determinism. The study of asymmetric relations between developed and developing states has not been at the forefront of social science concerns since the demise of dependency theory. Rather than subjecting the same themes to a different analytical apparatus the trend since the late 1970s is to neglect them in favour of what have been called the "West-West issues".[27] More recent attempts at thinking out dependent relations while eschewing the rigidities of dependency theory are to be found in the works of two French scholars, Jean-François Bayart and Bertrand Badie, whose approach informs this study. Taking individual and collective actors and their strategies seriously, both underline the

---

25　This goes beyond the remit of this book but certainly informs my outlook on the subject. In his great work on extreme violence in Colombia's Putumayo region, Michael Taussig seemed to accept these limits when stating that his goal was "to penetrate the veil while maintaining its hallucinatory quality". See Taussig (1987: 10). In a more recent contribution on Colombia, Robin Kirk seeks to make sense of her war experience in the following way. "I could follow the massacre stories and execution stories and "military target" stories like steps on a path. All were brutally logical. Yet once outside Colombia's extraplanetary atmosphere, I would read my notes with sick horror. It was all crazy." See Kirk (2003: 183).

26　In Andrew Apter's suggestive imagery. See Apter (1996) and Apter (2005).

27　Strange (1994: 12).

convergence of interest between "northern and southern princes" and the mutually beneficial elements that result. In contrast to dependency theory, which even at its most sophisticated still portrayed local rulers as "irresponsible puppets",[28] Badie and Bayart see them in empowered terms, and focus on their skilful management of the relationship. Neither underestimates the jagged nature of relations at the "financial, economic, military and technological levels [but also] symbolically".[29] Their goal is to bring politics back into the picture, perceiving the latter in terms of "a matrix for action"[30] amenable to political strategies by (constrained) agents, rather than a fixed, economically determined structure peopled by cardboard natives and arch-imperialists.

A large-scale inquiry. This study aims to understand a set of individual state trajectories as part of a wider process. Therefore, it adopts a broad spatial and thematic span. In this regard, I follow Charles Tilly's view that "work on comparative and interdisciplinary research in the social sciences" is greatly aided by "huge (but not stupendous) comparisons",[31] a view that mirrors historical sociology's penchant for "big questions".[32] The concern with large-scale processes also means that the structures that shape the political economy of oil and what is politically possible in these societies play a pivotal role in this study. The outstanding debt here is to political economists of oil, who have understood the similarities shared by oil producers and realised that "petro-states are not like other states".[33] The point about the structuring effect of oil on institutions, the agendas of empowered political actors, international relations and decision-making patterns is not meant to substitute determinism or "analogy" in the pejorative sense that Weber gave to it for serious inquiry.[34] But there is little doubt that the consequences of oil work in remarkably similar ways

---

28 Badie (2000: 10).

29 Ibid., p. 25.

30 Bayart, Mbembe and Toulabor (1992: 31-2); and also Bayart (2000: 218-19, 236).

31 Tilly (1984: 74).

32 Skocpol (1984: 7-12).

33 See Karl (1999) and section two of Chapter 1.

34 Weber (1978: xxvii, lii-liv).

across space and time. My goal is to acknowledge the preeminence of politics over economic determinism, but also the very structuring impact of oil, within the same analytical framework.

Empirical grounding and the attention to difference. The pursuit of common patterns that animates this book suggests neither a belief in identical historical experiences nor a desire to deal away with plural backgrounds for argument's sake. It takes as given that each country has "its own political history, a state of a different nature, its own power configurations" and unique "economic, social and political structures, geographic situations, economic conjunctures".[35] More importantly, the study accepts that "comparative analysis is no substitute for detailed investigation of specific cases".[36] This attention to local detail also holds important implications for the normative thrust, or lack thereof, of the book. I attempt to understand the subject without etching onto it personal expectations of historical progress, good governance, dysfunctionality and normalcy. This is not standard practice when looking at African politics, which is mostly assessed in terms of abnormality and what it fails to be rather than what it actually is.[37] Alternatively, I want to pursue Albert Hirschman's injunction of showing, as he put it, "a little more respect for life", for the accumulation of "heterodoxies" in the historical paths of these states.[38] In his analysis of Foucault's historiography, Paul Veyne brought out Foucault's interest in the "rarity" and "arbitrariness" of human facts, their spatial and temporal uniqueness.[39] This region-wide study will have worked if the wealth of local facts comes out enhanced rather than reduced to a common denominator. From this viewpoint, the trajectory of each state comes across not as a source of resented complexity thwarting a linear tale, but as corroboration of my reading.

The ubiquity of the political. It goes without saying that this wider significance, and the structural elements that a macro-perspective reveals, are

---

35 Hibou (2002: 4).

36 Moore (1993: xi).

37 Mbembe (2001).

38 See Hirschman (1971). Underlying this is a commitment not to underestimate the "description of political life itself", as noted in Vallée (1999).

39 Veyne (1978: 389-ff).

in turn indecipherable outside the immediate political context. External-ly, one finds that petroleum has suffocated most other forms of outside engagement, bilateral, multilateral and private sector alike, with the Gulf of Guinea, becoming its leitmotif and reducing it to the "place where oil comes from"[40]. Domestically, one finds petroleum to be everywhere. Oil penetrates every element of public and private life and can be understood only through "an analysis of which law, economics, society, culture, reli-gion, are but chapters of its significance".[41] Hence, rather than perceiving politics as a pigeonhole distinct from other categories, this book assumes that the political is to be found in "each and every domain".[42] To perceive the political as ubiquitous is also, as John Dunn wrote,

to repudiate categorically the possibility of a non-political [...] mode of expla-nation in the last instance, lurking reassuringly behind the unnerving flurry of political engagement, and to vindicate instead the logical and analytical priority of the categories of humane understanding over all superstitious apings of the magic of the scientific revolution in the attempt to understand politics.[43]

A note on research and sources

Most conventional bodies of evidence have not been made available for this book. Long-term access to some of the countries in the re-gion can be very arduous, which partly explains the small and uneven secondary literature on the politics of oil. Places such as the Cabinda enclave, Bioko Island or the Niger Delta have over the years been only intermittently accessible and most visits are short term, so that the majority of accounts remain anecdotal: ethnographic work in some of them ranks amongst the most dangerous worldwide. The local press is on occasion helpful, as in Nigeria, but it can also not exist, as in São

---

40 Author interview with oil company executive, London, May 2000.
41 Veyne (1976: 22). I am also following the work of Karl Polanyi and Max Weber, among others, who perceived economic history as built upon "the idea of the interdependence of the political, the social and the economic". See Bruhns (1997: 1274).
42 Navaro-Yashin (2002: 2).
43 Dunn (1989: 191).

Tomé and Príncipe. Local archives tend to be unavailable or chaotic; sometimes, there aren't any. A number of countries have a history of good access for scholars, such as Cameroon; others (Angola, Equatorial Guinea) actively discourage it. For their part, oil companies are extremely protective of the minutest details on local activities and relations with the host-state, and rarely publish separate annual reports for their subsidiaries. Payments from companies to states are not routinely revealed. Nor do oil firms facilitate their archives beyond the limiting and compromised grasp of company historians.

Faced with such paucity of authoritative information, a qualitative approach to research must be embraced, both because of its analytical strengths and because there is really no quantitative alternative. In recent years, scholars have adopted such an approach towards the study of Africa's unrecorded trade, the volume and character of which remains the subject of broad speculation.[44] Yet this should also apply to legal economic activity, where statistics are rarely more than guesses.[45] In the case of a legally traded commodity such as oil, even while there is a notion of how much is produced and how much it is worth, one is only now coming to an understanding of the character of oil money flows in the Gulf of Guinea region and beyond. The research for this book was thus based on the following: my own observations and experience of short- and medium-term multi-sited fieldwork; interviews;[46]

---

44 See, e.g., Ellis and MacGaffey (1996).

45 A fact observed by *The Economist* in a very critical survey of Nigeria (23 January 1982): "This is the first survey published by *The Economist* in which every single number is probably wrong". It goes on to quote the annual report of the Central Bank of Nigeria: "At the time of writing this report (April 1981) no actual data on federal government revenue and expenditure existed".

46 Most interviews were conducted from January 2000 onwards, with a small number emanating from earlier research work for Soares de Oliveira (1998) and (1999). With few exceptions, they were granted in confidence so that the names of most of my interviewees are undisclosed. In the absence of solid knowledge about much that concerns the study, the main purpose of the interviews was that of trying to ascertain realities on the ground (or at any rate, the perceptions of reality held by those with a personal involvement). They are used parsimoniously in the text and rarely if ever deployed to prove an otherwise unsubstantiated claim.

the local and international press; the generalist secondary literature on the region, even while only obliquely touching on the present subject; and the approximate economic data that is available.[47]

---

47 The main data sources used in this book are the US government's Energy Information Administration, the World Bank, and the Economist Intelligence Unit. In addition, I found Katz *et al.* (2004) and PFC Energy (2004: 151-61) very useful.

1

PETROLEUM AND THE SUCCESSFUL FAILED
STATE IN THE GULF OF GUINEA

In the global geopolitics of hydrocarbons, this has become one of the zones in which transnational and local factors are interlaced [...] the new oil frontiers coincide, paradoxically, with one of the most clearly marked boundaries of state dissolution in Africa

> Achille Mbembe, "At the Edge of the World: Boundaries, Territoriality and Sovereignty in Africa", *Public Culture*

After an hour of the closest attention, he had been able to verify for himself that the momentary cessation of breathing was not death, and that the regular breathing by which it was followed was not a prelude to health

> Italo Svevo, *As a Man Grows Older*

For the past decade, the continued relevance of the state, particularly but not only in sub-Saharan Africa, has been called into question. Superseded from above by a merciless international economy, challenged from below by agile social movements, in collapse or in oblivion, the state is often broached in the guise of obituary rather than as a fertile area of inquiry.[1] In the African context, this erosion is perceived with little of the ambiguity that characterises debates on the state in industrialised countries, where some processes of sovereign unbundling such as European integration are

---

1 The question of the continued relevance of the state at the internal and external levels figures prominently in the output of a number of disciplines; it is impossible to retrace the relevant bodies of literature. For the gloomy prognosis see Creveld (2002). For the opposite view, see Mann (1997).

seen as the state's own voluntary pursuits. Alternatively, it is the "failure" or "collapse" of the state and its incapacity to fulfil presumed sovereign obligations and deliver public goods to the population, as well as the need for its resuscitation, that are more frequently underlined.

This study offers an alternative scenario: the state in the Gulf of Guinea remains the "master noun of political discourse".[2] First and foremost, the state in the Gulf of Guinea counts as the institutional site that enables a certain form of engagement with the international economy—the extraction and sale of petroleum—to take place in a sustained, predictable and legal fashion. Secondly, the state remains the allocation terminus for rent-seekers, and therefore the ultimate prize of domestic political struggle. That this continuing centrality exists in tandem with, and does not disprove, many of the types of state deterioration more frequently emphasised is but one of the paradoxes I will seek to underline throughout.

This chapter provides the framework for analysing the paradox of continuing relevance—indeed, success—of the state amidst seemingly inauspicious circumstances. Section one begins by revisiting the theory of the modern state on the basis of which the fortunes of the state in the developing world are often studied, and stresses the high hopes it inspired and the practical fiasco of most postcolonial state-building efforts. The following sections discuss two postcolonial state experiences that together afford the best entrance point to the subject of oil dependence in the Gulf of Guinea. Though matching real state trajectories, they are presented as heuristic devices to help the reader make sense of the experiences of Gulf of Guinea oil exporters in a regional and commodity-specific framework. Section two focuses on the oil-producing state and its dismal development record in the twenty years that followed the momentary daze of petrodollars in 1973-74. Section three looks at the almost universally poor performance, and occasional implosion, of the state in postcolonial Africa. Section four looks at the oil state in the Gulf of Guinea which, while "failed" because sharing the worse elements of the petro-state and the African postcolonial state, is also "successful" because it is preserved from actual demise, supported by allies and sur-

---

2 Clifford Geertz, quoted in Crawford Young, *The African Colonial State in Comparative Perspective* (New Haven: Yale University Press, 1994), p. 13.

prisingly aloof from domestic and external pressures.[3] I conclude with
the paradoxical acknowledgement that this rapport between the international economy and the oil state, marked by violence, state decline and
human destitution, is viable in the medium-term and rewarding for those
partaking in it.

I. THE IDEA OF THE STATE AND ITS "CRISIS"
IN THE DEVELOPING WORLD

As Theda Skocpol succinctly pointed out, "states matter".[4] Yet interest
in the state on the part of the social sciences has varied widely in the last
century,[5] with periods of great concern interspersed with its falling from
view, neutered by pluralists or reduced to a tool of dominant class interests by Marxists. Despite a belated recognition and attempts at bringing
it "back in",[6] the state suffered further assaults on its analytical pertinence
from scholars of interdependence and, in the 1990s, of globalisation.
Ever since, claims of a "state-centric" approach, as in this chapter, have
often been conflated with an obstinate "realist" stance that fails to gauge
new trends and the emergence of non-state actors in lieu of the state. Yet
the concept of the state is everywhere foundational to debates in political
theory, to the extent that the most trenchant critique, or awkward avoidance, amounts to "a reproduction of that constitutive authority".[7] In this
section, I start by briefly reviewing the historically and geographically
circumscribed birth of the modern state, the universal assumptions it
spawned, and its export to all corners of the globe. I do not aim to enter
sprawling debates on the state that are peripheral to my purposes here,
but to provide the reader with the normative canvas within which the
postcolonial state is frequently assessed, and from which I will diverge

---

3 The idea of a "failed state" that can be "successful" because it continues to provide for
the prosperity of elites and because it benefits from resource flows that allow it to go on
standing is derived from Prunier and Gisselquist (2003).

4 Skocpol (1985: 21).

5 For the *parcours* of the state in twentieth-century social theory see Birnbaum and Badie
(1978).

6 Skocpol (1985).

7 Bartelson (2001: 7).

later in the chapter. I then confront the disappointing results of what Clapham described as "the Third World State",[8] including the ever more frequent instances in which the state has stopped functioning.

The state

The very centrality of the state makes it difficult to arrive at a convenient view on what it amounts to: careless usage labels so any form of moderately organised social life of the past millennia, but this is clearly insufficient. In the more restricted Weberian formulation that still orients much thought on the modern state, it is a monopoly of force within a bounded territorial unit holding a stable population. Weber's ideal-type is premised upon an institutional dimension ("a compulsory association which organises domination") and a relational one: he writes of "men dominating men, a relation supported by means of legitimate violence".[9] This overarching authority emanates not from traditional loyalties but from an impersonal bureaucracy administering the polity in a professional and "rational-legal" manner. Weber's state thus implies "institutional differentiation, centrality and verticality of political relations, spatial demarcation [...] and collection of authorised taxation".[10]

While interested in the general patterns of historical development, Weber was aware of substantial local variations and revealed a concern for the differing historicity of societies.[11] He understood the specificity and comparative newness of the developments that had led to the creation of the modern state form and its unprecedented assembling of "substantial military, extractive, administrative and productive organisations in a relatively central structure".[12] Weber's stance on the state was functional rather than teleological.[13] He saw the state as that which successfully fulfils the tasks enumerated above; he did not perceive it in reference to some "inherent goal, purpose, or aim".[14] Furthermore, while firm in the

---

8 Clapham (1985).
9 Weber (1948: 78 -82).
10 Mbembe (2001: 86).
11 Turner (1990: 195).
12 Tilly (1992: 12).
13 Geuss (2001: 42-ff).
14 Geuss (2001: 43).

belief that this was the most effective way of organising the polity, Weber was acutely aware of the drawbacks of the bureaucratic "iron cage" and scarcely an enthusiast for its export.

It is a matter of agreement that the modern state is a recent creation, shaped in and by the political experience of Western Europe from feudalism's last throes until the late nineteenth century,[15] when modern nationalism firmly cemented it around "imagined communities" of blood, soil and language.[16] It is also agreed that the functions now portrayed as the preserve of the state were actually fulfilled by countless non-state actors, sometimes until very recently. The "natural domain of the state" was in fact carved out through intense, long battles for monopoly over the deployment of violence, taxation, territorial control, etc.[17] But the idea of the modern state, while the embodiment of particular culture- and time-specific aspirations, is driven by an unambiguous teleology: that of the modern state as the ultimate, most civilised form of organising social and political life.[18] This normative drive was a powerful one, far more so as it coincided with the creation of the first truly global order that was forged by European expansion.

Indeed, the modern state's success in bringing about enhanced administrative and military efficiency, centralisation and the expansion of tax revenues, which, in turn, constituted the basis for increased domestic and international state power, greatly appealed to the battered empires

---

15 As Skinner argued, it is only with the Renaissance that the idea of a "civil authority which is wholly autonomous, which exists to regulate the public affairs of an independent community, and which brooks no rivals as a source of coercive power within its own *civitas* or *respublica*" emerges. See Skinner (1988: 107). One of the central concerns of historical sociology in the last three decades is the embedded-ness in western European history of the modern state and the way in which its formation and consolidation there "created the distinctions between legitimate and illegitimate, legal and illegal, that exist today". See Tilly (1984: 59).

16 Anderson (1983). In fact, and originally for the literature on nationalism, Anderson posits that Creole communities in America were equally important for the development of modern nationalism.

17 See, for instance, Spruyt (1994) and Thomson (1994).

18 Burckhardt (1995: 4) had already signalled "a new fact [which] appears in history—the State as the outcome of reflection and calculation, the State as a work of art".

of the non-Western world facing European prowess with dismay.[19] Starting with the Ottoman Empire, several non-Western states came to voluntarily implement many of the institutions and practices associated with European rising power, some successfully, most not.[20] The popularity of the modern state should be assessed as much from the prism of these modernising empires as from the more traditional perspective that sees it as an alien body, grafted onto societies by direct occupation and destruction of previous political systems. The idea of the modern state was thus a compelling one that would have guaranteed widespread attempts at adoption in the best of cases. The absence of credible organisational alternatives to stem the tide of modern economic, military and political challenges, together with the homogenising drive of the international system's imperial shapers, further strengthened its position.

Finally, the wave of independent ex-colonies, firstly in Latin America and more than a century later in Asia, Oceania and Africa, which ascended to the international system in the guise of the now-conventional state format, meant that the family of states became broadly coterminous with the world's landmass.[21] In the second half of the twentieth century, there was no other way one could conceive of self-government other than through the apparatus of a modern state. In this sense, scholars of different persuasions rightly emphasise the centrality of Weber's views for, however misread, his approach grew to constitute the assumption of what a state is like, looks like, and acts like. For a long time, it was the central discourse of state-led postcolonial advancement. This explains why prior to the unpredictable political developments in the postcolonial

---

19 The best general study on the adoption of the modern state apparatus is Badie (2000).

20 Some of the most enthusiastic importers of the idea and institutions of the Western state were thus to be found not in locations directly colonised by European powers but in formally independent non-Western polities such as Persia, the Ottoman Empire, Siam, Egypt, Ethiopia, China and Japan. With the exception of Japan, "importer" societies fared disappointingly in their attempts at emulation. But even where a degree of "success" was to be found, it rarely amounted to mimicry of the Western blueprint.

21 The very fact that these states were accepted into the sphere of sovereign statehood was due to substantial modifications in expectations of what sovereignty meant, as many amongst them were not viable according to pre-existing parameters. See Jackson (1990) and Mayall (1990).

world that I describe in the next sections, these were the expectations associated with the edifice of the state,[22] and why its unravelling has been so misunderstood. Conservative modernisation or socialist revolution, the path was there to be followed. Colonial administrators and nationalist political leaders of all persuasions were in broad agreement in regard to the modern state and its technology of organisation as the prerequisite for political capacity.[23] It nonetheless failed to materialise as expected, i.e. as an approximate of Weber's rational-legal ideal-type.

State performance in the developing world

That the "logics of importation" of individual societies would lead to the appearance of new practices and a reinvention of what the state means is in retrospect unsurprising. A great number of postcolonial states have fared sufficiently well and a smaller number have managed a qualitative developmental step forward, but my concern here is with those that have not. The relevant and very large body of literature highlights the following characteristics of the "Third World State".

Firstly, lack of institutionalisation.[24] With some exceptions, the state in the developing world never reflected the separation between private and public affairs that Weber had suggested characterised the modern state. Below the veneer of a modern bureaucracy, clientelism and corruption came to predominate and public office was abused for private ends. Personal politics played a central role and influential men collected a host of dependents. This intricate presence of patrimonial trends nestling within modern state structures has been called neo-patrimonial-

---

22 Although even at this stage, a contradiction was already apparent: between, on the one hand, the state as the outcome of historical progress, as seen in the West, and on the other, its role as the instrument of such a process in the developing world. See Birnbaum and Badie (1978: 59).

23 In Appadurai's paraphrase of Marx, "your present is their future [and] their future is your past" was the mantra of early analysts of postcolonial statehood. See Appadurai (1990: 4).

24 The next paragraphs draw substantially from Clapham (1985: 39-59).

ism.[25] In the postcolonial state, there is hardly a boundary between state and society and parochial interests penetrate the state itself.

Secondly, the rentier character of the state. The idea that "states based on external sources of income are substantially different from states based on domestic taxation" has led to the proposition of the concept of the "rentier state" by Hossein Mahdavy.[26] According to Beblawi and Luciani, the development of a modern fiscal system and reliance on direct taxation have often not taken place in the developing world.[27] Instead, the state has come to depend on external resource flows of varying sizes and types, be it in the form of aid or revenues from the sale of export commodities.[28]

The third characteristic is the state's legitimacy deficit. As noted by many authors, the paradox at the heart of the postcolonial state is that it is both strong and weak.[29] It is strong because it has at its disposal a mighty security apparatus, a bureaucratic structure, financial means and the monopoly of sovereign relations with the outside world that is by far the key source of sustenance in the postcolonial era. But the state is weak because it is associated with the few that control it and their core constituencies and is not recognised by the majority of the population as

---

25 This subject figures prominently in Chapter 3, so I will not discuss it here. I would simply add that the dualism, and perhaps viability, implicit in the theory of neo-patrimonialism may be short-lived. This is because, in most countries, the passing of "the postcolonial moment" has led to such a marked erosion of bureaucratic standards that the hybrid character of this mode of governance can no longer be taken for granted. To put it simply, "neo-patrimonialism" as an explanatory device withers away with the demise of the modern sector of the economy and state apparatus. See, for instance, Chabal and Daloz (1999).

26 Beblawi and Luciani (1987: 10). The notion of the "rentier state" has been faulted for being too vague: it is a statement of what is missing but lacks accuracy in terms of what is there. As discussed in the next section, the specific source of state income, and not merely its exogeneity, matters greatly.

27 Ibid.

28 Badie (2000: 25) mentions that such strategies are today "outrageously diversified" and include not only aid and export revenues, but also the importation of nuclear and toxic waste and the sale of a state's vote in multilateral organisations.

29 This paradox is best captured in the idea of the postcolonial African state as a "lame Leviathan", according to Callaghy (1987).

standing for their interests. In turn, this leads to it being seen as the "prize in political competition"[30] for aspiring political actors, to be won through cut-throat competition, warfare and coups d'état, and the insecurity this breeds feeds the lack of institutionalisation already referred to. In this context, legitimacy is constructed through the disbursement of material and ideal goods to clients, which necessarily excludes a wide number of constituencies on political, ethnic or religious grounds, thus hampering the establishment of any broad-based legitimacy.

Fourthly, there is what Clifford Geertz in his analysis of identity in newly-independent states called the political salience of "primordial feelings":[31] national loyalty towards the modern state was supposed to displace wider or lower affiliations such as ethnicity, religion, etc. but in many cases, it has not. These and other non-state sources of belonging have fired up political passions and led to the breakdown of many states into civil conflict and enduring suspicion between constituent communities.

This discussion highlights what the imported state failed to be more than what it amounts to. Others look instead at the extent to which the imported state has been "re-appropriated" by its subjects in a "logic of hybridisation" that would have finally made it "local".[32] It is clear that "the creation of a state tends, to a degree at least, to be a self-fulfilling prophecy of the development of a nation"[33] and that the state as "an imagined force"[34] exists externally to its functioning as a machine of administration. Nonetheless, most observers continue to point out the fundamental tension between the particularistic practices of the state and its supposedly rational-legal legitimacy[35] as well as its basic failure in providing public goods to the population. One can dismiss particular demands on the state (say, welfare provision or democratic accountability) as extrapolations of recent Western experiences that are not essential prerequisites for statehood. The fact remains that some modes of state governance in the

---

30 Clapham (1985: 41).
31 See Geertz (1973).
32 Bayart (1996 : 2ff).
33 Hannerz (1987) is referring here to Nigeria which, despite debt, civil strife and the lack of a strong sense of shared national identity, he sees as "a reality, of a certain kind".
34 Callaghy, Kassimir and Latham (2001: 19).
35 Badie (2000).

developing world not only clash with the loosest interpretation of the "modern state" but have mostly failed to bring about economic growth and human security as well. Analysts who emphasise the strength of the "national idea" over that of the materiality of politics misconstrue the importance of the demise of the "empirical" state.[36] Yet there are material shortcomings that the study of the political imaginary of the state cannot address.[37]

Towards state failure

State failure is, within the spectrum of disappointing postcolonial state trajectories, the worst possible outcome and an increasingly widespread one in the past decade,[38] particularly in Africa: the histrionics of state crisis are almost as old as the modern state itself,[39] but this time it is anything but metaphorical. Pointing out a continuum of state enfeeblement that verges from weak statehood to the rarer "collapse of the state", Rotberg notes that states have failed when they are "consumed by internal violence and cease delivering positive political goods to their inhabitants. Their governments lose credibility, and the continuing nature of the particular nation-state itself becomes questionable and illegitimate in the hearts and minds of its citizens."[40] In short, failed come about through the unrestrained and unsustainable acceleration of the "Third World State" trends outlined above, such as the exclusion and repression of segments of the population and the neo-patrimonial running of the state apparatus.

---

36 Patrick Chabal has stressed that if the state is taken to be a transcendental concept to be found in every instance, it is hopelessly diluted as a result. See Chabal (2000: 828).

37 Bartelson (2001: 34) notes that the state concept is both "empirical" and "transcendental", but this is lost on most scholars, who proceed according to either prosaically materialist or "imaginary" lines of inquiry. The injunction by two prominent students of colonialism that complex phenomena must be addressed in both dimensions should be heeded far beyond their immediate areas of concern. See Cooper and Stoler, (1997: 18).

38 This section draws from the useful contributions to Rotberg (2004). The participants in a seminar on state failure organised by the Cambridge Security Programme, Corpus Christi College, Cambridge, 22 June 2004, provided additional insights.

39 See Hont (1994).

40 Rotberg (2004: 1).

Failed states provide nothing in the way of public goods to their citizenry, cannot cover their nominal territory and are frequently assaulted by insurgent groups which may prioritise plunder over the more classical goal of regime change. Most observers agree that failed states appear more easily in contexts such as Sub-Saharan Africa, where the tradition of statehood is shallow, the state never had at its disposal the means to run a complex administrative structure, and the national economy inherited from colonialism was very fragile. More significantly, in view of the survival of states with unimpressive prospects and the foundering of many a credible state, such as Lebanon or Somalia, is the presence of a "destructive leadership".[41]

It is important to underline the role of human agency lest one accept the idea that state failure is a natural phenomenon like "famine" was for long assumed to be. This is noteworthy because discussions of the "failed state" are rife with claims of failure that reflect the expectations of observers much more than the capacities or limitations of states themselves. In turn, these normative expectations have evolved over time and are still the subject of intense discussions and, at any rate, may simply not be shared by local elites. Therefore, in addition to the quantifiable criteria used by Rotberg, Zartman[42] and other analysts to measure how failed a state is, one must pay attention to what rulers are actually trying to do and the manner in which they explain decisions to themselves.[43] In failed states, corruption and criminality flourish to the extent that "unparalleled economic opportunity" is created for elites, who may even have a vested interest in state collapse.[44] Often enough, these are short-termist strategies that thrive on the confusion and lack of regulation of failed states to make quick gains in contraband, plunder or illegal activities such as slavery, narcotics or arms. While these are rational strategies for self-enrichment, they are usually not sustainable strategies for political survival, as is shown by the astonishing turnover in political leadership in failed states like Sierra Leone, Liberia and Afghanistan.

---

41 Rotberg (2004: 11).
42 Zartman (1995).
43 See Chapter 3.
44 This is the argument of Reno (1995).

The degree of failure one ascribes to the oil-producing states in the Gulf of Guinea is greatly dependent upon not only the particular history and context of each state but the criteria one prioritises. Nonetheless, and taking account of the differences between, say, Angola coming out of decades of total war, the misruled but still stable Gabon and the paranoia prevalent in Equatorial Guinea, all present what can be described as variants of failure or extreme weakness. As I will later argue, though, the presence of petroleum means that economic depredation, while leading to state failure according to the conventional criteria, is not politically irrational because the root of viability—the oil economy—is not threatened as it is in the case of non-oil states. For this reason, and though other strands of the historical texture of these states also matter, the following two sections examine the legacies of postcolonial African statehood and petroleum wealth and the readings these have evoked. This is followed by a discussion of the type of state experiences and institutional configurations that have resulted and are being reinvented across the Gulf of Guinea today.

II. THE PETRO-STATE[45]

Though petroleum was known and occasionally used as an energy source for thousands of years, its modern history started quite recently, with the successful drilling of an oil field in the Pennsylvania countryside of the late 1850s.[46] It soon witnessed a growth so spectacular that by the beginning of the twentieth century it had displaced coal as the fuel of the modern economy. The oil business was by then already controlled by a handful of major companies with a tight grip over production, transport, refining and distribution.[47] In addition, a cartel of major producers created in 1960, the Organisation of Petroleum Exporting Countries (OPEC), would also come to exercise power over pricing and the struc-

---

45 "Petro-state" is meant to convey not only a state that is a major producer of oil, but one in which oil constitutes a predominant source of revenue for the government and is an important part of the domestic economy. In this sense, major producers like the United States and Canada do not qualify as petro-states.

46 The reference work on the history of oil remains the celebratory Yergin (1991). The absence of an overview that is less in thrall of its subject is a real gap.

47 See Chapter 4.

ture of oil markets.[48] No other commodity since has approached its value
or overriding significance. The developed world's growing dependence
on petroleum led to steady collaboration between the firms and imperial
governments and ever-closer involvement in the politics of colonial or
formally independent locations in possession of it. Considerable strife
resulted, with forceful external entanglement in local power struggles
(Iran, Venezuela) and early gestures of economic nationalism (Mexico,
USSR, Bolivia) premised on the ownership of oil.

The unwelcome attention that petroleum had brought to resource-
rich regions in the developing world did not impinge negatively on local
perceptions of it. Newly independent nations merely assumed that the
treasure so recently fought for by Western powers was now within their
reach, the lynchpin of postcolonial redemption and development to be
arrived at by national ownership of the oil and indigenisation or outright
expropriation of the industry. Odd early voices that thought otherwise
(famously, that of OPEC founder Juan Pablo Pérez Alfonzo[49]) were simply dismissed. And when the 1973-74 oil weapon was deployed by the
producers' cartel, it was the perceived shift in global power and wealth
that was underlined,[50] not the disruptive effects of petroleum on the oil-
rich states.[51] Most economists perceived oil wealth in equally unproblematic ways despite the fact that dependence on oil extraction created a
rentier context of the type decried since David Ricardo referred to rents
as the opposite of profits, i.e. as not the result of "ingenuity or work",
but of nature.[52] In retrospect, the economic and political consequences of

---

48 On OPEC see, among others, Mabro (1998) and Skeet (1988).

49 Conversation with Terry Lynn Karl, Kribi, 12 September 2003, and Karl (1997). For
Alfonzo's scepticism on the developmental merits of oil, see Alfonzo (1976).

50 See, for instance, Vernon (1976: 2): "The crisis that surfaced in 1973 [...] was signalled by an extraordinary event: the emerge of a group of unindustrialized oil-producing
countries with both the aspirations and the seeming power to control the international
oil market".

51 Two articles on the thirtieth anniversary of the embargo brought out both the unprecedented transfer of resources it afforded and the nefarious political and social consequences it left in its wake. See Fouad Ajami, "The Poisoned Well", *New York Times,* 17
October 2003, and Daniel Yergin, "Thirty Years of Petro-Politics", *Washington Post,* 17
October 2003.

52 Ricardo (1973).

the sudden influx of money into oil exporters were so unprecedented and visually captivating that most analysts prioritised the study of short-term frenzy over hypothetical scenarios of a future bust.

It was only with the 1980s slump in oil prices, the crippling debt that came to plague exporters, and the political straits they found themselves in that serious analysis of oil as, at best, an ambivalent source of wealth acquired academic and policy space within which to be pursued. The trend was set by World Bank economist Alan Gelb's edited volume on oil windfalls, the title of which was a then provocative "Blessing or Curse?".[53] As substantiated by later outstanding research, the appropriate title (without a question mark) should have read "Rarely a Blessing, Mostly a Curse".[54] The contrast between the heightened expectations of oil producers and their populations, on the one hand, and the dire impact of oil dependence on development, on the other, is one of the most dispiriting tales of postcolonial hope gone astray. On the basis of studies of oil dependence conducted in exporters of the five continents, there now exists a clear understanding of how it tends to impact on economies and societies the world over. Their cultural, demographic and geographical diversity notwithstanding, it is the similarity of choices and outcomes among exporters that stands out.

The economic consequences

While there seems to be nothing innate to petroleum that should warrant such widely shared negative consequences, the fact remains that most producers of this commodity have fared poorly in the last three decades, especially in comparison with analogous countries that are not oil-rich. Few states have had unambiguously positive experiences with petroleum,[55] which suggests that there are clear governance prerequisites for

---

53 Gelb and Associates (1988).

54 On the resource curse see Auty (1993). The landmark empirical study on the macroeconomic effects of mineral resource dependence is Sachs and Warner (1995). See also the very useful Stevens (2003).

55 The best examples are those of Norway and of a number of regions of the developed world such as the province of Alberta (Canada) and the state of Alaska (USA). Indonesia has also been pointed out as a country that for a time wisely managed its oil resources, if not the rest of the economy.

proper absorption of oil wealth, such as a strong democracy and the rule of law, that are not present in most developing countries. The following discussion briefly summarises the most important economic outcomes for oil exporters that have not been able to count with or create such an environment. While these are not "iron laws", they are nonetheless "stronglt recurrent tendencies".[56]

Firstly, macroeconomic instability. The price of oil is famously volatile, with swings that can take it from modest figures to $60-plus per barrel in a comparatively short period of time. This is a function of petroleum's status as a political commodity only obliquely dependent on the realities of supply and demand, making price variation difficult to predict. Such volatility should mean that the budgets of oil-rich states are thought out with unpredictability in mind but, in reality, what little economic planning there may be is normally pursued within the framework of unrealistically high revenues. Time and again, producers that face a season of windfalls spend extravagantly in oblivion of the boom-and-bust nature of oil markets. Added to this, while some states have the structure within which to (mis-)spend revenues, most are faced with money flows that far surpass their absorptive capacity, resulting in overheated economies and rampant inflation.

Secondly, the rentier context created or exacerbated by oil revenues. In the petro-economy, the state is the endpoint for oil revenues (whereas other types of capital inflows are less centralised) and it is accordingly to the state edifice that political actors coalesce for access to funds. Typically, this gives origin to an inner circle of main beneficiaries chosen on a religious, ethnic, family or political basis that profits disproportionately from the opportunities created by oil wealth. On an outer ring lie more indirect beneficiaries of state largesse, be it in the form of a grossly expanded, largely useless civil service or, even further from the core, in the form of subsidised consumer goods for the wider (normally urban) population. The state seeks to legitimatise itself and realise its vision of prosperity by investing in ambitious infrastructure projects. These respond to pressures to spend the newly acquired treasure, often through the distribution of contracts to insiders, and to do so in a visible, dramatic way. Even at the

---

56 Auty (1994: 12).

height of the oil boom the majority does not share the windfalls and macroeconomic growth does not lead to poverty alleviation.

Thirdly, the so-called "Dutch Disease", a widespread phenomenon in exporters of petroleum that is named after the negative impact upon the Dutch economy of natural gas-related revenues in the 1970s.[57] It describes the process by which hard currency revenues from a booming primary export sector usher in the appreciation of the local currency's exchange rate. Contrary to sound policy that could be deployed to prevent Dutch Disease, governments tend not to devalue for reasons of national pride, the result being an incentive for importation and the development of local non-tradable sectors. This leads to a delay in industrialisation and a decline of domestic agriculture in favour of cheaper imports, and provides a general disincentive for local productive activity. In many oil exporters this impacts on rural dwellers, comprising the bulk of the pre-boom population, who abandon the neglected countryside in hope of a better life in urban centres. Because it is an enclave activity with skilled and very limited labour needs and few linkages to the non-oil domestic economy,[58] however, the petroleum industry fails to provide alternative employment for the newly urbanised masses. The result of the Dutch Disease is thus a steady decrease in non-oil productive activity and, in the face of the eclipse or lack of a non-oil fiscal base, the growing dependence of state coffers on petroleum rents. In fact, the OPEC members' average of oil revenues as part of total government revenue is an astonishing 42 per cent.[59]

Lastly, the contracting of unmanageable debt burdens. As the 1970s wore on, oil producers placed a substantial proportion of their revenue windfalls in western banks. This boon gave the latter an extraordinary incentive for lending to developing countries as their own available resources exceeded the needs of traditional private sector industrial world borrowers, in what became known as the "petrodollar recycling process": petroleum producers were, in effect, borrowing their own money,

---

57 Fardmanesh (1991).

58 Linkages are minimal, even at the level of infrastructure that, however unsuited to local needs, did feature highly in colonial era investment in mining or agricultural development.

59 2001 World Bank estimate.

at interest. Banks peddled all types of loans to all types of questionable sovereign entities, which contracted them to satiate internal demands that even then exceeded available resources, in the illusion that contemporaneous revenue flows were sustainable or increasing. They were proved wrong when oil prices took a dive in the 1980s, leaving many saddled with heavy debts and unbearable arrears to this day. This has not hampered the rapport between the banking industry and oil exporters, though. It is credit, rather than capital, that hurls the world economy onward.[60] And credit, as Polanyi noted, is in turn based on the credibility of the borrower and the extent to which he can inspire trust in the lender.[61] Proven oil reserves do this, so that heavily indebted economies low on creditworthiness and high on investment risk are constantly bailed out by banks eager to access the oil (now used as collateral) and the onerous interest rates to be squeezed out of rogue debtors. The continued presence of oil wealth means that indebted producers constantly seek out new loans, and these are readily available at ever more unfavourable rates. The petro-state thus hovers on the edge of bankruptcy but always finds a manner, in collaboration with the international system and with the aid of occasional high oil prices, to sustain itself this side of functionality and viability.

The political consequences

The political consequences of oil wealth have been shown to be equally negative and, indeed, it is primarily its impact on institutions, mentalities and the quality of governance rather than macro-economic trends that one must look at in order to account for the predicament of the petro-state.[62] As pointed out, the idea of unbounded wealth and limitless power is a common feature in states experiencing a boom in production or massive financial windfalls as the result of increased oil prices. The surreal nature of proclamations and the spending frenzy that ensues have had observers gaze in fascination at what can only be described as the delusion of it all. Visions of an "El Dorado" are invoked by many authors

---

60 Strange (1994: 28-30).
61 Polanyi (1957: 12).
62 For this argument see, among others, Leite and Weidmann (1999).

in an attempt to convey the perception of oil windfalls: they appear, seemingly from nowhere; they are not the fruit of labour, or indeed, of any visible effort. They are God-sent, bestowed on a nation for no reason other than its manifest destiny to greatness. But petroleum refuses to live up to these prospects, and when a state enters the bust phase of the cycle, pent-up political energies explode with unanticipated angst.

An immediate consequence of oil revenues is the increase in state power and the absolute social, economic and political centrality it acquires. It can suddenly afford expensive bids at changing the physical and social landscape, and it often seeks to do so: the boom phase characteristically includes a growth in construction and proliferation of employment in the state sector and the services sector. As the key actor of the domestic economy and main supporter of the lingering private sector through handouts and public contracts, the state is everywhere and in constant expansion. The institutional pace and formats this takes are shared by most oil exporters. They comprise a bloated civil service, mushrooming of parastatals and soaring expenditure on security forces. They also include the creation of an institution that will rapidly become the key entity within the petro-economy, as employer, political actor and cash cow: the national oil company (NOC), which is discussed in Chapter 2. The oil economy is meanwhile managed under a shroud of secrecy, kept in place by the mostly authoritarian petro-state. Production levels and financial amounts transacted are matters of speculation; budgets are fictitious; and money flows are the object of unheard-of detours and misappropriations in a context of no public accountability. In the popular mind, the rapport between oil and the state takes on a mysterious character.

This has to do with the fact that oil emancipates the state from dependence on society. Easy oil money means that the state does not need to expand the domestic fiscal basis; in cases where the building of state institutions runs parallel with the availability of oil wealth, no fiscal administrations are created at all.[63] Often, the state does not seek to pursue this option anyway for it faces articulate national constituencies that see themselves as legitimate beneficiaries of the national wealth rather than net contributors to it. And even those furthest from the "national cake"

---

63 This is the case with the Persian Gulf economies studied in Chaudhry (1997), where instruments for internal taxation were negligible or non-existent.

will fiercely protect the one privilege (e.g. food or fuel subsidies) they perceive to be benefiting from. This fiscal aloofness strengthens the executive but also cuts its link with society as well as the minimal social contract the link might imply.[64] The point is obscured in the prosperous early years, when there is still enough money flowing to keep even the poorest under the illusion that development is near at hand. Once the bubble bursts, though, the hoax is revealed. In difficult times of readjustment, the inner circle goes on benefiting from petroleum wealth, but the rest of society is brutally severed from it.[65] By then, as Apter writes, "oil scarcely circulates" in the body politic any more.[66] The pauper masses of the oil producer, so often labelled acquiescent by those underlining the fiscal autonomy of the state[67] but less the discontent it can brew, then revolt against the rulers of unresponsive state structures.

The petro-state at the brink of failure is highly indebted, with a vast unemployed urban working class and restive youth, a large and intermittently paid civil service, a neglected countryside and an inequitable pattern of wealth distribution. Recent research shows that, faced with a powder keg of distorted economies and dissatisfied populations, oil states are far more likely than non-oil states to suffer from civil war or separatist bids from resource-rich regions.[68] This body of literature has been criticised for the underplaying of other variables and the mono-causal stress on oil as explanation for all misgivings. But some of its conclusions should be taken seriously, and anecdotal evidence from many of the major producers broadly conforms to it. This said, it is noteworthy that instability in the petro-state is state-directed, i.e. the goal of insurrectionists is the

---

64 Vieille (1984 : 28).

65 See Ross (2001) for evidence that the populations of mineral-rich countries are poorer than their mineral-poor counterparts.

66 Apter (1999: 319).

67 Okruhlik (1999). As the example of the Iranian Revolution shows, the social transformation and ostentatious wealth of the boom years can be highly disruptive.

68 Fearon and Laitin (2003: 75-90) suggest that "oil-exporting states have had more than twice the annual odds of civil war onset" than other states. For other analyses of oil's role in impeding democracy and promoting war see Ross (2001) and Collier and Hoeffler (2000).

capture of the state apparatus in order to benefit from its resources[69] rather than its implosion. While oil remains, the state is the wellspring of prosperity and the aspiration of dissenters.[70] Oil threatens the quality of state governance but is, while it lasts, the paradoxical underpinning of state survival and viability and its mark of international importance. An oil state is always taken seriously by the international system.

The state oil builds

On the basis of the evidence, there is broad agreement as to the structuring effects of oil wealth, a commodity that is without parallel, financial or strategic, in the contemporary global economy. As Karl writes,

> it matters whether a state relies on taxes from extractive industries, agricultural production, foreign aid, remittances, or international borrowing because these different sources of revenues [...] have a powerful (and quite different) impact on the state's social and economic programs, create new organizations, and direct the activities of private interests. Simply stated, the revenues a state collects, how it collects them, and the uses to which it puts them define its nature. Thus it should not be surprising that states dependent on the same revenue source resemble each other in specific ways.[71]

Oil affects states in a way that is dissimilar from that of dependence on foreign aid or other sources of rent, even if some forms of rentier behaviours are comparable. In turn, while malign consequences are due to the character of boom-era policies (themselves deeply related to the pre-oil political context) rather than to an unwavering property of oil economies, it is undeniable that the latter promote institutional patterns and incentive structures that consistently yield poor results in the developing world. The impact of oil is apparent in recurrent features of oil producers, whether in the development of the ubiquitous NOC or in the virtual

---

69 This excludes the secessionist attempts that are typically centred on the oil region and therefore are not allowed to succeed, for this would be suicidal for the viability of the state.

70 The case of Algeria is paradigmatic: the Islamic insurgents of the 1990s left the oil infrastructure broadly intact because they too aspired to access the state's oil wealth in the event of a successful takeover. I thank Professor George Joffe for highlighting this to me.

71 Karl (1997: 13).

demise of pre-existing domestic taxation structures. "The capital flows of the 1970s", writes Kiren Chaudhry, "reshaped the domestic institutions and economies of each constituent country: whole classes rose, fell, or migrated: finance, property rights, law and economy were changed beyond recognition."[72] This is also the case with the type of economic actors it fosters: the entrance costs to the oil sector mean that only states, as owners of the mineral resources, and oil companies, with access to appropriate technology and finance, can go into it.

The problem with petro-state structures is that they are not freestanding: they are premised on oil money. In the boom years, institutional and decision-making patterns reflected "soaring external capital inflows, whereas the bust pattern [after 1986] demonstrates how institutional and political relationships forged in the boom break down in times of crisis, conditioning institutions, organizations, and policy outcomes in unexpected ways."[73] In short, the same political economy of oil that shapes class interests, institutions and expectations ends up undermining them. The petro-state provides an ideal illustration of what was referred to in the previous section as the paradoxical nature of the postcolonial or Third World State, namely, that of its strength and weakness. Oil means that its powers are "simultaneously amplified and destabilised".[74] On the one hand, it has enormous resources at its disposal, can pursue ambitious developmental schemes and stand aloof from many of the constraints (vis-à-vis donors, for instance) that other developing countries face. On the other hand, these are offset by the boom-and-bust pattern, debt, clientelism and the impossible demands made upon it by claimants to the "national cake".

III. THE AFRICAN POSTCOLONIAL STATE

"When we speak of the state in tropical Africa today", we are told by two prominent academics of the field, "we are apt to create an illusion."[75] If it were not the beneficiary of a particular sovereignty regime that priori-

---

72 Chaudhry (1997: 3).

73 Ibid., pp. 5-6.

74 Watts (1994: 418).

75 Jackson and Rosberg (1986).

tises external recognition over internal consistence, and which therefore affords sovereign status to entities that may not have empirical standing,[76] it would either no longer exist or would have gone through a substantial measure of reconfiguration. For the path of the African state in the last three decades has become synonymous with economic decline, grievous human suffering, armed strife and incapacity to correspond to expectations of what a state should be able to accomplish. These expectations include not just the post-1945 Western social contract on welfare provision to the population but also the more classic requirement of state command and self-sustainability. The following section is a brief look at this trajectory and the interpretative contributions it has elicited.[77]

The African postcolonial state is the result of multiple "repertories", including different pre-colonial linkages with the international economy, cultural and religious traditions, resource endowment and geographical location, etc.[78] Prominent among these structuring experiences is European direct rule between the late nineteenth century and, in most cases, the early 1960s. The scramble for Africa substituted a fairly arbitrary set of boundaries for a multifaceted history of socio-political organisation that included different types of societies but was mostly characterised by low degrees of "stateness". These artificial demarcations and the political units they created would not become a substantial source of conflict in the postcolonial era: instead, they became the basis for African postcolonial statehood and a source of conservative anxiety among power-holders eager to prevent endless territorial claims and fragmentation.[79] Other than territorial limits, the colonial enterprise also created the administrative edifice, the patterns of social and regional division, the economic base and the external economic linkages that were inherited by independent Africa.

---

76 See footnote 136 in this chapter.

77 This account draws mainly on Clapham (1996a), Bayart (1989), Cooper (2002), Lonsdale (1981), Villalon and. Huxtable (1998), Van de Walle (2001) and Joseph (1999).

78 Bayart (1996). Two of the best works on the relevance of the colonial experience for present-day politics are Mamdani (1996) and Young (1994).

79 From the 1990s, though, and especially in the Great Lakes region, states routinely got together to bring down neighbouring governments they disliked. But outright annexation remains a taboo.

In a recent contribution, Fred Cooper perceives the African state as a "gatekeeper" between the external and internal realms: a state with limited and uneven capacity to influence the latter but able to mediate between it and the international economy. The relevance of the concept stems from the fact that, unlike many other academic epithets born of the contingent turns of the state in the last decades, it applies to both the postcolonial state and its colonial predecessor[80] and is pivotal for the much-neglected analysis of institutional continuity in Africa. Like its postcolonial successor, the colonial state was incapable of optimal internal taxation—which it never developed despite metropolitan unwillingness to subsidise imperial rule—and instead derived its fiscal revenues from customs duties on imports and exports and the mostly extractive companies operating in the territory.

Within this framework, the colonial state developed a minimal infrastructure geared towards the export of commodities for sale in developed economies and, correspondingly, mostly failed to develop integrated national or regional markets, strong institutions, or capable civil services. Neither could it ensure, for most of the colonial era, an effective administrative coverage beyond key urban centres and circumscribed spaces of vital economic importance, although variations existed between, say, France's Sahel provinces and British Nigeria, where the European presence was scant, and the tightly controlled Southern African settler colonies. Alternatively, the presence of the colonial state was felt in a roving manner.[81] While the colonial state was willing to deploy its coercive might across the vast hinterland to affirm its authority in the event of rebellion, it mostly tolerated and even fostered "native" power-yielding entities. Although the dissimilar histories of imperial powers brought with them different emphases, strategies and (at least discursively) goals, these traits were common to the colonial experience

---

80 Cooper (2002: 156ff). Cooper's book is deliberately pitched against the standard approach to twentieth-century African history that severs the late colonial period from a meaningful explanatory role in the understanding of postcolonial politics. From a different perspective, Jeffrey Herbst attempts a similar long-term view that underlines the importance of, in his case, pre-colonial forms of deploying power across space, for an understanding of today's approach to territoriality. See Herbst (2000).

81 Cooper (2002: 157).

in most of sub-Saharan Africa. In the last, development-oriented years of colonialism, efforts were made to expand the bureaucracy and infrastructure of colonies and create a measure of socio-political space for Africans to envision a self-governing future.[82] But no lasting efforts were made to diversify economies away from over-reliance on primary commodities of fluctuating value in international markets or leave behind durable, competent institutions.

Early post-independence years

State experiences in the first years of independence were varied, mirroring each former colony's constraints and opportunities and the myopia or dynamism of political leadership. But despite a low starting point and some early slippage—most famously in Congo—the early path of the African state can in retrospect be seen as a moderately positive one.[83] The social mobilisation of the late colonial years as well as the democratic trappings left in the wake of British and French decolonisation were soon discarded and single party structures or personal rule were put in place instead. But an enabling international environment, the opportunities created by the appropriation of the administrative apparatus by the late colonial non-white bourgeoisie, and still-favourable terms of trade for the commodities Africa exported meant that the state could cultivate a system of broad-based patronage around itself. On the basis of liberal distribution of export revenues and development aid, civil service and parastatal jobs, educational opportunities, public contracts and access to health, and other state-controlled material and ideal goods, the state was able to satisfy a number of internal constituencies. Such policies were pursued both in an attempt to build a degree of legitimacy around the state and as a patent consequence of the absence of that legitimacy. Even at its most encompassing, though, a patronage system was premised on insiders and outsiders, whose access was allowed or precluded on ethnic, religious, regional or political grounds. While resources were available

---

82 This was of course, not so in the Belgian and Portuguese possessions, where no such efforts were made at all.

83 Dani Rodrik lists fifteen African countries with growth rates of at least 2.5 per cent until the 1970s oil shocks. See Rodrik (1999), table 4.1.

to keep it going, the system was already producing its marginalised; but when resources became scarcer and the circle of those benefiting from them was much reduced, the potential for social conflict extended to previous beneficiaries as well.

This centrality of external resources for the construction of internal politics, a process Bayart termed "extraversion",[84] meant that access to the revenues of the international system was the condition for state survival in the early postcolonial era.[85] The diminution of these flows could challenge the whole setup. The quick transformation of the international economy from an embracing to a hostile one in the early 1970s was the first trend to question the early choices and structures of the African state. While the oil-exporting states that are the subject of this study celebrated the sudden quadrupling of available resources as a result of the 1973-74 OPEC price hike, the majority of oil-poor developing nations were faced with a major impediment to the strategies hitherto pursued:[86] firstly, because of the heightened price of petroleum, which ate away at their foreign exchange reserves: secondly, because of the economic downturn this caused in the developed economies, which led to lower demand for Africa's export commodities and a general decline in already dwindling foreign direct investment. The African state was thus confronted with a shortfall in foreign exchange earnings needed to keep up the expensive administrative and clientelistic edifice it had erected around itself in the previous years. This shortfall was provisionally resolved by massive borrowing that private and public lenders were only too eager to indulge. By the early 1980s, however, the return to unfavourable interest rates confronted developing nations with huge debts.

The state unravels

Structural adjustment programmes (SAP) were the response of the international financial institutions to the economic straits of indebted de-

---

84 Bayart further argued that this is the key mode of interaction between Africa and the global economy. See Miller (1988) for a landmark study of "extraversion" in the pre-colonial coastal enclaves.
85 Clapham (1996a: 55).
86 See Penrose (1976).

veloping states.[87] Although put forward in this context, their creation coincided with paradigm shifts in international economic policy (away from the post-war consensus and towards neo-liberal precepts) and in the analysis of African economic decline (away from an emphasis on the international economy and towards African misrule).[88] The average SAP was premised on the privatisation of state companies, the end to state monopolies, the end to subsidies for consumer goods, the devaluation of the local currency and a general commitment to budgetary restraint and macroeconomic stability. In short, the goal was to debilitate the institutions and strategies that had thus far enabled the survival and prosperity of local elites and provided the structure for existing, if no longer viable, patronage systems.

Yet SAP designers were unwilling to recognise the political character of many of the issues at stake, preferring to phrase them in "economistic" terms.[89] This unwillingness is particularly clear in the studious avoidance of the fact that those supposed to implement reform (local elites) were the main targets of a reform effort aimed at the very heart of political and economic control, and therefore the key losers, were reform to take place. Most programmes failed, remained unimplemented or had disappointing results, often because governments were no longer able to enact economic policies. But they soon became an indelible part of Africa's political landscape and contributed much to the acceleration of state decay by speeding up the predatory drive of elites best placed to benefit from the wholesale, and externally approved, privatisation of the economy. The scant leeway left to Africans states in the wake of heavy indebtedness meant that some version of reform had to be pursued; the fact that IFI plans could be given domestic uses that never crossed the minds of their Washington architects explains the record of partial and perverse implementation.

Although Sub-Saharan Africa had previously had its share of armed conflict, the 1980s and 1990s witnessed a host of many-sided, seemingly intrac-

---

87 There is an exhaustive, if uneven, bibliography on SAPs. For this account I have mainly drawn on Hoeven and Kraaj (1994), Mosley *et al.* (1991), Gordon (1993) and Walle (2001).

88 See Clapham (1996b) for a succinct analysis of this intellectual shift.

89 The point about "depoliticization" is most famously put forth in Ferguson (1994).

table wars. The Horn of Africa experienced interstate war, civil conflict, great power involvement, famine and secession between 1977 and 1993. Southern Africa was confronted with white South Africa's destabilisation efforts: Mozambique, for instance, faced debt, famine and a particularly vicious insurgency until 1992;[90] Angola's civil war dragged on mercilessly until 2002. Starting in 1989, Liberia became the epicentre of a regional economy of war and plunder. There was a marked increase in insurgency, often staffed by excluded social actors who had been either marginalised from the start (like Burundi's Hutus and Rwanda's Tutsis[91]) or "down-sized" as part of the shrinkage of the patronage circle in time of economic decline.[92] Throughout, the resources available to the African state via the formal channels of the world economy (aid, commerce, credit) decreased and the number of conditions placed on assistance by ever more demanding donors mounted.[93] The end of the Cold War and the consequent loss of two important external patrons, which was coupled with the slow but inexorable shifting of French foreign policy priorities away from France's former African colonies, exacerbated these trends.

Collapsed or privatised? Crisis as lifestyle

By the early 1990s, and if one excepts the brief and quickly disillusioned enthusiasm for a "third wave" of democratisation that was to have in-cluded Africa,[94] images of the continent were as bleak in academia as in the customarily negative media treatment. Africa's formal economy was in free-fall and its share of international registered trade was dwindling into insignificance. Despite the existence of a number of state trajectories

---

90 See among others Geffray (1990) and Vines (1996).

91 On Rwanda see Prunier (1995); on Burundi see Malkki (1994).

92 As illustrated in Reno (1995).

93 The OECD's *African Economic Outlook of 2003/2004* (OECD, 2003) claims that aid to Africa ($22 billion in 2002) has rebounded in the last two years from the all-time low of the late 1990s but "remains far below the levels reached in the early 1990s".

94 A decade later, the number of established African democracies is disappointing, bearing in mind the number of "national conferences" and "transition processes" that marked the beginning of the 1990s. The "authoritarian re-establishment" on the basis of often-rigged elections is a much more prevalent outcome. See especially Boulaga (1993) and Bratton and Walle (1997).

that do not fully fit this profile,[95] most African "states" were patently no longer states according to any conventional criteria, and for the first time, the farce of sovereign statehood found many unwilling to play it. In the context of humanitarian intervention, or that of development aid, the state was now often ignored or bypassed by outside actors. Vast swathes of territory not only ceased to be administered from the centre, a common enough occurrence in early post-independence and even colonial days, but actually fell under the informal but long-term writ of non-state actors.[96] The Africa analyst thus discovered that the "warlord", a historical footnote of China's republican chaos, was alive and well in the continent's emerging politics.[97] State implosion was everywhere accompanied by communal violence in which ethnic or religious themes were given centrality. This was compounded by particular post-Cold War developments such as the forms warfare took in Liberia, Sierra Leone and Northern Uganda: they seemed positively inexplicable.[98] The human costs of these developments, and of the epidemic spread of new and old diseases, were such that it was difficult for observers not to read African politics solely in terms of what had gone wrong and what was missing. The equation of "Africa" with "crisis" became received wisdom.

More recently, a body of scholarship that seeks to separate the empirical from the normative and primarily engage with the former has counteracted these face-value readings of the African crisis.[99] There are perils

---

95 Tanzania, Kenya and Senegal (despite the Casamance rebellion) come to mind. There are also success stories of failed or weak states whose fortunes have improved in the past decade such as Mozambique, Ghana and Uganda.

96 See Roitman (1998) for a look at the new "regulatory entities" that control illicit trade around the Lake Chad basin, an area formally shared by Cameroon, Nigeria, Niger and Chad but in no certain possession.

97 The key contributions here are William Reno's.

98 The trendsetter is the much derided but influential Robert Kaplan, "The Coming Anarchy: How Scarcity, Crime, Overpopulation and Disease are Rapidly Destroying the Social Fabric of Our Planet", *Atlantic Monthly* (February 1994) on Sierra Leone. A noted counter-argument is Richards (1995). See Ellis (2001) and Behrend (1998) for important studies on Liberia and Uganda respectively. See also the *Politique Africaine* issue on "Liberia, Sierra Leone, Guinée: la régionalisation de la guerre" (2002).

99 Prominent amongst these are works by Reno, Bayart, Hibou and Mbembe, as well as of a number of mostly Francophone writers associated with the journal *Politique Afric-*

to this approach. It is right in not dismissing apparently self-destructive or stupid actions as such, and instead finding identifiable agendas and hidden beneficiaries in Africa's politics. But sometimes this has led to an over-estimation of the viability of resulting arrangements and the explaining away of extreme violence and the breakdown of spaces for human exist-ence in terms of conscious strategies of political domination[100] or, more insidiously, of the growth pains of African "state formation".[101] Never-theless, this has in essence been a very positive turn. While it is evident that an analysis of the real collapse of living conditions must be central to a reading of present-day Africa,[102] to limit it at that would amount to a great impoverishment of our understanding of contemporary politics. Noting the eminently rational character of actor strategies in the face of economic decline, these authors speak instead of the "privatisation of the state", a process encompassing not just divestment of state assets, but of key sovereign functions such as the monopoly on coercion and taxation.

According to this view, the "new organisational solutions"[103] are pur-sued by state actors themselves. These are characterised by rapid de-bureaucratisation of the state, the retreat of decision-making power to shadow entities and heightened deployment of private violence for eco-nomic ends.[104] Public institutions that remain are hollow, inconsequential shells designed to keep a sovereign facade that satisfies the external gaze and ensures the trappings of statehood, with "the gatekeepers of the inner state" hiding from view. Unable to compete in legal areas of the global economy, including those in the hitherto-dominant primary commodi-ties whose prices are now much less favourable,[105] African economies have swerved decidedly towards the unrecorded and/or criminal niches in which they hold a competitive edge.[106] Smuggling, human trafficking,

---

aine.

100 Perhaps the obvious example here is Chabal and Daloz (1999).

101 Bayart, Ellis and Hibou (1999: 44).

102 Clapham (1996a: 164).

103 Mbembe (2001: 78).

104 Hibou (1998: 23).

105 The current exception is the type of commodities craved by China and other Asian industrialising economies, the price of which has considerably increased over the past years.

106 Bayart, Hibou and Ellis (1999).

the importation of toxic waste, the drugs trade and international fraud have compounded the incomes local power holders still derive from more traditional sources such as development aid. Control of the state becomes less central than in the first postcolonial decades, but state actors are still at the forefront of economic activity, and an important role in the informal economy is normally premised on political connections. In this context, "partial reform and the actual implementation process" provide state elites with "new kinds of rents, as well as with discretion over the evolution of rents within the economy".[107] Thus the state remains an arbiter in the economy, although its influence is indirectly exercised.[108]

At the moment of writing, none of these trends is new, nor are the explanations. As Nicolas van de Walle points out, Africa has entered a third decade of what can be labelled a "permanent crisis",[109] with its time-honoured rituals of debt rescheduling, half-implemented reforms, humanitarian and development aid and the trumpeting of occasional progresses as a return from the brink. Some changes have indeed taken place, mostly for the worse. States, for instance, are now much less capable of implementing macro-economic or infra-structural reforms, even if they choose to do it. In most places, populations have created alternative welfare and educational arrangements that do not presuppose the existence of the state; its absence from whole dimensions of the life of "their" peoples seems well nigh immutable. The NGO sector has grown from insignificance to be a major provider of basic amenities to the population as well as the site for the type of clientelistic activity formerly afforded by the state alone.[110]

This outline of the trajectory of the African postcolonial state can lead to two conflicting readings. The first is that the bundle of processes I described amounts to "creative destruction", with new formats replacing imported models of institutional organisation unfit for the po-

---

107 Walle (2001: 159).

108 Hibou (1998 : 67).

109 Walle (2001).

110 Walle (2001: 157). The author further notes that "the NGO sector seems in large part an emanation of the state elite". This conforms to my fieldwork experience but, as Walle writes, the process and its implications remain under-researched.

litical realities of the continent. Validation for this view can be sought in analogies with early modern European state-building that uncover plunder and organised crime as the root of the absolutist state and ultimately, of Western modernity.[111] A second, alternative reading would be that the "African crisis" suggests dismal prospects, not only for sustained economic growth and less-than-feeble institutions, but also for any broadly shared positive outcomes at all if the current framework is not fundamentally transformed. Both readings are however compatible with Crawford Young's claim that "the postcolonial moment has passed [...] deeper continuities with pre-colonial social and political patterns, and novel experiences of coping with the realities of state decline in recent decades, combine to close a set of parentheses".[112] The economic and political arrangements of those first postcolonial decades have definitely been subverted and are becoming unrecognisable.

IV. THE SUCCESSFUL FAILED STATE IN THE GULF OF GUINEA

The outline of two of the most common and disappointing experiences of statehood in the postcolonial era provided by the previous sections is essential for understanding the oil state in the Gulf of Guinea, its institutional patterns and its very survival. For its nature is paradoxical. The Gulf of Guinea oil state shares many of the elements that are common to the trajectory of the African postcolonial state and can be described in terms of what the previous section referred to as a "failed" or very weak state. But it is also the beneficiary of an engagement with the international economy guaranteeing that it is respected and indulged and that it can access foreign exchange with ease.[113] Most importantly, the international system actively promotes its survival. The oil state in the Gulf of Guinea is therefore both failed and successful.

---

111 See especially Tilly (1985).

112 Young (2004: 49).

113 Clapham (1996a: 165) notes the "critical importance of foreign exchange in maintaining the state itself, and the lifestyles of those who were most closely associated with it" and the extent to which the "crisis in foreign exchange" impacted on African states. This has been much less of a concern for oil states.

Failure

The dimension of failure can be dealt with very briefly because it is empirically undeniable, as the overview of the domestic governance of these states in the next chapter shows. To start with, the pre-oil context of Gulf of Guinea states was amongst the least likely in the world to foster oil-based development.[114] Every structural prerequisite was missing for sound use of oil revenues. Severe pathologies already characterised the politics of many states before the arrival of oil rents: their economies were fragile, volatile, with a weak fiscal base and badly run, and institutions were always inefficient and feeble. Neither were the quality of pre-oil political life and, more specifically, the agendas of empowered political actors inspiring of much hope. Long before oil wealth arrived, authoritarianism, lack of accountability and factional competition for the state and its spoils were common currency. In none of the states in the region can one claim that a "developmental elite" was ever at work for the public good. These traits, which are common to the African postcolonial state experience, have everywhere led to disappointing outcomes. It is therefore unsurprising that the manner in which Gulf of Guinea states dealt with their wealth was path-dependent.

The importance of a bad legacy, however, should not lead one to think that oil-era political and economic developments are simply the playing out of pre-existing dynamics. On the contrary, the creation of an economy dependent on oil revenues, as well as deepening the dysfunctional trends of the African postcolonial state that already haunted the Gulf of Guinea, has brought with it new problems and noxious incentive structures that fostered even poorer and more irresponsible decision-making.[115] I will not discuss these pathologies in detail here for they are the focus of Chapters 2 and 3. Suffice it to say that they have resulted in a pattern of state dereliction that is deep, multilevel, and durable.

This demise of the state in the Gulf of Guinea, as in most of Africa, has resulted in such a loss of life and such a decrease in the living conditions

---

114 The next two paragraphs are based on Soares de Oliveira (2006).

115 As Chapter 3 argues, poor and self-serving performance by decision-makers plays a key role in explaining current outcomes and must be placed alongside (but not replace) structural factors in accounting for them. See also Soares de Oliveira (2006).

and dignity of those who survive that it would be callous not to accord it great relevance. Simply to denounce the normative underpinnings of the discourse on state failure is not enough.[116] As Hedley Bull wrote, human societies may be plural in terms of the disparate goals they prioritise, but there are nonetheless "elementary goals" that all seek to secure not only or their own sake, but as the precondition for the "realisation [...] of any other goals".[117] I want to keep the empirical element of state deterioration in full view and, in this context, the language of state failure is not entirely inappropriate. For I am writing under the assumption that:

People and states must be secure from the fear of violence at the local, national, regional and international levels if an enabling environment for sustainable political and economic development is to be created. This means that states must be adequately protected against external aggression and internal subversion, and that the lives of ordinary citizens must not be crippled by state repression, violent conflict, or rampant criminality.[118]

There is no doubt that lack of capacity or unwillingness of states to pursue such goals has resulted in tragedy.[119] This said, the success of those in control of the state apparatus, and the legitimacy and support they have achieved internationally, put a question mark over explanations that do not give enough emphasis to the rewards encompassed by this kind of politics.

---

116 See, for instance, Hibou (1998: 12). Hibou and others argue that the normative view is skewed, and that the "failed state" is actually performing a set of tasks distinct from those that critics, steeped in expectations of the liberal-democratic modern state, blame it for not performing. In light of this, they argue, one should revisit what "failure" means. I agree with the view that there are no *a priori* attributes of the state or tasks it naturally fulfils but only historically contingent expectations. The "anti-normative" approach is nonetheless limiting because if fully accepted, it would ultimately dismiss tremendous declines in infrastructures, healthcare, education, life expectancies, etc. as valid criteria to assess the state. A convincing analysis is surely that which is capable of explaining the contemporary politics of the African state while giving great centrality to the human suffering it abets and oftentimes creates.

117 Bull is referring here to the right to life, the right to contractual predictability, and the right to property (2002: 3-5).

118 Ball *et al.* (2003: 263).

119 See Gary and Karl (2003) and also the brief discussion on the end of public goods provision in section two of the next chapter.

Success

This is where the "successful" dimension comes in. For oil ensures that the state is preserved from actual demise, that allies and cohorts uphold it and that it can afford weaponry to fend off armed challenges.[120] The fiscal independence and foreign exchange abundance provided by oil allow it to ignore external pressures and domestic society outside elite circles. Because of decay, the tasks the state can perform are much reduced, but this occurs within a reconfigured field of political action in which empowered actors have new expectations for the state apparatus and discard many of its old obligations vis-à-vis the citizenry. This means that, more frequently than not, the lingering benefits of statehood are privatised while its drawbacks and failures are "public", to be shouldered by the masses. In such new circumstances, success and failure may not be contradictory at all because each pertains to different segments of the overall political outcome. This does not imply a revisionist take on failure but rather that an analysis of failure must allow for the fact that certain forms of governance can be "successful as measured against its own parameters and judged by the standards of its political program".[121] Some instances of failure, such as the abandonment of territory and the end of public goods provision, should be seen as policy choices as well.

The success of oil states is premised on outside actors needing them to perform a specific role. While most external relations with Africa have been "de-stated" over the last two decades,[122] the state remains at the centre of the oil business; political power derives not so much from control over informal market, as elsewhere, as from formal market channels.[123] On account of the capital-intensive nature of their invest-

---

120 The Angolan government's long and successful war against UNITA is a case in point. But perhaps the most extraordinary example of the last decade is that of Algeria. Despite a bloody civil war that claimed the lives of 150,000 people, the country's oil sector remained insulated from the fighting and continued to provide the military government with the financial resources to pursue its military effort and in time claim the upper hand. See Martinez (2000) for an influential study of the conflict.

121 Prunier and Gisselquist (2004: 104).

122 Clapham (1996a: 256).

123 See Reno (1995) for the argument that politicians can exercise political power through the control of informal markets.

ment and the need to secure it, oil companies are absolutely dependent on the existence of a sovereign entity with which they can sign contracts and strike partnerships enabling the long-term business of oil to achieve its ends. If things go wrong, it is impossible to sue a non-state entity such as an insurgent group.[124] Because "the whole framework of international commercial law depends on the courts of national systems and or systems of arbitration, both national and international, that exist at the will of states and require national systems for the execution of their decisions",[125] no exploitation of oil resources by firms could occur outside a state-firm partnership. This means that oil companies not only do not hamper the state; they increase its power through diplomatic leverage exerted on its behalf, assistance in obtaining loans and, most importantly, by allowing it to profit from resources it would otherwise be incapable of accessing. For a state that is beset by structural weaknesses and poor governance, the oil relationship is an asset and an opportunity. Perhaps the most striking element in the specificity of the oil states in the Gulf of Guinea is the amount of money poured into them by foreign investors and the continued access to international banking more than two decades after most other African states lost it.[126] These are capital-intensive, foreign-led economies, very different from the majority of the many corners of Africa where "investors do not go to".[127] This is the Africa that counts. Externally, this means that

---

124 Even if that type of partnership worked, the firm could always be sued in turn by the state for having paid taxes or royalties that are legally its own. The cautionary tale here is the many business transactions gone wrong with Liberia's Charles Taylor prior to his assumption of the Presidency in 1997.

125 Foreword by Anthony K. Appiah in Sassen (1998: xiii). As Reno (2001: 207) notes, "firms seek to limit risk through international certification of their political partners' sovereignty so that agreements can be enforced in foreign courts, operations can be indemnified, credit on attractive terms can be procured from multilateral and bilateral agencies [...] and capital markets can be tapped".

126 Of course, any country with a valuable resource can arrange for its mortgaging, as Guinea-Bissau did in 2003 with its future fishing production, but the amounts of money are simply not comparable with the mammoth oil-backed loans—sometimes in the billions of dollars—that are negotiated by oil states, as discussed in Chapter 2.

127 Cooper (2001: 207).

the state possesses considerable autonomy from postcolonial traps such as pressure from donors and multilateral organisations.

But how can an economic partnership of such complexity be operated by a state that is facing infrastructural, bureaucratic and political decline? Surely the international partners of the state also suffer from its weakness or failure? The answer to this lies in two enabling factors. The first is the creation of a parallel economic system that insulates oil companies from the unreliability of local conditions, with its own acceptable legal framework and logistical efficiency. In such enclave contexts, companies can operate freely and do not face the rent-seeking, contractual uncertainty or threat of expropriation that are widespread outside the sector. (Rent-seeking will in turn reach epidemic proportions around the state institutions that are the terminus for the oil producer's share of the revenue.) In this sense, the character of oil as an enclave sector without linkages to the non-oil economy, everywhere seen as one of the limitations of oil-geared development, may be the condition that permits the oil rapport to develop at all. This privatisation, or discharge, of state functions onto non-state actors is a mode of political action studied by Weber, particularly in medieval Europe and the Ottoman Empire,[128] and has already been briefly described in section three. It is characterised by a state that is unwilling or unable to perform tasks such as tax collection and the administration of justice, which proceeds to sub-contract them in the broadest terms in search of positive outcomes like growth in revenue in the postcolonial period or *mise en valeur* and budgetary self-sufficiency in the colonial era. The second enabling factor is the creation of state-owned organisations, the national oil companies, which articulate state interests in the oil sector with comparative prowess and are spared the administrative decline evident elsewhere. As shown in Chapter 2, they bring together the scarce human resources of the oil state and are its enabling arm in negotiations, in joint ventures and, ever more often, in the contraction of oil-backed loans.

Over the years, the myth of oil riches has become ingrained in the political culture of the Gulf of Guinea states. While economically and

---

128 See Hibou (1998) for an influential usage in the context of "privatisation of the state"; see also footnote 100 of Chapter 2.

politically exclusionary, oil enfranchises the population around expectations of wealth and perceived membership of something blessed (or cursed). It is difficult to conceive an idea of the Nigerian state, or of an Angolan state, that does not revolve around the ownership of oil. The same applies to many of the ethnic nationalisms that have grown around the desire to benefit from the oil wealth. Oil has been grafted upon "the conflicts and political cleavages" of each society and has participated in the shaping of "how public life [is] organised, how the contours of rival groups were created, and in which sites a new political culture eventually emerged".[129] Oil is an essential part of the continuous historical process of conflict and interaction through which a shared state mythology and national space of politics are created, even where state institutions no longer operate.

Limits to success

The success of oil states is highly constrained, however. Firstly, they are in a position of great dependence. Their reliance on oil revenues is almost unprecedented: with the exception of Cameroon, oil is the overwhelming source of revenue, as table 2 shows. On account of their limited technical capacity to explore, transport and refine oil (and in some cases, in the field of negotiations as well) oil states are also dependent on foreign technology and finance. The state's reliance on a handful of tax-paying entities confers great centrality to the firms involved in oil extraction. Secondly, and tied to the first point, the weakness and lack of indigenous capacity to run the oil sector mean that, if compared with international standards, the contractual terms are some of the most unfavourable world-wide. This is so not only in terms of the host-state's take but also in the way oil contracts are structured. Shaxson noted that the structure of PSAs in the Gulf of Guinea tends to create an oil company revenue stream that is much more stable than that of the host-state, which has to cope with an unusual degree of volatility.[130]

Thirdly, oil contributes to the deepening of an extensive legacy of institutional and economic fragility at most conceivable levels. In the Gulf of Guinea, the resources available to the state heighten the high levels of rent

---

129 Bernault (1996: 26).
130 N. Shaxson (2005).

seeking to be found in the rest of Africa. This does not *per se* distinguish the oil state from states without oil, but it deepens the trend of political actors converging on it as well as the political centrality of the state for the internal power game. As Cooper notes in regard to Nigeria, "oil can turn a gatekeeper state into a caricature of itself":[131] familiar political dynamics take on unheard-of proportions in the scramble for the generous spoils. Whether or not one accepts the notion of a resource curse, dependence on oil revenues has contributed dramatically to the unpredictability of economic and political life in the Gulf of Guinea. In short, the presence of a legal commodity that needs to be transacted via a state, far from having stemmed the trends present in resource-poor Africa, has instead precipitated some of the worst cases of brutality, war and near-implosion of the state.

	Angola	Cameroon	Congo Brazzaville	Equatorial Guinea	Nigeria	Gabon
Oil revenues as % of GDP	45	4.9	67	86	40	73
Oil revenues as % of government Revenues	90	20	80	61	83	60

Table 2. Crude oil production 2002 per country as percentage of GDP and share of government revenue[132]

The impact of oil in the material world of the Gulf of Guinea is as backhanded, and as pregnant with unintended consequences, as the indelible mark it has left in the popular imagination. From an institutional viewpoint, oil ensures that the state does not disappear fully. But it does not foster genuine state building, instead promoting the creation of commodity-based enclave states, in much the same way as long-distance trade led to the creation of coastal states until the end of the nineteenth century.[133] These are ramshackle, provisional structures, if internationally

---

131 Cooper (2002: 172). See p. 41 for Cooper's concept of the "gatekeeper state".
132 From Gary and Karl (2003) various sources.
133 For a discussion of the very different impacts of foreign trade on state-building drives across Sub-Saharan Africa prior to the Scramble, see Freund (1998: 51-71).

recognised ones, organised around the export of a prized substance: they are not the seeds of a future conventional state and they are not viable without oil. Thus the "success" of the oil state can only be understood in its context of tenuous survival and from the viewpoint of the select number of actors that benefit from it. Elsewhere, oil revenues would be studied solely as the source of system-wide failure and civil conflict, the poison enfeebling otherwise competitive economies.[134] Yet in the Gulf of Guinea, all things being equal, oil has played the role of lifeline for very weak governments. It must be assumed that, in the absence of oil, many of the states under review would have fallen into other governance traps that have led non-oil states in sub-Saharan Africa down the path of institutional implosion. If anything, oil is responsible for the lingering "stateness" of many of the oil states, and for those at the helm it is not a curse but a blessing.

Sovereign credibility

I have previously described how the tenuous grasp over internal sovereignty by oil states does not fulfil Weberian criteria. Yet their external sovereignty (i.e. recognition by other states) is actively constructed by a supportive international setting that detaches the survival of the oil state from its domestic wreck, making a seemingly unsustainable scenario of economic decline and institutional decay viable because of the oil link with the world economy. At a time when the "failed state" has become the key concern for strategists of industrialised nations as providing breeding grounds for terrorism, it is ironic that some of the most decrepit states around are remarkably immune from external sanctioning. But then, as Badie writes, "even when it is unmasked [sovereignty] is not abolished".[135] It should not come as a surprise that the new discourse on state failure (and the controversial discussion on sovereign de-certification) is highly selective and detached from internal circumstances, for it

---

134 See section two of this chapter and much of the literature on natural resource endowments in the developing world.

135 Badie goes on to note that "it is not necessary for the concept to be clear in order to be abundantly employed [...] the fiction is useful and provides a lot of services". Badie (1999: 111).

merely echoes the "organised hypocrisy" that has characterised the history of sovereignty.[136]

This foundational hypocrisy was, if anything, accentuated by decolonisation's creation of essentially different "sovereign games"[137] for different (Northern and Southern) players, and the international recognition of "quasi-states"[138] that might not exist empirically but could count on a UN seat and the support of the international system. This unlikely arrangement worked throughout the early postcolonial period, as there were plenty of rewards and next to no obligations attached to it for the quasi-states. Those who believed it was a sustainable arrangement were wrong, however.[139] Following the end of the Cold War, it became clear

---

136 Amongst many see Krasner (1999). US President Wilson's view of sovereignty as premised on national self-determination was the first meaningful intellectual departure from the association of sovereignty with capacity to sustain it. It was in the aftermath of the Second World War, however, that nationalist claims finally became the index for attribution of statehood. This was so even where these claims had little local support and pertained to territorial units with spurious demographic relevance and no hope of economic viability. The right of self-determination was thus separated from the empirical capacity for self-government. This does not mean that the "practice of sovereignty" was not highly ambiguous beforehand: Krasner demonstrates that it has always been honoured in the breach. Nonetheless, before 1945, this pertained to the realm of external recognition rather than capacity for domestic survival. In fact, amongst the hundreds of states that disappeared through annexation or voluntary union from the European political map in the last centuries, one can find many viable states with good credentials, like Bavaria or the Kingdom of Naples. Conversely, there are many examples of European states being dependent for survival on a supportive international system rather than on the capacity to fight off challenges such as Luxembourg, Belgium and the successor states to the fallen Central and East European empires. But their capacity to hold internal sovereignty was never open to question because, until the postcolonial era, states that could not function were simply gobbled up by expanding rivals. Quasi-states are truly a creation of postcolonialism.

137 Sorensen (1999: 591ff).

138 Jackson (1990).

139 If I understand him correctly, Robert Jackson (writing in 1990) expected the "quasi-state" to survive because the international system had accepted that a certain category of states need not conform to previous criteria. Empirical decay was therefore immaterial to its deployment in the world stage. It seems that Jackson's otherwise excellent contribution greatly overestimated this postcolonial sovereignty paradox as a stable form.

that little political advantage was to be gained from propping up every weak state in the developing world. While not yet accepting de-certification[140] or re-colonisation, which would entail "the death of the principle of universal equality between states",[141] Northern states are no longer willing to spend the necessary resources and time to prop up unimportant, peripheral states.[142] (Conversely, their willingness to invade and occupy them on human rights and bad governance grounds has grown.) Public anxiety in the West about failed states may have increased, but so has the acknowledgment that they are part of the international landscape and that not all can be saved or treated back into functioning statehood. For the time being, this momentous alteration—the fact that the "quasi-state" regime of tolerance, generosity and support is no longer available to all—goes unrecognised because political innovation and conceptual change are different things and because the history of sovereignty has been able to support gross contradictions. And yet it is there. In turn, weak and failed states have pursued heterogeneous strategies in response to this novel insufficiency of juridical recognition for sustainable statehood. This may at times actually include the implosion of the state, as in the case of Sierra Leone.[143]

The Gulf of Guinea states, however, are a meaningful region-wide example of sovereign units that are continuing to be not only passively tolerated, but also actively nursed by powerful actors of the international system. They survive not because of the conservatism of an international society unwilling to acknowledge their dereliction, but because of the networks of international complicity that underpin them. The mainte-

---

140 An early call for de-certification of failed states was Herbst (1996).

141 Bain (2003: 59).

142 The post-9/11 stress on failed states has not sufficiently changed this deficit of attention and resources. This is so despite a number of surprising conversions to the urgency of addressing state failure. Witness, for instance, Francis Fukuyama's recent acknowledgement that "many of the problems of our current age, from poverty to refugees to human rights to HIV and Aids to terrorism, are caused by states in the developing world that are too weak. [...] State building [...] is something we will have to think seriously about". Fukuyama, "Bring Back the State", *The Observer,* 4 July 2004.

143 See Reno (1995).

nance of some weak states by the international system is thus more and more a function of their individual importance and not of a generic desire to uphold sovereign statehood in the developing world. How else can one explain that many such efforts come not from other states or international organisations, but from self-interested private sector actors?

Despite its invaluable descriptive use, then, the "failed state" category remains at heart a deeply political one. Mobutu's Zaire was never a "failed state" according to its external patrons; nor is Colombia, whose government has no access to large parts of the country; nor are the worthy allies in the war against terrorism in Central Asia. This "embryonic language of the new imperialism", as Robert Skidelsky called it,[144] is never marshalled against "our" failed states, but merely against enemies or extraneous states. This is the case with the oil states of the Gulf of Guinea and other resource-rich, "useful" states that demand capital-intensive, long-term investment. Their "failure", because it is unrelated to international terrorism or other externally damaging developments, has caused little anxiety except when it imperils the oil flow, as in Nigeria. The international system is ready to shore up their existence through credit, diplomatic support, tolerance of poor governance, and enhanced external visibility, because it needs them to exist. Even when grievous crimes are committed, as in the recent Darfur massacres in Sudan, the international consensus needed to vilify the culprit oil state simply does not exist.[145] The international system thus plays a key role in the construction of oil states as credible international actors, in what can only be described as a form of sovereignty-enhancing intervention designed to enable legal resource extraction rather than improve internal sovereignty gaps.[146]

---

144 Skidelsky (2003: 31).

145 In the case of Sudan, China, which holds a considerable stake in the country's oil industry, has been instrumental in watering down Security Council moves to tackle the issue. See the discussion on Chinese companies in section one of Chapter 6.

146 Though pursued in the context of oil states, this argument is conceivably relevant for broader discussions of the external politics of failed states that "still matter". As already pointed out, the role of empowered actors of the international system in shoring up failed states is particularly clear in the context of "war on terrorism" alliances, though the reasoning in those cases is primarily political rather than economic.

Conclusion

In describing the medical condition of one of his characters,[147] Svevo got the point precisely right: one hovers in permanent crisis, a crisis that is neither terminal nor benign. The most loudly asserted consequence of the "African crisis" is the deterioration of state sovereignty across the continent, measured as economic decay, declining terms of trade, debt, donor conditionality, increase in external intervention, etc. Yet as one enters the third decade of unstoppable decline, the condition of the African state is increasingly perceived not as an extraordinary one (in the literal sense of the word) but as a permanent feature. Political theory, notes Istvan Hont, is simply not equipped to confront an understanding of crisis that bears all its ills but not the expected outcome.[148] This is particularly evident in the case of the Gulf of Guinea where the known ailments of failed statehood do not lead the state to collapse and even allow for the steady prosperity of those at its helm.

This chapter has suggested that, on the one hand, the oil wealth of the Gulf of Guinea does not mean that the states in possession of it will escape Africa-wide social and political trends; indeed, some may be exacerbated by its presence. Warfare, ethnic strife, poverty, population growth, the AIDS pandemic and much more will be shared by them. Oil will not provide for insulation and the Gulf of Guinea's politics will only be understandable in the wider African context. On the other hand, *oil will change the calculus of state survival*. It will guarantee that whatever the domestic political conditions, multiple external and internal actors will have an interest in maintaining a notional central structure, and that enough resources will be available to coerce or co-opt enemies. This will frequently work in favour of incumbents and allow them to build a political order that is violent, arbitrary, exploitative but fairly reliable. The political process is and will be unstable and fragmentary, but the

---

147 See the quote at the start of this chapter.
148 Hont (1994: 169) identifies two common scenarios in political discussions of crisis: "bad crisis" and "good crisis" that leads to "something more positive and perfectly healthy". Yet he notes that a third scenario consisting of "a happy escape from death, which falls short of achieving a utopian return to real health [...] a non-revolutionary model of permanent crisis, without the decisive outcome of genuine crisis theories" is conspicuously absent.

structure of politics itself will be stable and viable—while oil lasts. Yet far from leading the oil state down the route of institution building, oil provides a paradoxical contribution to the accelerating deterioration of institutions and lives prevalent across Africa.

My argument accepts the compatibility between the dysfunctional trends of the petro-state and the African failed state, on the one hand, and the astonishingly successful strategies of political survival amidst decay, on the other. Other states in the region that have no oil do not possess such an obvious source of sustenance and viability. There are many alternatives to oil, including foreign aid, the ever-decreasing agricultural export revenues, mining, the sale of fishing rights and many other legal and illegal engagements with the international economy.[149] But these options do not originate revenue streams comparable to the oil sector and cannot provide the basis for long-term, sustainable political arrangements. Likewise, some of the more widespread business opportunities in sub-Saharan Africa today do not presuppose a role for the state, which is increasingly sidelined. Conversely, petroleum investment is long-term, involves unparalleled revenue, and necessitates the state. How this puzzling, seemingly unsustainable mode of governance is articulated at the domestic level is the subject of the next chapter.

---

149 Internal revenue based on fiscal extraction, as discussed in the Chapter 2, is nowhere a real option.

2

THE GOVERNANCE OF THE OIL STATE

During the five years that it implements the giant Third Plan (1975-1980) [Nigeria] will also lay the foundation of a modern industrial state and so begin to develop the power commensurate with its size and influence.

Guy Arnold, *Modern Nigeria*

[During the oil boom] the ethics of business penetrated politics, the ethics of politics penetrated business; the ethics of the gangster penetrated both.

Gavin Williams, *State and Society in Nigeria*

The trains may not run on time... but the army and Sonangol [Angola's national oil company] work and that's all that's necessary!

Interview with European diplomat, Luanda, 26 January 2004

Turning to key domestic arenas, this chapter seeks to understand the role of oil in structuring the prevalent forms of governance in the Gulf of Guinea, the state goals that are pursued under the shadow of the oil economy, and their implications for life in the oil state. My argument is twofold. Firstly, I argue that during the boom years there were fitful attempts at exercising power in a way that Michael Mann has termed "infrastructural",[1] and at garnering a modicum of popular legitimacy. Nonetheless, these were always trumped by rent seeking, the yielding of despotic power, and the

---

1 In Michael Mann's helpful definition state power can have two meanings (Mann, 1984). The first is *infrastructural power*, or "the capacity of the state to actually penetrate civil society, and to implement logistically political decisions throughout the realm". The second is *despotic power*, which relates to "the range of actions the elite are empowered to undertake without routine, institutionalized negotiation with civil society groups".

neglect of the political and economic measures that could underpin the institutionalisation of the state. Secondly, in the subsequent period, the weak and privatised state has striven to endure through arbitrariness, despotism and resource control alone. Always primarily a ruler of things rather than people, the oil state progressively abandoned attempts at governing the economically useless population and became exclusively concerned with mastering oil enclaves and money flows. While many trends described here are not unique to the oil state, the sum of its approaches to violence, bureaucracy, taxation, territory, economy and population constitutes a particular set of attitudes towards the exercise of political authority [2] that is determinative for the politics of the region.

This chapter proceeds as follows. Section one looks at the centralisation brought about by oil and the ways in which the boom-era state was managed. It also discusses the policy areas that were neglected by this apparently ambitious state, such as domestic taxation and the agricultural sector, as well as the consequences of this neglect. Section two notes that an ensuing pattern of institutional decline, privatisation and shedding of welfare responsibilities that is outwardly similar to that of other African countries has, in fact, unique aspects. In particular, national oil companies—the pivotal actors in the rapport with investors—have not only been spared the fate of other public institutions but acquired expertise and organisational capacity of a type assumed to be defunct in Africa. Geared up to allow the failed state access to oil revenues, bank loans, and opaque financial strategies, NOCs

---

2 The hypothesis of a "governmentality"—a term coined by Michel Foucault in his 1977-78 Collège de France lectures to denote a "set of attitudes towards political authority", brought into African studies by Jean-François Bayart—that is peculiar to the oil experience of the Gulf of Guinea cannot be investigated here. It is not clear at this stage of my research whether there is one to be contrasted with the generic sub-Saharan African "governmentality" which Bayart succinctly summarised in terms of the "politics of the belly". Nor is it obvious that an inquiry based on a "petro-state versus the rest" approach is the best way to go about understanding the specificity of the oil experience in the Gulf of Guinea. See, amongst others, Foucault (2001) and Dean (1999) as well as Bayart (1989) and Clapham (1996b).

are the lynchpin in its successful quest for external resources. Section three discusses how the state perceives space and exercises authority across its nominal territory. The focus is, firstly, on the geography of the oil state, its pattern of concentration on "useful" industry enclaves and urban centres, and the non-state "useless" spaces that proliferate where no economic imperative exists; secondly, on how, and through which coercive vehicles, state control is enacted across this patchwork of alternatively neglected and occupied areas.

I. GOVERNING THE BOOM-ERA STATE

There was a brief decade—the oil boom years of the 1970s and early 1980s—before the current age of state decay set in, during which the goals of the oil state were exceedingly ambitious, especially when compared with its subsequent, much more limited scope of action.[3] This was an era of high spending oil capitalism that saw the state metamorphose from a minimalist apparatus into an all-encompassing structure with powers seemingly emanating "from itself".[4] Centralisation, authoritarianism, a disregard for all things rural, the non-pursuit of internal taxation, and broad-based patronage as well as rent seeking, which are discussed in Chapter 3, characterised the state-building pattern of this period. Economically, the abandonment of productive activity in non-oil sectors, large-scale infrastructure development, macroeconomic instability and (towards the end of the period) heavy external borrowing were the norm. Infrastructural power was seriously, if incompetently, deployed across state territory during these years as an integral part of the brash and vocal oil-geared nationalist state-building drive. This section looks at the actions of the state in areas then thought to be its preserve before examining areas that were neglected even amidst claims of determined interference, and the momentous consequences of that neglect.

---

3 Although this section focuses on the experience of the older producers during the 1970s and early 1980s, it has obvious implications for the study of the new producers, and especially Equatorial Guinea, which are repeating in great detail many of the foolhardy policies of old.

4 Coronil (1997: 2).

Oil boom and the expansion of the state

The advent of oil was the single most important factor in the early postcolonial history of the oil-rich states of the Gulf of Guinea.[5] Culminating in the enormous 1974 increase in revenues, the oil boom informed their outlook on the international system, their ephemeral and optimistic nationalism and the domains that petro-elites defined as fit for state involvement. It took place at a time when the weight of the state was taken for granted in both the socialist and the capitalist models[6] and provided the means for bold state action beyond the wildest dreams of non-oil exporters. At this point, states saw virtually the whole of society and the economy as their direct or indirect area of intervention and proceeded to create or expand an enormous array of tools to accomplish that.

In some cases, the oil-driven expansion of the state prospered on fertile ground. For Congo-Brazzaville, formerly the administrative seat of French Equatorial Africa (AEF) and inheritor of a bloated bureaucratic apparatus, oil and socialism were mere justifications for the maintenance and reproduction of older big-state tendencies.[7] Angola's legacy, too, combined the cumbersome tradition of Portuguese bureaucracy with a late colonial civil service geared up for the administration of a settler society of about 400,000 Europeans; its socialist post-independence successor could but emulate it.[8] Yet stating the existence of a public sector bias is not enough to explain the region-wide expansive trend. For even states that had had a late colonial or even postcolonial measure of prudence in this area suddenly reacted in much the same manner. Everywhere the number of state employees swelled exponentially, as did their salaries. The Gabonese state's employment of a fifth

---

5 The numbers are staggering. For instance: between 1960 and 1984, the value of Gabon's exports grew 76-fold and the state budget 124-fold. See Pourtier (1989: 189, vol. 2).

6 See section three of Chapter 1.

7 Bazenguissa-Ganga (1996a).

8 In other respects, there was little continuity with the colonial period. Certainly the settler project of an economy based on both industrial import substitution and the export of primary commodities was aborted at independence.

of the work force is perhaps the clearest example,[9] though Cameroon's post-1978 doubling of the civil service was nothing short of spectacular.[10] This was also the case in Nigeria, which by 1980 had an estimated 850 parastatals[11] and a grossly overpaid civil service.[12] This expansion, and the "indigenisation" policies that accompanied it, were often *trompe l'oeil*, because many key technical tasks were still performed by expatriates[13] and real (oil-related) policy-making was increasingly located in parallel structures or the presidency, rather than in the bureaucracy. The growth in the number of state employees, state institutions, and areas of engagement was thus rarely accompanied by increased capacity for competent execution.

The centralist and unitary state tradition of the non-Anglophone states was ready-made for absolute central control over state resources.[14] Where this was not so, rules were quickly changed. This was the case of Nigeria, where the pre-oil decentralised character of revenue allocation was promptly modified by the military government of General Gowon to a new formula that brought the financial might of oil revenues to the hands of the federal state. Decree 13 of 1970, Osaghae writes, "strengthened the federal government's role as the sole allocative and distributive authority and gave it a domineering fiscal position"[15] at a time when the end of the civil war and heightened oil production were turning Nigeria into one of the world's key produc-

---

9 This is a 1984 estimate. By then, civil service numbers had grown tenfold since independence according to Pourtier (1989: 205-7, vol. 2).

10 See Jua (1993: 144-9) and Jackson and Rosberg (1994: 301).

11 Lubeck and Watts (1994: 216).

12 Public sector salaries were often unrealistically increased, as in the case of the Udoji awards of 1975 that saw Nigerian civil service salaries go up 30-100 per cent.

13 This was obvious in the smaller producers such as Gabon, as noted by Pourtier (189: 261, vol. 2), but even in Nigeria, "indigenisation" focused primarily on managerial, rather than technical, tasks.

14 By the time oil revenues started to flow in the late 1970s, Cameroon's President Ahidjo had already revoked the federal structure of Cameroon and merged the Francophone and Anglophone segments into a unitary republic in the French mould. Angola's provincial governors have substantial responsibilities but are directly appointed by the president.

15 Osaghae (1998: 73).

ers. Beyond the excessive concentration of power it encompassed, the strengthening of federal control over revenue had three other results. Firstly, it sidelined the principle of derivation that would allow the oil-producing regions to keep a significant share of the oil money, and thereby created the conditions for the Niger Delta conflicts of the last decade. Secondly, it created incentives for federally nurtured states with no need for a local fiscal base—in effect, rentier states of a rentier federal state—which led to calls for state creation across the country, feeding ethnic claims on oil money and further bureaucratic growth.[16] Thirdly, it led to the ubiquitous presence of the military-run federal government in all corners of the Nigerian economy, a trend furthered during the 1979-83 Second Republic.[17]

In these circumstances, petroleum revenues were the reason for "the high degree of centralisation of authority on the executive",[18] with presidential cliques taking control of the levers of the state to an extent that rendered formal institutions irrelevant.[19] Secrecy surrounded everything that was oil-related. While the ownership of oil and spending of oil revenues came to define the self-images of the oil exporters, the details of the business and related practices of the state and its international partners and friends remained absolutely confidential. Annual reports of national oil companies and local subsidiaries of foreign companies were not published. Amounts of money transacted and oil production levels remained subjects of speculation, and audits were unheard of. According to a journalist working in the region in the 1980s, neither the operators nor the NOC even disclosed production locations.[20] In Cameroon, secrecy was justified in terms of the threat of oil revenues to "the agricultural economy and peasant development"[21] and, more generally, to prevent the exodus of youth to the cities and

---

16 Suberu (1991).
17 The key work on this period is Joseph (1987).
18 Karl (1997: 90).
19 See Chapter 3.
20 Author interview, Cambridge, 16 July 2004.
21 Ndzana (1987).

a "boom mentality".[22] In Nigeria, oil crept into politics unannounced, years after it had become an important economic factor: until then, political life had gone past it, oblivious to the underground wealth. It was only in the mid-1960s that the sudden realisation of it struck the country like a storm. In no time, "political feelings about petroleum changed from apathy to euphoria"[23] but factual knowledge of the sector remained scarce. This was but one dimension of the wider shroud of mystery deliberately bestowed upon petroleum: as Coronil notes, secrecy was key to the dramaturgy of political power as exercised by the oil state.[24]

The grandiose and unrestrained economic policies of the oil state are another important dimension of its conduct. In a recent review of the Gulf of Guinea's oil producers, the IMF noted that "the challenge of macroeconomic policy [there] is to stabilize budgetary expenditures and sterilize excess revenue inflows in the context of medium- to long-term sustainability considerations".[25] The established producers have not heeded this advice for the past three decades, and as in many other areas, it seems unlikely that the newcomers will follow the straight and narrow path of balanced budgets and revenue-stabilisation funds. Generally speaking, economic policies during the boom era were shortsighted, politically driven, economically self-defeating and characterised by unproductive and ostentatious public and private spending as well as theft.

Since resources were apparently limitless, all budgetary discipline was abandoned. States spent copiously, especially on capital-intensive infrastructure projects with low or at least very long-term returns on investment, in energy provision, mining, heavy industry, transport and defence. They also favoured prestige projects with barely any productive

---

22 Ngu (1988: 112). See also "Pétrole: une production et des réserves 'modestes' selon le président Ahidjo", *Marchés Tropicaux et Meditérrannens,* 4 June 1982. Claiming a commitment to forestalling the "Nigerianisation" of Cameroon, President Ahidjo kept oil revenues on a off-budget foreign bank account, and his successor Paul Biya kept it off-budget until 1993.

23 Pearson (1970: 137).

24 For this argument see Coronil (1997) and specifically on Nigeria, Apter (1996) and Watts (1994).

25 Katz *et al* (2004)

rationale. The study of the lavish amounts of money spent in building (and importing wholesale) a "modern", industrial economy goes beyond the scope of this chapter. There are endless anecdotes of boom-era white elephant projects. But Nigeria surely tops the league with its new capital, Abuja—which, after a decade and a half of slow growth, has recently seen a veritable explosion in public works—and the multibillion dollars-worth, Russian-built Ajaokuta steel complex that never worked. In addition to his megalomaniac redesigning of central Libreville ahead of the 1977 OAU conference, President Bongo insisted on building the hugely expensive Transgabonais railway to Franceville. (Bongo's attempts to have France build a nuclear plant in Gabon did not bear fruit.)[26] Ambitious five-year plans were designed to lift the newly rich economies from timeless poverty. Though everywhere profligate, the spending strategies of the oil state were "conditioned by the country's distinctive social and political makeup"[27] as well as the resources available.

Import substitution models influenced policy in Cameroon and late colonial Angola,[28] but nowhere was the modernising frenzy more evident than in Nigeria, which saw itself "on the threshold of an industrial revolution",[29] the leader of independent Africa.[30] In a class of its own amongst the region's oil producers on account of its size, population and oil endowment, Nigeria proceeded to make major investments designed to create Africa's first industrialised power. Yet even during the boom years, spending far exceeded available resources. Most oil states did not

---

26 It seems that the injection of oil money continues to give birth to pompous dreams worthy of the 1970s: witness, for instance, President Obiang of Equatorial Guinea's recent enthusiasm for the building of a new capital, "Malabo 2".

27 Bienen (1988: 228).

28 Roque et al. (1991).

29 Watts (1994). The Nigerian experience with state-led industrialisation in the 1970s is unique in the region and better seen in the context of other major oil producers that adopted similar visions, like Algeria, Venezuela, Iraq and Iran. A comparison of Nigerian federal revenues of 1978 with those of a decade earlier shows a 22-fold increase. See Panter-Brick (1978: xiii).

30 In the 1970s, Nigeria provided aid for varied purposes to Guinea-Bissau, São Tomé and Príncipe, Mozambique, Mali, Senegal, Upper Volta, Chad, Mauritania, Niger, Ethiopia, Sierra Leone, Somalia, Zambia and Sudan. See Osaghae (1998: 72) and also Mayall (1976: 30).

immediately engage in borrowing sprees, and sought to deal with budget deficits by drawing on accumulated reserves or by printing money, none of which was sustainable. Gabon, which did start borrowing at this early stage, found itself in a major crisis in 1977 that led to "a sharp contraction of [GDP], which fell by an annual rate of –12.6 percent"[31] and had to call in the IMF—an early warning of what was to come. By the time of the 1979 oil price hike, though, a belief in permanently high oil prices was coupled with the willingness of foreign banks to lend,[32] and restraints were cast aside.

Social investment during the period was substantial in most states, especially if compared with the neglect of social expenditure in the subsequent phase discussed in section two. Yet this was so in quantitative rather than qualitative terms and not only failed to achieve stated goals but amounted to "an exercise in massive waste".[33] This was because development was fundamentally equated with the building of "modern" physical structures. Thinking much beyond the material dimension to include the human investment needed to man those structures, however, was scarcely a priority for governments. Schools and hospitals were built, but the human resources to staff them were always insufficient. (It is noteworthy that enthusiasm for public works had as much to do with a misguided state-building drive as with the immense opportunities for siphoning off resources that these afforded.) Despite the developmental rhetoric of the boom years, the oil state was, and remains to this day, fundamentally a machine for resource allocation, whether through actual disbursements or opportunities to make further profits.[34]

Moreover, even if one ignores the misspending, corruption and incompetence, it is not surprising that these states did not substantially improve living standards even when elites seemed to be concerned with

---

31 World Bank (1988: 1).

32 See the discussion of debt in the next section.

33 Athul Kohli reports that Nigeria's educational expenditure increased eight fold in the 1970s alone. See Kohli (2004: 323-4, 358). At the educational level, Congo's commitment during the years of single party rule was probably more substantial than in the other states of the region.

34 The schemes were the same as elsewhere—import licensing and the granting of public contracts to insiders, for instance— but the scale was wholly different.

the popular welfare. This is because though enormous in absolute terms, the money was in actuality never enough to single-handedly bring about "development",[35] although there was of course great room for considerable improvement on the basis of sound revenue management. To top it all, oil states spent substantial amounts of their export earnings on consumer subsidies, perhaps the only tangible benefit to reach the population. Predictably, petrol and other refined products were mostly sold at a fraction of their real cost, especially outside the Francophone countries. In Nigeria, this created a huge incentive for crude and refined oil smuggling, as the price in neighbouring non-oil countries was often much higher. Other heavily subsidised areas were electricity, water, and basic consumer goods such as food staples. These subsidies were exceedingly costly and ultimately detrimental.

To sum up this discussion, the statist discourse of the boom era was characterised by claims of state ubiquity in all economic sectors and most dimensions of public and private life. The state did indeed sporadically meddle with practically every conceivable domestic domain. But there were clear biases to its attention and many policy dimensions were under-emphasised or neglected with far reaching consequences for the post-boom era. Nowhere is this clearer than in the cases of agriculture and taxation.

Neglect of the countryside and the decline of agriculture

The decline of agriculture during the boom years is the direct consequence of state policies and non-policies that, through neglectful or punitive actions, ensured its virtual demise across most of the region. In her important discussion of food supply, Jane Guyer supports the view that

---

35 Although the sums are monumental in absolute terms, the oil endowments of some of the Gulf of Guinea states are not sufficient on their own "to deliver the $10,000+ GDP level of prosperity similar to Kuwait or Brunei", with the recent exception of Equatorial Guinea and perhaps also Gabon. In Nigeria in particular, the large oil rents must be seen in the context of an estimated 130 million citizens, whose per capita oil endowment is accordingly very low. See Keith Myers, "Petroleum and Poverty Reduction in Sub-Saharan Africa", paper presented at the conference on "Oil and Gas in Africa", Royal Institute of International Affairs, London, 24-25 May 2004.

African agriculture "is characterized by a particularly detrimental combination of backward techniques and predatory state policy".[36] Earlier research by Robert Bates had already placed the responsibility of agricultural decline on the shoulders of exploitative states using marketing boards to extract a surplus from the pauperised peasantry, thus promoting not just a decline in output but a rollback of producers' engagement with the formal economy.[37] But overall, most African cities have been fed from their hinterlands for most of the past century.[38] In today's Gulf of Guinea, this is no longer so. The region's under-performing agricultural sectors, together with the lack of incentives provided by the availability of oil money, have led to very low production and acute dependency upon imports. While only referring to cereals, Table 3 shows that all states have become net importers and that dependency on imports in the cases of Angola, Gabon and Congo is very high indeed.

Throughout the 1970s, export-oriented agriculture suffered a decline as steep as that of food production. This was partly due to the general deterioration of the terms of trade for agricultural commodities and the malign policies by marketing boards, but the simultaneous growth of the oil sector meant that agriculture—already the subject of ambivalence by urban elites—simply stopped being a priority. Most crops were thus relegated to the quiet neglect accorded to the rural world more generally. Gabon spearheaded this move more than three decades ago:[39] always an agricultural under-performer, it did not hesitate to earmark oil revenues for importation of produce it could conceivably

---

36 Guyer (1987: 3ff). Reflecting these shortcomings the food supply of cities in sub-Saharan Africa has never been taken for granted by colonial and postcolonial governments and has always been the subject of intense intervention. Guyer writes of "market surveillance, expansion of agricultural research, road development, extension of co-operatives, and the institution of marketing boards" as widespread policies (p. 37). In regard to the backwardness of African agriculture, Goody (1971) had already noted that it precluded even the use of a "feudal comparative framework" for assessment, as "the so-called feudal systems of Africa lacked a feudal technology".

37 Bates (1981). See also the influential World Bank (1981), commonly known as the "Berg Report" after its main author.

38 Guyer (1987: 234).

39 See Lebigre (1980), Pourtier (1989) and Gaulme (1987), among others.

find or promote internally. It also indulged its urban population with French and other foreign luxury products that were soon embraced as the locals' own. Gabon remains a caricature of food dependency, but the experience of other states in the region is that of gradual convergence with it. Angola, a major agricultural exporter until independence, has devoted little attention either to its export-oriented agriculture[40] or to food production; the government subjected both to experiments in collectivisation it neither believed in nor strenuously pursued, and then abandoned them. Equatorial Guinea's oil boom has not led to the revitalisation of the coffee and cocoa sectors or an increase in food production, but rather to importation. São Tomé's much-neglected cocoa sector seems bound for oblivion when oil rents reach the country in the next years, despite statements to the contrary.[41]

Cereals (excluding beer) Millions Tons	Angola	Cameroon	Congo Brazzaville	Gabon	Equatorial Guinea	Nigeria
Production	560190	1434717	7678	31667	—	21027064
Imports	779380	426715	275127	197369	—	3818747

Table 3. Estimates of Cereal Production and Imports, 2002[42]

Experts debate over the role of oil on the major agricultural shifts of the last decades,[43] especially that of the so-called Dutch Disease.[44] The appreciation of the local currency and the repeated unwillingness to depreciate it made imports a much cheaper option to local produce,

---

40 Angola ranked amongst the world's top five coffee exporters during the two decades before independence in 1975 and was self-sufficient in terms of food production. See Roque et al (1991) for a good study of the Angolan economy and its shift away from agriculture.

41 Frynas, Wood and Soares de Oliveira (2003).

42 Source: FAO Statistical Database (www.fao.org), accessed 24 November 2004).

43 The main Nigeria-focused contribution remains Watts (1987). The picture presented is a complex one, where broad decline in agriculture takes place in tandem with crop-specific successes and continued elite interest in land acquisition.

44 See section two of Chapter 1.

the insufficiencies of which are compounded by the structural problems I have alluded to. As Sara Berry observes, mounting food prices in the mid-1970s in Nigeria led not to an increase in agricultural production, but to the acquisition and subsequent sale of imported products, often by erstwhile farmers who, on account of inflation, could sell it at a profit in the space of weeks.[45] Although intuitively apparent, the impact of the Dutch Disease has been disputed because it is so hard to measure and because many countries experienced it differently. In the case of Cameroon, for instance, the agricultural sector actually grew in the first years of the boom;[46] in Nigeria, agriculture contracted, but manufacturing swelled owing to the huge investment of successive military governments. Angola's economy shrank and production outside the oil sector virtually ended in 1975, but this was due to civil war, foreign invasion and the departure of the colonial professional class much more than to the economic impact of oil, although it is clear that post-independence policy choices deepened these limitations. Therefore, even while recognising that Dutch Disease is present, most analysts have sought additional explanations[47] such as urbanisation and the dwindling of rural labour, the growth of the informal economy and the changing dietary requirements in the urban sphere.[48]

Food imports deepen the decline of a crucial labour-intensive sector that is also the only alternative to mass unemployment. Moreover, food imports are a drain on foreign exchange reserves, whereas a healthy export-oriented agricultural sector could have provided an additional source of revenue. There is thus a speedy path between food self-sufficiency, the onset of importation of foreign products, and their transmutation into

---

45 Berry (1993: 263).

46 Aerts *et al* (2000).

47 While Nigerian food imports rose 700 per cent between 1970 and 1980, the authors in Watts (1987) emphasise that because of unrecorded trade, self-consumption, smuggling, etc., this may not necessarily mean a real decline in production. Nonetheless, it does mean a relative decline in terms of the state's agriculturally derived fiscal revenue.

48 As a recent survey of Nigeria's political economy notes, the main imported foodstuffs have been wheat, sugar, fish, milk and rice, none of which are traditional staples. The authors observe that it "appears to reflect an inability [by national agriculture] to diversify to meet urban tastes rather than a failure to match demand for [traditional staples]". See Bevan, Collier and Gunning (1999: 43).

staples of local diet and social life.[49] Of the cases under review, only Cameroon, where food self-sufficiency is a widely shared postcolonial goal, maintains anything close to a vibrant agricultural sector; and even there, relative decline and the creeping in of imports have formed an important preoccupation of recent research.[50] In keeping with the interventionist, centralising logic of the boom days, most other states paid lip service to the idea of rekindling agriculture or attaining self-sufficiency, yet in cases such as Angola and Gabon or, more recently, Equatorial Guinea, this never amounted to much.[51] Nigeria saw expensive attempts at reversing the decline of agricultural output such as President Obasanjo's late 1970s' "Operation Feed the Nation" and President Shagari's "Green Revolution Programme" during the Second Republic.[52] All these efforts failed dismally[53] and their main impact was in the proliferation of rentier opportunities in the form of contracts (e.g. for the importation of fertilisers[54]) and donor assistance (e.g. the World Bank support for the Green Revolution Programme). In short, those states which have channelled some of the oil monies to agricultural development "simply converted a formerly independent engine of economic development into an oil-subsidized activity".[55]

The decline of agriculture happened because there was no pressure to invest in it and because such a decision would have contradicted the

---

49 This is argued persuasively in Sidney Mintz's (1985) work on the history of sugar.

50 See, for example Walle (1989), Courade (2000), and Aerts (2003).

51 In 1971, Congo saw a high-profile Operation Cassava but it too fell short of its goals.

52 The preference was for public sector intervention in ambitious agribusiness schemes, i.e. in consonance with the then-popular statist approach. Bienen (1988: 245-7) mentions the creation of, and/or massive investment in, government food companies like the National Livestock Production Company, the National Grains Production Company and the National Rootcrop Production Company as well as related projects like the River Basin Development Authorities (RBDAs) and the integrated Agricultural Development Projects (ADPs).

53 Osagahe (1998: 94) mentions that during "Operation Feed the Nation", imports actually rose and production declined.

54 See Soyinka (1996: 74-ff) for a good description of this and other notorious boom-era scams.

55 Karl (1997: 81).

modernist state aspirations of the times. Oil elites are either of urban origin or firmly entrenched in the cities and often are disinterested in things rural, while disenfranchised rural constituencies are more likely to crave town life and opportunities.[56] A reduction in agricultural output has thus proceeded in tandem with the displacement of the rural population away from the countryside and towards the cities, making agricultural decline and urbanisation mutually enforcing trends.[57] In the boom period, scores of mostly informal tertiary sector jobs proliferate[58] and oil states become a magnet not only for rural dwellers but for foreigners as well.[59] Yet the oil industry itself is capital- rather than labour-intensive, and part of the human resources it requires is highly skilled and unavailable in the local economy. It may employ a large number of workers in a project's construction phase but once that is over it needs comparatively few people to man oil production and exports.[60] The oil sector certainly creates nothing in terms of job opportunities remotely close to what manufacturing or agriculture does.[61]

The prospect of employment in the oil-fuelled modern economy was nonetheless a widespread expectation leading to very rapid urbanisation.

---

56 Indonesia is perhaps the only case where oil revenues were invested primarily in agriculture and the countryside rather than in the cities. See Booth (1992) for a good survey and Bevan, Collier and Gunning (1999) for a comparative analysis with Nigeria.

57 In cases such as Congo, this deepened a pre-existing urban bias. See Vennetier (1963) for a study of pre-oil urbanisation.

58 Berry (1993b: 276, footnote 54) cites as an example the huge motor repair industry that resulted from the flood of imported cars into Nigeria in the 1970s and the "mechanics, panel beaters, vulcanizers, battery chargers and other specialities, whose buoyancy and low barriers to entry attracted many farmers' sons".

59 These jobs are easily wiped out by an oil slump, and those previously holding them can fall into dire poverty or, if they are non-nationals, be expelled from the country. The more notorious such expulsions were in Gabon and Nigeria, but most oil states have resorted to similar measures.

60 This is the case of the Chad-Cameroon pipeline, which mobilised thousands of workers from all across Chad. Now that it is complete, the pipeline and related activities will only employ an estimated 300 people (Chad's population is about 8 million).

61 Furthermore, many oil companies in the Gulf of Guinea, used to subcontracting tasks to foreign service companies because of the lack of local alternatives, also hire foreigners to do unskilled or easily trainable jobs that could be done by natives, such as cooking, security and cleaning.

Apter is on target when he describes social perceptions of the bonanza: "oil meant money and modernity; it was revitalizing and glamorous".[62] It is not necessarily a mirage, at least not always: despite slump reversals that may upset urban-rural wage ratios to the detriment of the former, city-ward migration may at times be a rational decision.[63] Peasants seldom enjoy their condition, and the choice is clear when modern life and supposed upward mobility are contrasted with a stagnating or declining menial occupation, with few rewards at the best of times. In turn, the state's lack of interest in the rural setting provides further instigation to the exodus. Even during the boom years, the few public goods provided by the state were located in the cities.[64] Such population shifts in a labour-intensive sector such as agriculture cannot fail to have a major impact on production.[65]

A complex phenomenon such as urbanisation can seldom be accounted for by any one cause.[66] Certainly, in the case of Angola, a civil war pursued in various guises throughout 41 years must be given centrality in explaining why the capital alone accounts for a third of the country's population.[67] Paul Collier notes that city-countryside wage differentials in Nigeria were already quite substantial before the oil boom, thereby giving rural dwellers a good incentive to decamp.[68] That said, oil-related

---

62 Apter (2005: 6).

63 The point is made by both rational choice theorists and Marxist analysts, who commonly stress that the fiscal exploitation of the countryside by the city means that migration makes sense. Nonetheless, Coquery-Vidrovitch (1991: 45-7) asserts that one cannot disregard the role of "fanciful and largely erroneous ideas about urban opportunities and urban life" in bringing forth mass migration.

64 The Nigerian government's decision to provide amenities only to communities of more than 20,000 people is a good example of this urban bias (Bienen, 1988: 243).

65 Morgan (1977: 176).

66 See Coquery-Vidrovitch (1991) for a seminal review of African urbanisation. Berry (1993a) notes that the oil boom in Nigeria did give rise to exponential growth in tertiary activities that figured high in the aspirations of rural migrants.

67 Likewise the fact that the MPLA government soon lost interest in the countryside and did not make any effort at controlling it.

68 In the same work, the authors note that by the 1980s oil slump, and the massive urban pauperisation that ensued, the wage differential had eroded or even been reversed. See Bevan, Collier and Gunning (1999: 87).

urban bias and state neglect of the countryside account for unusual mass shifts of populations from rural areas to cities. A survey of Congolese politics labels the Congo a "suburb of Brazzaville"[69]—no exaggeration, as the capital and other urban areas account for an estimated 62 per cent of the population, as opposed to the Sub-Saharan African average of 34.5 per cent.[70] In the same way that the boom-era state moved capital away from productive activities to unproductive consumption, the move to the city meant a shift of the labour force away from economically productive roles.[71] This would further limit options when the oil boom turned to bust.

Taxation

In the oil state one finds two mutually enforcing fiscal trends: that of the traditional under-taxation of subsistence economies with weak state structures, and that of the fiscally aloof state reliant on externally derived rents. Direct taxation has traditionally been lax in sub-Saharan Africa and much of the developing world, the extraction of a surplus from unwilling subjects ever the nightmare of resource-hungry rulers and administrations.[72] As elsewhere, predatory behaviour has often led to short-term results, but always at the cost of the integrity or even survival of the fiscal base, as a brief exploration of concessionaires in early twentieth-century

---

69 Several articles in this special issue are relevant but see especially Achikbache and Anglade (1988: 7), which notes that the urban population of Congo grew by an astonishing 6 per cent a year between 1974 and 1985.

70 World Bank *African Development Indicators*. (2000). Gabon is in a similar situation: according to the OECD's *African Development Outlook 2003/2004*, 82 per cent of the country's population in 2001 was urban.

71 Fed primarily through imports, the petro-city evades the equation of a city being inseparable from its countryside, as commonly agreed by most students of urban history. It is not really "parasitic" of the hinterland, as Weber regarded the pre-capitalist city to be, despite sharing the latter's nonproductive and consumerist character, because the exploitation of the peasantry is not pursued by the petro-state with the vigour of agrarian states. But neither is it a "generative" city in any sense of the word.

72 There is no comprehensive history of tax or labour evasion in sub-Saharan Africa, but the point is made in numerous overviews and monographs. For a discussion of current fiscal issues see Sindzingre (2001).

AEF shows.[73] As in other colonial contexts, fiscal earnings throughout the Gulf of Guinea mostly derived from customs duties and were never enough in terms of the needs of a strong state apparatus.[74] This continued throughout the early postcolonial era and acted as a clear limit on the necessarily costly expansion of the state. With the advent of the oil boom, the emphasis shifted to oil taxation, but in both instances there was never a meaningful domestic dimension to taxation. The gradual and difficult task of constructing a tax administration was happily dropped by the state when it found what seemed to be uncomplicated prosperity in oil rents.

	Angola	Cameroon	Chad	Congo Brazzaville	Equatorial Guinea	Gabon	Nigeria
% of non-oil revenue in total government revenue	10.6	68.3	—	22.6	16.5	32.5	17.8

Table 4. Non-oil taxation per country, 2000[75]

The "enormous influence" of the fiscal element is a frequently underrated dimension of the history of states,[76] and in the context of the oil state, a key explanation for "its form".[77] Once the petro-economy was put into place, the state's administrative incapacity to tax was compounded by the creation of social classes whose interest ran counter to domestic taxation.

---

73 See section one of Chapter 4. In this regard, Gabriel Ardant (1971-2: 13; 399-436) notes in his landmark history of taxation that fiscal pillage on the part of state authorities, which he equates with the destruction of the tree "together with its fruits", is as recurrent as revolts by those subject to taxation.

74 The exception to mostly exogenous sources of state income was the surplus extracted from the peasantry by the state through means of marketing boards, which also proved unsustainable.

75 Source: Katz *et al* (2004).

76 "The fiscal history of a people is above all an essential element part of its general history": see the classic essay, J. Schumpeter, "The Crisis of the Tax State", in Schumpeter (1991: 99-140).

77 Schumpeter (1991: 108).

The avoidance of taxation of domestic groups thus became state policy.[78] As Ardant notes, the notion of tax exemption as a sign of social privilege is an integral part of the history of fiscal extraction.[79] But rather than shifting the tax burden onto the poor or the rural population—a common experience in postcolonial Africa—the oil state simply abandoned most attempts at direct taxation (see Table 4). In Nigeria, for instance, taxes were "essentially eliminated with the coming of the oil revenues".[80]

The oil state, by generalising the exemption to the totality of the population, both pursued the interests of the elite and ensured the acquiescence of the masses, if only through the "negative" liberty of tax avoidance. While the latter's active support within a circumscribed, much-reduced patronage system of the post-boom years is no longer a priority, rulers are still careful not to deploy policies that will bring about the wrath of the population, especially in the cities.[81] This is evident, for instance, in repeated Nigerian government attempts to phase out petrol subsidies that cost the state billions of dollars a year as well as feeding the smuggling of fuel to neighbouring countries where prices are up to 90 per cent higher. Invariably, these have led to the mobilisation of millions, the halting of important sectors of the economy, and the government having to backtrack, in a demonstration of something resembling taxation's "insurrectionary power".[82] The fiscal conditions that have underpinned the petro-state are thus paradoxically supported from different corners of society, for different reasons, and are difficult to modify. This holds two sets of consequences.

The first set of consequences concerns the domestic realm. The lack of taxation means that there is no reciprocity in the state-society rapport. Taxation may not automatically mean representation, as analysts who forget the complex history of fiscal extraction in the West tend to argue,[83] but it is surely a prerequisite. From an institutional perspec-

---

78 Karl (1997: 87).

79 Ardant (1971_2: 68).

80 Kohli (2004: 350-53). See also Frynas (2001: 29).

81 See section one of Chapter 3.

82 Ardant (1971-2: 401).

83 Current discussions on taxation tend to establish direct linkages between taxation, expenditure and accountability, and ignore the fact that, in early modern Europe, in-

tive, the result of non-taxation is the creation of a state apparatus that simply does not depend upon or want to pursue one of the key tasks that textbook accounts of state-building emphasise: the extraction of a surplus from the population. For the latter, one consequence of the state's oil wealth is that it is not needed. As Okruhlik notes, the citizenry is unnecessary in both Weberian (not the target for tax collection) and Marxist terms (not the subject of class exploitation).[84] Veille points out that the state is thus "released from the social formation", not needing a domestic fiscal base to underpin its financial, and therefore coercive and interventionist, potency.[85] In addition, there is the institutional impact of not having a domestic tax base. Fiscal institutions are often the scaffolding for a vast array of state institutions that rise around it: it is the will to tax that creates the need for information about citizens, territory, resources, etc., i.e. that leads a state to "map out" its surface and know its realm. Without the need to tax, the state is "blind", not only spatially, as mentioned in section three of this chapter, but also statistically, topographically, economically, etc. The "competent" institutions that are in place, such as the NOCs, exist for the management of the state's external relations; their impact on domestic governance is so limited that it does not compensate for the institutional thinness of the state.

The second set of consequences pertains to the external implications of the absence of domestic taxation. Dependence on oil rents to the detriment of taxing domestic economic activity implies that a few foreign companies account for a large share of tax revenues, and this puts immense political power in their hands. There are great contrasts between dependence on only a few taxpayers, as in the case of Congo, Equatorial Guinea and Chad, and the diversified oil landscape of Nigeria or Angola, but even in these cases one is still talking of a handful of tax-paying firms. This means that companies can extract unusually favourable terms for their investment. The actual state-firm sharing of profits depends on the bargaining power of states and is very different across the region, but

---

creased taxation led to absolutism in the short run. Nonetheless, while there may be disagreements on whether increased fiscal revenues make the state *better*, they certainly make it *stronger*.

84 Okruhlik (1999: 295).

85 Vieille (1984: 25ff).

the rule is that the Gulf of Guinea offers companies a fiscal setting that is amongst the more generous worldwide.[86] Overwhelming dependence on one source of fiscal revenue also means that whatever political order is premised on it is radically questioned when the money evaporates[87] either through a fall in oil prices or, more ominously, through the depletion of oil resources. Although some of the key producers such as Angola and Nigeria have reserves that will last for many decades, others such as Gabon and Congo have rapidly dwindling amounts of oil. Especially in the case of Gabon, whose very identity is indiscernible from the structuring role of oil since the 1960s, it is nothing less than the survival of the state as we know it that is at stake.[88]

II. GOVERNING DECLINE AND "PRIVATISATION"

Beyond the continuity of mismanagement that characterises the whole post-independence period, much has changed since the heady boom days: what the state sees as its business, where it wants to invest and act, and what (and who) it wants to relinquish. The decline of oil producers set in with the fall in oil prices and worldwide demand of the 1980s, for which their high-spending economies, nursed by the sky-high prices of the previous decade, were unprepared. When the debt crisis ensued, the state's presence across the economy, along with the heavy bureaucracy and state-owned companies and the white elephant project commitments, were the obvious target for the newly influential IFIs, which pressed for reform as the prerequisite for debt relief. The subsequent implementation of deregulatory and privatising policies and the festering economic decline meant that elements of the state's overbearing presence were rolled back. The boom-era interventionist ambitions were abandoned, and so were most attempts, though not claims, at exercising infrastruc-

---

86 As a recent IMF publication notes, "governments [in the Gulf of Guinea] collected about 50 percent of the total export value of oil [in] 2001", but this includes variation that go from "90 percent in Nigeria to 21 percent for Equatorial Guinea" (Katz *et al.*, 2004: 5). Chad's share of the oil revenue is reportedly 12.5 percent, i.e. uniquely low ("Chad and Cameroon: Country Analysis Brief", EIA, December 2003). It is likely that the state share of future oil production from São Tomé and Príncipe will be lower still.

87 Chaudhry (1997: 34-5).

88 See section two of Chapter 6.

tural power. But fiscal and executive control was not abandoned. The era of reform created not the conditions for an end to the crisis—that, despite subsequent oil price rises, continues in one form or the other till this day—and sustainable growth, but rather a reconfiguration of politics allowing the privatisation of power and economic opportunities. Rather than having been eroded wholesale, the oil state in the Gulf of Guinea seems to have retreated to a series of key sites wherefrom it can affect the arenas the elite remains interested in.[89]This is the case of one of today's more prevalent institutional survivors from the boom era, the national oil company, which I look at in some detail.

Contraction and non-reform

The economic involution of the bust period cannot be overstated. Nigerian revenues slumped from $27.4 billion in 1980 to $11 billion in 1983 and $7 billion in 1986. Between 1985 and 1987 the Gabonese budget was cut by half. The triple shock of a higher exchange rate, lower oil prices and lower production sent Cameroon into a ten-year downward spiral.[90] States in the Gulf of Guinea failed to adjust quickly enough to new circumstances for four reasons. Firstly, there was a lack of budget discipline and tools to smooth revenue flows and plan for the contingency of low oil prices. Secondly, the necessary measures were politically too costly and were therefore delayed. Thirdly, there were misplaced expectations that the oil market prices would rebound quickly, so that no reform measures were immediately taken: the next boom would pay for the current problems. Fourthly, because Gulf of Guinea states are even more dependent on oil exports than other petro-states,[91] oil price volatility disrupted them in a manner without parallel.

---

89 O'Leary, Lustick and Callaghy (2002) discusses the institutional redesign, including state contraction, necessary across the developing world for "responding effectively to destabilizing and often violence-laden conflicts". The processes described in this section are certainly an example of reduction in the "size, scope or ambitions of organisations" but their goal is hardly that of enhancing the state's "future prospects".

90 Aerts (2003). Cameroon's real GDP growth rates in 1988 and 1989 were negative, 15.7 and 11.4 per cent respectively, according to the Economist Intelligence Unit (quoted in Walle (1991: 392).

91 See Table 2 in section four of Chapter 1.

IFI structural adjustment proposals for reform were fundamentally the same in oil economies as elsewhere:[92] balance budgets, devalue the currency and liberalise exchange rate regimes, end the licensing system, terminate all subsidies to consumer goods, including fuel, and retreat as much as possible from the economy. There was some initial resistance to reform in all the oil states, though eventually only Angola evaded the SAP straightjacket. After bringing down the Second Republic, General Buhari led Nigeria for two years of reduced public expenditure and cuts in the civil service, but the naira remained grossly overvalued and discussions on reform with the IMF and major creditors stalled.[93] Gabon and Cameroon accepted SAPs but felt cushioned into non-compliance by French support, both bilateral and through the IMF.[94] Equatorial Guinea prevaricated on even mild attempts at reform and went off the radar of the IFIs since oil rents insulated it from external pressures.[95] Nonetheless, in due time, the oil producers accepted some intrusion of the IFIs and proceeded to reshape their economies. Nigeria in the Babangida years saw an attempt at local "ownership" of the reform agenda by the introduction of changes—including a 60 per cent devaluation of the naira in 1986— that were initially lauded, though they too proved disappointing.[96] Even Angola had to introduce something akin to a reform agenda

---

92 For a summary of SAPs, see section three of Chapter 1 and the bibliography on the subject provided in the footnotes.

93 Olukoshi and Abdulraheem (1985).

94 This was shared at the time by other French-speaking states in Africa. "France has often been instrumental in softening the conditionality of [the IMF and the World Bank] by interceding on behalf of African states and by providing infusions of capital into their economies at key times". See Walle (1991: 401), which credits the Franc Zone arrangements for having "delayed significant reform after the second oil crisis". Clapham (1996a: 178) provides the example of President Biya of Cameroon, who benefited from a last-minute extension of new aid facilities by the IMF and France on the eve of the 1992 presidential election against John Fru Ndi. Biya won, although electoral fraud probably played a more meaningful role.

95 Relations between the IMF and Equatorial Guinea have much improved since 2003.

96 Lewis (1996) is a good account of "the politics of attempted reform" between 1985 and 1994. The reformist bent of the Babangida presidency was revealingly abandoned during the brief 1990-91 oil boom that was caused by the Gulf War. By 1993, the budget deficit was again over 10 per cent of GDP (*Nigeria Country Profile 2003*, EIU).

to tackle the serious and recurrent economic crises it suffered from the late 1980s, although these homegrown solutions were self-serving, ineffectual, and quickly discarded.[97]

At any one time in the last twenty years, the Gulf of Guinea states have been implementing reforms, reneging on reforms, or both, sometimes simultaneously. Despite a variety of state-specific tempos and policies, the upshot of the 1980s crisis and the ensuing decay was the curtailment of the formal domain of the state.[98] As elsewhere, this encompassed the privatisation of some state property, especially after an initial period when the more statist elites feared irrevocable loss of power.[99] But its chief outcomes were the renunciation by the state of putative state obligations, the decline of public institutions, and the *de facto* abandonment of vast swathes of territory. The end of welfare provision was made easier by the fact that many of the ambitious goals of old had already foundered completely or were no longer accepted by elites as the state's own, but the overall retrenchment had to do with declining capabilities of the state. Reform thus deepened into the farming out of core state tasks to non-state entities, the use of remaining levers of the state for the pursuit of private goals, and the abandonment of public goods provision to non-state actors such as NGOs, international organisations, companies or churches. Weber termed this process "discharge" and argued that it occurred in instances when the state was unable or unwilling to pursue alleged "state" goals.[100] In such circumstances, the state relinquishes its

---

97 See Hodges (2001: 89-122) for the "limits of reform" in Angola. In particular, see pages 104-5 for a useful table that summarises the nine economic reform programmes pursued by the Angolan government from 1989 to 2000.

98 This is the general trend, which is not challenged by the numerous exceptions. In Angola, for instance, the 1990s actually saw an increase in the size of the civil service, but this did not detract from the abandonment by the state of public goods provision. As Hodges notes, this increase took place almost exclusively in Luanda.

99 This was the case with the Angolan elite. See Ennes Ferreira (1995), for the initial hesitation about, (and subsequent embracing) the privatisation agenda by the Angolan elite. Privatisation in the oil states did not in fact go as far as elsewhere, and at the present time, substantial remnants of the parastatal sector are still in state hands.

100 The clearest discussion on the subject can be found in Max Weber (1991: 85-91). As far as I am aware, the concept of "discharge" was brought into African studies and current debates in the comparative politics of privatization more generally by Béatrice

involvement in some arenas but holds on to "state authority" and the capacity to regulate the activities that remain of interest to it.

The paradoxical result of the reform effort for political authority was that, far from unravelling the centralisation of the boom years, it de-bureaucratised, furthered, and personalised the hold of the leadership over oil resources and the "real" state. Peter Lewis refers to Nigeria's transition during the reform years from "decentralized patrimonial rule" to "predatory control".[101] This led to the acceleration of plunder, afforded by late 1980s and early 1990s developments such as financial deregulation in Nigeria, the end of Marxism in Angola, and the *fin-de-royaume* atmosphere in Congo and Cameroon. As later chapters show, domestic political developments did not much affect the flow of oil.[102] The minimal local role in the fundamentals of oil exploration and production meant that decline of the host-state was immaterial for the oil sector.

National oil companies as a tool

Although the experience of decline of the oil state shares most features of the broader crisis of the state in Africa, it diverges from that of the non-oil exporters in its pattern of institutional decline. Starting with Nigeria, oil states during the boom years realised that, even if the technology was

---

Hibou. See in particular Hibou (1999). Hibou wants to think of "new modes of power and government" and look at state strategies that seem to point to retreat or decomposition but which are really part "of the continuous process of state formation, as a new way of producing the political" (Hibou, 1998: 13). While fundamentally agreeing with this notable contribution, I find that Hibou partly obscures the thrust of Weber's views. For Weber, "discharge" might be "historically banal" but it was also a "mediocre" alternative, while Hibou attacks normative views that see the "privatised state" in dysfunctional terms. But Weber himself had no compunction about using value-laden terms such as "moral degeneracy" to describe the process and saw it as the result of "the degeneration of the monetary economy" and the "absence of fiscal instruments".

101 Lewis (1996: 79-80).

102 In the case of Nigeria, where the state had taken up much more ambitious responsibilities, the impact of bad governance was felt in oil production, although not enough to discourage investors away. I am referring here to the recurrent shortfalls in NNPC's contribution to the joint ventures with the oil majors. The terrible management of the country's oil refineries, which led it to become a net importer of refined products despite being one of the world's leading producers, had no discernible impact on the upstream.

beyond their reach, the ability to "collect and interpret information that affected their negotiating position" was essential and that "a technical elite, even if very small, could play a powerful role in negotiations".[103] The oil economy necessitates institutions capable of speaking the language of high finance, business contracts and oil technology. Therefore, it had to be insulated from, initially, the incompetence of the bureaucracy and, later on, the inexorable decay of the state apparatus that is as decrepit in the oil state as in the rest of Africa. National oil companies—responsible for the managerial and regulatory demands of the oil business—not only have survived the onslaught, but also remain islands of relative bureaucratic and technical capacity.

Most states take equity interest in the oil sector, indirectly or via the setting up of a NOC.[104] Some of these are fully integrated companies present in refinery and marketing as well as in research and development of oil resources. Others that are less experienced or endowed perform the role of repositories of the state's petroleum rights but are not involved in most other aspects of the oil industry,[105] which continues to be operated by the private sector. As a result of the resources available to it, the NOC typically developed into the biggest industrial conglomerate in a national economy, as well as the government's most important tax gathering agent. Yet in many instances[106] NOCs have been pernicious economic actors.[107] As McPherson emphasises, NOCs across the world

---

103 Vernon (1976: 8).

104 For a detailed study that includes profiles of the five regional state companies, see Soares de Oliveira (2003).

105 Khan (1987: 185). See also the useful section on NOCs in Oxford Centre for Energy Studies (2004), especially the contributions by Robert Mabro and Giacomo Luciani.

106 There are, of course, exceptions. NOCs presently control a majority of world oil reserves, and their performance, in countries such as Brazil, Indonesia and Saudi Arabia, has often been positive, despite more recent neo-liberal antipathy towards the idea of the nationally owned oil player. Some, such as Malaysia's Petronas and Norway's Statoil, have developed into extremely competitive commercial entities.

107 A non-exhaustive list of shortcomings includes the following. Firstly, a near monopoly of local technical and commercial talent to the detriment of all other institutions and firms. Secondly, the prevalence of opaque accounting, lack of auditing and generally questionable financial practices. Thirdly, the frequent incapacity to provide the state's share of investment in joint ventures (when these exist) or to invest in research and

became "self-styled emporiums, states-within-states",[108] mired in corruption, inefficiency, non-commercial obligations and incompetence.[109] This said, one should not dismiss NOCs as simply "mismanaged", for the crux of their mission (political patronage and control and elite empowerment) is only secondarily of an economic nature. In this sense, there may be no contradiction between politicised, "incompetent" and even criminal activities pursued by the NOC and the regime benefits expected.

At the present time, five of the seven regional oil exporters have a NOC.[110] They are Sonangol (Angola), NNPC (Nigeria), both in a class of their own, and the smaller SNPC (Congo-Brazzaville), SNH (Cam-

---

development, in the case of autarkic NOCs. Fourthly, the deleterious impact on the sector—especially but not exclusively in distribution—due to an absence of competition, corruption, and recurrent smuggling and shortages. Lastly, the fundamental contradiction between regulatory and commercial roles, on the one hand, and commercial and non-commercial tasks, on the other. For a helpful, brief analysis, see McPherson (2002).

108 McPherson (2002: 4-6).

109 As a recent anatomy of the Mexican NOC, the world's fifth largest oil company and worth about $50 billion a year, put it, PEMEX is "a cash machine for the government, a slush fund for politicians and a patronage mill for party loyalists". It is, in the words of an administrator, not really a company, not really a ministry, but "something weird". Indeed, from the conventional viewpoint, it is a textbook failure with losses of about $1 billion to corruption *per year*, millions of stolen gallons of oil, thousands of ghost jobs and contracts, widespread corruption and the political and physical intimidation of would-be reformers. See Tim Weiner, "Mexico's corrupt oil lifeline", *New York Times,* 21 January 2003.

110 Gabon had a NOC, Petrogab, until the late 1980s, by which time it was dissolved following bankruptcy, its responsibilities reverting to the relevant ministry. Petrogab had been created in 1979 "as a vehicle for the government's equity holding in the local affiliates of foreign oil companies as well as the state's interests in new production sharing agreements [...] In addition, Petrogab was granted the authority to sell directly on the market Gabon's share of oil". See *West Africa*, 30 August 1982. In Gabon, a NOC is not essential because many of its functions are fulfilled by Elf Gabon, which, differently from Elf subsidiaries in Angola or Congo is a joint venture in which the Gabonese state holds 25 per cent and Gabonese "entrepreneurs" (all close to the presidency) hold 17 per cent. As discussed in section two of Chapter 4, Elf Gabon has had a business empire that goes far beyond the ambit of an oil firm for many decades.

eroon)[111] and Gepetrole (Equatorial Guinea). Their core mission includes the roles of sector regulator, revenue collector and sector monitor. Although NNPC and Sonangol are incipiently involved in upstream activities, the five NOCs are primarily supervisory bodies, with some presence in refining and distribution but no meaningful engagement either in exploration or production. Most of the available evidence questions the way the NOCs are managed, even when bearing in mind significant differences between, say, the NNPC maze and the more tightly run Sonangol.[112] In particular, the disparity between presupposed oil revenues and actual amounts reaching the state budget hints at massive leakage in the middle passage. Until very recently, none published internationally acceptable accounts and none was independently audited. Nonetheless, it is fallacious to assess them as primarily commercial entities. The character of the NOC as a "highly political body"[113] stems from three key tasks it performs in addition to the aforementioned oversight of the oil sector. Firstly, the handling of the government's share of oil production, which it sells onto the world market, directly or through oil traders; secondly, the conclusion of joint ventures with international players in oil and natural gas production, oil sector services, and a host of other tertiary activities in insurance, telecommunications, logistics, etc., with advantageous results for power-holders; and thirdly, the use of physical ownership of oil for the contraction of oil-backed loans.

Angola's Sonangol[114] has provided other oil-producers in the region with a model of commercial and political accomplishment.[115] Founded in 1976, the company immediately benefited from pragmatic partnerships

---

111   The best analysis of the Société Nationale des Hydrocarbures remains the two-volume Gaffney, Cline and Associates (2000).

112   A 1995 World Bank report, quoted in McPherson (2002: 4) estimated that annual NOC losses in the sub-region were in excess of $1.4 billion, roughly the same as the Bank's lending activities in the whole of Sub-Saharan Africa.

113   Congo Brazzaville Country Profile 2003, EIU, p. 32.

114   *Sociedade Nacional de Combustíveis de Angola*: http://www.sonangol.co.ao.

115   Interviews with Chevron and Total officials in Luanda in March 2002 were consistent in their portrait of Sonangol as a trustworthy and mature business partner, with some of the same faces presiding over investor relations for twenty years. The paradox that a country undergoing a civil war could be such a reliable investment location was often underlined by the interviewees.

with actors as diverse as Nigeria's NNPC, Algeria's Sonatrach, and the Arthur D. Little consulting firm. By the early 1990s, when the MPLA government shed the remnants of Marxism, the company was in an ideal position to ride high in the new capitalist wave—a move aided by its restructuring as a holding company with autonomous subsidiaries. In addition to its regulatory functions, Sonangol is involved in oil product distribution, oil industry support services, banks, the Luanda refinery, and many other areas of the economy, such as insurance and shipping. In the last five years, it has aggressively pursued a policy of joint ventures in oil services with major international players, a trend perceived to diminish competition and increase costs as Sonangol tends to oblige foreign firms to use these rather than possible alternatives.[116] Sonangol has crowded out all types of entrepreneurs (bar those connected with the presidential inner circle) at the domestic level by investing in lucrative businesses however unrelated to its main thrust. It also provides a vast array of benefits to insiders, including extensive social and welfare benefits, educational opportunities,[117] scholarship programmes, and well-endowed literary prizes. Furthermore, it is a disciplined and politically pliant NOC: similarly to the central bank and the oil ministry, Sonangol is run by presidential loyalists, with managerial positions jealously kept away from the more critical wings of the ruling party.[118]

---

116 A Western oil official is quoted as fearing a "decentralization of corruption" with the emergence of the new joint ventures—the likely hideout for opaque deals if NOC transparency ever materializes. See N. Shaxson, "Angola's Sonangol moves onto a peacetime footing", *African Energy*, 60 March 2003. There are now about 60 such joint ventures.

117 This includes Indiana University's "Sonangol Training Program", which has provided English language skills to Angola's chosen teenagers "until they are placed in undergraduate programs at US universities" (http://www.indiana.edu~ird/cieda/sonangol.htm, accessed 4 November 2004).

118 Examples are Joaquim David, Desidério Costa, and Manuel Vicente, the current head. Two articles provide the best analysis of the firm's dominant personalities: "Sonagol's New Masters", *Africa Energy Intelligence*, 5 February 2003, and "Uncovering Sonangol's Secrets", *Africa Energy Intelligence*, 30 January 2002. A counter-example is the presence in the company of some of Luanda's critical intellectuals such as Filomeno Vieira Lopes.

Sonangol is charged with selling Angola's share of petroleum production, which amounts to 40 per cent of the total.[119] The company is the centrepiece of what has aptly been called "the Bermuda triangle": the black hole of Angolan finances defined by the triangulation of Sonangol, National Bank and Finance Ministry responsibilities,[120] in a nexus of huge parallel financing involving large-scale diversion of resources. Major discrepancies between reported government revenues and those initially received by Sonangol have been uncovered,[121] as well as a number of widespread tactics for siphoning off revenue.[122] As revealed by the Angolagate trials in France, the company also played a prominent role in the secret procurement of weapons throughout the 1990s. *Africa Confidential* is on the mark when stating that Sonangol "is not a normal company in any sense of the word".[123] But in spite of its awkward practices, the company is impressively well networked,[124] taken seriously by business and an important instrument of Angolan ambitions in the region and elsewhere.[125]

---

119 Following sustained problems with oil traders in the early 1980s, and in the footsteps of its Algerian counterpart, Sonangol established London, Houston and, more recently, Singapore operations for that purpose.

120 See the two influential report by Global Witness (1999) and (2002).

121 "Angola: 'Slanderous' Report", *The Guardian*, 4 December 2002.

122 These included delaying the transfer of receipts to the government in order to take advantage of local high inflation or the perennial underestimation of oil prices in the state budget so that the additional revenue can be pocketed. See Angola Country Report, EIU, May 2002.

123 "An Edited Peace", *Africa Confidential,* 19 April 2002.

124 Sonangol invests heavily in lobbying. A recent Washington lobbying contract is with Patton Boggs.

125 Senior employees have been dispatched to São Tomé and Príncipe (an old ally), Equatorial Guinea, where a new Gepetrole/Sonangol joint venture has been recently struck, Congo-Brazzaville, Gabon, and refining interests in the Ivory Coast and elsewhere. Further to this, Sonangol is active as a distributor in STP, Cape Verde, the the DRC and Portugal. Sonangol has played a key role in the ongoing restructuring of national oil sectors in some of these countries and its involvement is likely to deepen. More recently, Sonangol has established a partnership with the Chinese government through a Hong Kong based subsidiary, China Sonangol International Holding, in a $5 billion bid to invest in oil exploration in Argentina, another example of the closer relations between

The case of Nigeria's NNPC, which is by far the biggest and most chaotic of all the NOCs, is much more complicated.[126] Created in 1977,[127] NNPC is the sector regulator, tax-gathering agent and the vehicle for joint ventures with the multinational oil companies.[128] It is also a mess: a conventional business analysis would effortlessly rank NNPC as one of the world's top mismanaged firms. It has failed to keep up with its joint venture obligations, maintain a fully functioning downstream sector, and account for billions of dollars on a yearly basis. More recently, with Nigeria's economic diversification into international fraud, NNPC has figured prominently in a string of "419" scams that are evidently run or abetted by senior employees of the company.[129] The role of NNPC, and that of other prominent institutions such as the army, in organised smuggling of crude and refined oil, which amounts to an estimated 20 per cent of the country's daily production, is also the subject of public speculation.[130] NNPC has been a key battleground of Nigerian politics, which results in no continuity of policies or leadership.[131] This said, if one looks at NNPC with less orthodox expectations, the picture is different.

China and Gulf of Guinea oil producers. See "Angola will be part of a Chinese-Argentine exploration deal", *Miami Herald,* 23 November 2004.

126 There is no space to discuss NNPC in depth here. See Soares de Oliveira (2003: 16-19) for a discussion of NNPC and a bibliography on the political economy of the company.

127 NNPC was the result of a merger between the Nigerian National Oil Company, created in 1971, and the then Ministry of Petroleum Resources, which has been recreated since. On the NNOC background see Turner (1980).

128 In addition to the majority stakes in the oil-producing consortiums with foreign companies, NNPC's activities include some oil exploration, production and refining, pipelines and storage terminals, marketing of oil, gas and refined products and petrochemicals. It also holds minority interests in companies engaged in various activities, including liquefied natural gas production and exportation (NLNG, a joint venture with Shell) and the wholesale and retail marketing of refined petroleum products. It performs badly at all these, especially downstream. See, for instance, World Bank (1993).

129 See Apter (1999: 277ff) and Hibou (1999: 69-113).

130 See Kalu Agbai, "The Aso Rock Oil Bunkering Mess", *Insider Weekly,* 24 November 2003 for an indictment of the highest reaches of Nigerian politics, including Vice-President Abubakar, in the theft of oil.

131 According to Osagahe (1998: 14), the company had eight CEOs between 1985 and 1996.

Employing an estimated 17,000 people, NNPC remains the leading actor in the economy. Although the post-1999 Obasanjo Administration has moved to curb, or at least centralise, the leakage potential of NNPC, it has not sought to fundamentally upset it: the goal is to turn it to its advantage rather than reform its ways.

Upstarts in the oil sector unsurprisingly want to emulate smooth operators such as Sonangol. Despite claims that they want to become technically able companies, it is the opportunities that the NOC allows, such as loans and joint ventures, that really move the newcomers. In the wake of his 1997 military victory, President Sassou created the SNPC[132] to replace the former NOC, Hydro-Congo. Besides selling the state's share of production through a London-based operation modelled on Sonangol's, SNPC plays a pivotal role in the government's strategy of contracting oil-backed loans.[133] Equatorial Guinea has also created a NOC, Gepetrole, which compensates for its lack of human resources by hiring Western consultants,[134] who have been carefully nurturing the company's public image through the diffusion of glossy brochures, CD-ROMs, a website[135] and a steady presence in international business fairs. These consultants, together with Lebanese affiliates of the regime and Angolan experts, have also promoted the setting up of a London branch to sell Equatorial Guinea's share of production on terms similar to those of Sonangol and

---

132 Societé Nationale des Pétroles du Congo.

133 See IMF (2003) for an extended analysis of oil in the Congolese economy that pays great attention to SNPC developments. The report notes with some concern that "while other state-owned enterprises are being privatized, the SNPC is expanding" (IMF, 2003: 23).

134 These include the Exploration Consultant Group, which had already organised Equatorial Guinea's 1998 bidding round for the country's offshore blocks (and is actually listed in the company's brochure as a key UK contact) and the Anglo-Norwegian company InSeis Terra. An oil marketing firm, Addax, which has won several concessions despite lack of experience in research and development, is also seen as close to the regime (*Equatorial Guinea Country Profile* 2003, EIU, p. 36). See also "Now a Power in own Right", *Africa Energy Intelligence,* 18 June 2003.

135 Gepetrol's website is http://www.equatorialoil-oil.com/pages/GEPetrol%20 page.htm (accessed 20 November 2004).

SNPC.[136] Most recently, Chad has announced the intention of creating a "national oil company", which is also likely to take the form of an oil trading and loan-contracting agency. The lessons of Sonangol are spreading, and the need to preserve or create the institutional capacity needed to enhance and better capture resource flows has never been greater.

NOCs are the "prime interlocutors for oil firms, traders and bankers".[137] They are the key policy makers, planners, tax collectors and negotiators in oil for the government, as well as the regulatory entity for the oil sector. They are the partners in joint ventures with international firms, the providers of politically sensitive, non-commercial services and goods, as well as significant local "entrepreneurial" forces. The fact that these five companies work in unconventional ways is not in dispute—nor is the amount of incompetence, inexperience, waste and bad timing that, ubiquitous here or occasional there, are features of all. It is important not to romanticise their capacity: most NOCs cannot even calculate if costs are over-reported by companies or whether governments are being paid what they should.[138] But whereas NOC capacity may often be

	Production	Exploration	Refinery	Involvement in non-oil economy	Oil-backed Loans	Oil Trading
Sonangol	Yes	Yes	Yes	Yes	Yes	Yes
SNH	No	No	Yes	Yes	Yes	Yes
Gepetrol	No	No	No	Yes	No	Yes
NNPC	Yes	Yes	Yes	Yes	No	Yes
SNPC	No	No	Yes	Yes	Yes	Yes

Table 5. Present and past activities of national oil companies in the Gulf of Guinea

---

136 The London office was reportedly established with the aid of oil traders Glencore and Stag. See "Building up Steam", *Africa Energy Intelligence,* 19 November 2002, "An Affiliate in London?" *Africa Energy Intelligence,* 24 September 2002.

137 "National Oil Companies", *Africa Energy Intelligence,* 29 March 2003.

138 Presentation by Menachen Katz, Assistant Director, African Department, IMF, Workshop on Petroleum Revenue Management, The World Bank Group, Washington D.C., 23-24 October 2002.

exaggerated in absolute terms, "it is probably correct in relative terms".[139] Despite notable flaws, they are essential to keep the downsized state and its end of the oil partnership running. Theirs is a strange marriage of the latest financial expertise and oil market savvy-ness with the narrow enrichment goals of a failed state leadership, and it works.

The path of the NOC is somewhat dissimilar from the vision of "indirect private government" that stresses the introduction of arbitrariness, the absence of clear boundaries and "constant re-negotiation" as features of the weak, privatised state.[140] Conversely, the NOC strives to appear predictable and competent in the eyes of oil multinationals, even when it does not manage to be so. As the state withered away, "the state" retreated into the NOC, the last refuge for educated, technically able personnel working in partnership with international companies and consultants. But while this takes place under the firm hand of the presidency, a careful distance between the NOC and the state is cultivated, lest the final implosion of the state take the NOC down with it. In situations where hostile creditors have tried to impound the assets of the NOC as repayments of state debts, these have argued in court that they are an entity different from the state. SNPC has taken this to an extreme, by de-linking its UK operation (the valuable part of SNPC) from the rest of the company.[141] While this strategy is not always successful, the fact that Sonangol's credit rating is better than that of the Angolan state shows that the NOC can, on many dimensions, lift itself from the smouldering ruins of the oil state.

Debt as opportunity

Perhaps the most crippling and lasting element of the post-boom crisis is debt, brought about by declining revenues and a rise of interest rates

---

139 McPherson (2002: 6).

140 Hibou (1998: 24-ff).

141 The now-defunct SNPC UK was owned by two entities, Excellet Investment Ltd and Quickness Ltd, which was owned by SNPC but managed by the Eversheds law firm. This meant that the London branch was not a direct affiliate of SNPC. See "Congo-B: The Real Oil Bosses", *Africa Energy Intelligence,* 13 March 2002. This did not convince French courts, which (following complaints by creditors) declared SNPC UK to be an asset of the Congolese state and, therefore, "freeze-able", thus forcing the closing of its London trading operation in 2005.

that caused arrears to accumulate. For the past two decades, with the partial exception of Equatorial Guinea, all the states in the region were heavily indebted.[142] Some, like Congo, were amongst the most indebted in the world. As an extensive body of literature on the debt crisis in Africa and elsewhere shows, this has obvious implications for the types of economic policy that can be pursued. In the Gulf of Guinea, it guaranteed that outmoded big-state thinking decreased and that the recurrent expenditure to cover previous infrastructure commitments could not be fulfilled. Yet, crucially, the debt of oil states did not lead to the total withdrawal of international lending which debt burdens of a similar caliber have encompassed elsewhere, despite the deterioration of relations with Paris Club creditors and the closing of "conventional" doors at normal lending rates. Even in such desperate situations, the oil state possesses a source of great credibility that other states do not. This is the state's present and future share of oil production (in most instances, owned and sold by the NOC) to which it has resorted when it no longer has access to international capital markets, i.e., very frequently.

Loans contracted on the basis of oil production, or oil-backed loans, are invariably defined by the IMF as unfavorable to the borrower because of their short maturity and interest rates that are often extortionate. The approach of the Fund, while correct in its assessment of the burden of oil-backed loans for public finances and the public interest, neglects the fact that they also amount to an important source of cash for the ruling clique. In effect, there is a straightforward divorce between public and private interests, with "the state" shouldering the weight of irresponsible contraction of debt while the elite basks in the rewards of easy money.[143] What is irrational from a standard economic viewpoint makes a lot of sense when seen from that of empowered state actors.

---

142 The timings were different: Cameroon's debt mounted at around 1986, while Congo's and Nigeria's date from the early 1980s. Nigeria quickly went from a small debtor to an enormous one because of a borrowing spree by the Second Republic at the time of the second oil price shock. Gabon found itself bankrupt and in need of IMF intervention as early as 1977 as the result of a particularly ostentatious OAU meeting in Libreville. Angola's debt was small until about 1986, when the government abandoned its borrowing restraints, and mounted exponentially in the 1990s.
143 Vallée (1999: 56).

Equatorial Guinea, for instance, borrowed money from oil companies at very high interest rates in the first years of its oil bonanza.[144] And though it is now awash in dollars it can barely spend, it has more than once flirted with the idea of an oil-backed loan.[145]

	Angola	Cameroon	Chad	Congo Brazzaville	Equatorial Guinea	Gabon	Nigeria
Debt burden as % of GNP (year)	22.1 (2001)	108.3 (1999)	5.5 (1999)	219 (2000)	49.7 (2000)	4.2 (2000)	93 (1999)

Table 6. External Public Debt[146]

That is why some form of oil mortgaging has been pursued by all states under review, although it is Angola—with plenty of wartime experience with oil-for-arms deals—and Congo-Brazzaville that have focused the recent attention of IFIs because oil-backed loans constituted a disproportionately large share of their total debt.[147] Nigeria started the trend in the early 1980s when the Buhari regime got a $1 billion loan from BCCI in exchange for oil.[148] In 1992, Cameroon got a $180 million loan from the Canadian Imperial Bank of Commerce through its national oil

---

144 See Shaxson (2007), especially chapter 2.

145 See "Equatorial Guinea to join oil-backed loan bonanza", *African Energy* 68, November 2003 for details on the loan, which was abandoned by the government at the last minute, probably for internal political reasons. The loan, which was being arranged by Deutsche Bank, was reportedly to pay for Equatorial Guinea's equity participation in Marathon Oil's LNG project. See "Malabo backs away from $400m pre-export deal amid claims about clan conflict", *African Energy* 73, April 2004.

146 Source: Economist Intelligence Unit.

147 Angola is the most enthusiastic at using its oil as collateral. In 2004 alone, it negotiated two separate loans: a $2 billion loan from China and a $2.5 billion loan from a consortium of UK banks headed by Standard Chartered. See N. Shaxson and David White, "Angola's oil group seeks dollars 2.5 bn syndicated loan", *Financial Times,* 21 April 2004 and "Sonangol's secret plans", *African Energy Intelligence,* 25 February 2004. Congo contracted such loans from both commercial banks and the oil companies for about two decades accounting for around half the country's total debt, which is thus unusually hard to manage.

148 Graf (1988).

company.[149] In the intervening years, oil-backed loans have grown in sophistication and size, as they became the lifelines for a number of states.
These loans mostly take place through "special purpose vehicles", which
are financial tools that allow lenders to benefit from very safe repayment
structures through oil price protection, debt service reserve accounts
and an accelerated repayment mechanism. They are also syndicated (i.e.,
shared by a number of banks), which further decreases risk for the creditor, and their short maturity of three to five years as well as inordinately
high interest rates mean that there is never a lack of lending interest.[150]
This appetite "for asset-backed lending by [countries] with a poor governance record but big oil and mineral resources",[151] explains why, in
addition to commercial banks, oil traders, multinational firms, and other
states (through the creation of oil-backed credit lines) are keen to get
involved. In addition, private interests have acquired public debts, as in
the case of Angola's debt to Russia.[152] This amounts to both the privatization of debt[153] and the construction of a politically appealing, if financially
dire, avenue of autonomy from the demands of the IFIs.

The current debt burden of Gulf of Guinea oil states can thus be seen
as the result of a short-term policy choice as much as a structural con-
straint. Contrary to non-oil states such as Guinea-Bissau, whose debt is
virtually impossible to pay, some of the oil states have the cash flow to

---

149 "Enquête Cameroun: les secrets de l'or noir", *Lettre du continent* 362, 19 October
2000.
150 As the precedent of South Africa's apartheid state shows, there is no contradiction: it was precisely the status of nominal international pariah that made the latter such
an attractive business partner, especially through the inevitable contraction of unusually
high-interest loans. See Nicol del Innocenti and John Reed, "Apartheid lawsuit targets 34
companies", *Financial Times,* 19 May 2003.
151 "Angola sticks with oil-backed loans, but promises fiscal probity", *African Energy* 74,
May 2004.
152 According to Swiss investigators, Angolan politicians personally benefited from the
payment of Angola's debt to Russia, which was handled by Pierre Falcone. See Global
Witness (2004: 36-52), Global Witness (2002) and Verschave (2001: 117-76).
153 Because they were never very attractive to bankers, most of the debt of most African
countries is public debt owed to bilateral donors and international financial institutions.
Although this is true of the oil states as well, a growing part of their debt is contracted
from the private sector.

cope with their debts, if not erase them. However, a proactive approach to debt reduction may presuppose some of the "good governance" of an IFI straightjacket and an end to opaque money deals. This is in the interest of the state as an abstract entity and of the population as a concrete entity, but hardly of those who hold the levers of the state. States with little leeway such as Cameroon may have been cornered into accepting some form of engagement with the IFIs.[154] But those that can avoid such course of action have found it more fruitful to "stumble from loan to loan".[155]

The logic is not necessarily different when oil states experience a boom, as is again the case since 2003. Despite unheard-of revenues, Angola preferred to contract numerous oil-backed loans with western banks and China rather than sorting out its debt in a sustained way. Exceptionally, reformers may be able to carry the day and pay off outstanding debts. This was the case with Nigeria's Finance Minister in 2005 and her sorting out of an estimated $30 billion debt to the Paris Club through the payment of $12.5 billion—an astonishing deal than earned Nigeria an improbable investment grade from credit-rating agencies for the first time in two decades.[156] Minister Okonjo-Iweala's feat, however, was received with great anger by Nigerian legislators and state officials who had made claims on the windfall revenues;[157] a year later, she was no longer in government, and the pre-election frenzy bode ill for any sort of budgetary constraints on oil-fuelled spending and borrowing. From the perspective of many empowered actors in Nigeria, the goal of ending the country's debt burden clearly paled in comparison with access to ready cash. Other states in the region are finding out that, more often than not, such painful choices do not even have to be made anymore. Congo, for instance, saw an important part of its debt to IFIs forgiven with only the

---

154 Cameroon's ldeclining oil sector opportunities, together with the declining importance of oil revenues as source of patronage, have created the context for significant efforts at reform. Paradoxically, the chances for success are closely intertwined with the growing irrelevance of the sector under reform.

155 N. Shaxson, "Angola: the New Frontier", *Business in Africa*, May 2000.

156 See the discussion on reformers in section two of Chapter 3.

157 Several author interviews, Abuja, March 2005.

flimsiest commitment to reform and in the face of numerous oil-related scandals.[158]

The end of public goods

Success for the few during the years of decline could thus continue to be pursued through the instruments left to the state and allowed by oil, and the lingering rewards remain impressive. But what of the population? The era of privatisation, I suggested above, brought about a pragmatic re-adjustment with the state now seeking "less the control of the population and more that of the sources of wealth (and the possibilities of wealth)".[159] Oil money has stopped circulating fully through the economy: the links between the state and its close circle of immediate beneficiaries, on the one hand, and the majority of the population, on the other, have been severed. Within a continuum that verges from Gabon's dwindling pa-tronage to Angola's free-for-all, all states have to a large extent aban-doned their people, ideologically driven projects of social improvement, and attempts at buying off broad-based legitimacy by fulfilling some of the needs of the majority.[160] The results can be gauged from the tragic human indicators of the Gulf of Guinea. "In general", the IMF notes, "oil-producing countries in Africa have not achieved better social indica-tors than other Africa countries",[161] a staggering outcome after, in some cases, four decades of oil revenues. For the state, people no longer count in most circumstances and in most locations. Public goods are no longer

---

158 See section one of Chapter 6, especially footnotes 112 and 113.

159 Hibou (1998: 56).

160 This affects *everything*, including the provision of water and electricity in urban areas where the state is supposedly still "in business". The Nigerian electrical provider NEPA is an infamous case, as documented in Olukoju (2004) and also World Bank (1993: 1-12). In Angola, the World Bank noted that "the Empresa Provincial de Água de Luanda [EPAL, the water provider] is unable to perform the normal functions of a company, such as providing regular maintenance, issuing bills, making connections, and maintaining ac-counts and records" (World Bank, 1999).

161 See Katz *et al* (2004). Bienen (1988: 228) notes that Nigeria is the only case in a series of petro-state case-studies in the same volume (none of the others are in Sub-Saha-ran Africa) where the standard of living was lower at the time of writing than in pre-oil boom days.

a right, but a privilege bestowed by private entities, often on a capricious, rapidly subtracted basis. The state may be weak in many aspects, but it is remarkably autonomous from popular pressure outside the elite and a number of urban constituencies still able to exercise leverage.[162] Mostly the population has been deemed expendable, and no resources or attention are lavished, even fruitlessly, on it. While the recent oil boom seems to have resurrected elements of the Gulf of Guinea love affair with prestige buildings and needless infrastructure, it has not led to any re-assumption of old state responsibilities towards the population.

As often as they could, governments delegated to other parties the fulfilment of popular needs; those not picked up by a non-state actor simply went unfulfilled. States are more than glad to have oil companies, churches, civil society groups or the UN—all intrinsic to the present governance of the oil state—perform tasks that are well beyond the moral sphere of intervention, and the practical means, of the state elite. Across the Gulf of Guinea, NGOs permanently fulfil basic tasks in public health and food security, while presidential "private foundations" provide services or handouts to selected constituencies.[163] In Angola, social spending is a fraction of state spending on defence or luxurious consumption. The rural setting and its inhabitants in particular have been utterly forgotten,

---

162 The size of this politically influential part of the population is constantly shrinking and its character changing: previously protected groups such as the bureaucracy have suffered considerably from the economic decline of the last two decades, occasionally through redundancies and most often through steep falls in real income. The benefits of being a low strata insider are less tangible by the day, even where the number of civil servants has grown in absolute numbers over the past decades, as in the case of Angola.

163 For the Angolan case see Messiant (1999). The *Fundação Eduardo dos Santos* website can be found at at http://www.fesa.og.ao/. FESA has recently been joined by a Fundação Angolana de Solidariedade Nacional that works in similar terms. The idea has been picked up in Congo as well, with a "Fondation SNPC" one of the pet projects of SNPC director Denis Okana (various interviews, Congo-Brazzaville, 6-13 November 2005). See also "Une fondation pétrolière Sassou?", *La Lettre du Continent* 447, 20 May 2004.

ejected, as it were, from the realm of state obligations.[164] Everywhere epidemics are expected to be fought with external resources.[165]

Although in most locations other foreign actors have taken the brunt of this process of state divestment of social responsibilities, it has also thrown a growing number of state functions into the arms of oil firms ill prepared to conduct them. Oil companies in the Niger Delta and Cabinda, to their annoyance, are called upon to satisfy the needs of populations abandoned by the central government and, in the case of Nigeria, the victims of neglect by rivals who have captured local government. They must do this while also providing for the security and infrastructure which they need and which are no longer proffered by the state. Companies are not actually replacing the state; to assert this would assume a human and financial commitment that is simply not taking place.[166] But perhaps the most important consequence of state failure in oil-producing regions is not the actual swelling involvement of companies, but the shifting of popular expectations away from the state towards the private sector. This occurrence is inevitable, for it is not only at the level of specific tasks that the state has retreated: it is also at the level of whole swathes of the "national" territory, where its presence is no longer felt.

III. GOVERNING SPACE AND VIOLENCE

The notion of the bounded, exclusionary and directly administered territory may be at the heart of the modern state but, as John Ruggie writes,

---

164 Two years after the end of the Angolan civil war, the World Food Programme was scrambling for funds to feed 1.4 million people on a regular basis. See e.g. "UN Agency Appeals for Urgent Aid to Feed 1.4 Million Hungry Angolans", *IRIN*, 10 August 2004. By 2006, an astonishing 98 percent of Angola's medical doctors were in Luanda. See "Angola: Neglected Provinces Need Share of New Wealth", *IRIN*, 27 June 2006.

165 See e.g. Sharon LaFraniere, "In oil-rich Angola, cholera preys upon the poorest", *New York Times*, 16 June 2006.

166 The possible exception is the important, if ineffectual, oil company engagement across the Niger Delta discussed in section three of Chapter 5. Civil strife in the Delta over the past decade has resulted in companies, and especially Shell, making some meaningful efforts at local development or, at least, at the co-optation of potentially belligerent locals. It has not worked.

"it appears unique in human history".[167] Most past systems of rule have lacked an element of territoriality, perceived it in non-monopoly terms or approached it in ways that presently seem unorthodox. There was an earlier, and brief, moment during which the Gulf of Guinea oil state attempted more or less conventional approaches to statehood, including territorial coverage. But today's oil state presents one such historically unexceptional context in which, for lack of bureaucratic capacity, political willingness and economic imperative, only a fraction of the nominal territory counts—that enclosing the centres of decision-making and production of oil.[168] The rest, though claimed, is expendable and often depopulated by war or economic migration. This geographically uneven pattern of state presence, which Catherine Boone has labelled "political topographies",[169] is shared, to a degree, by many states in Africa that have

---

167 Ruggie (1993: 149-52).

168 I do not assume a previous golden era of absolute territorial coverage. The examples of Chad, which never effectively controlled its territory, or Angola, which spent twenty-seven years enmeshed in an extremely violent contest for control of the state, should caution against such reading. But the early post-independence years were nonetheless characterised by attempts at constructing formal ("Weberian") state structures that have by and by been abandoned. This brief striving for territorial hegemony by the early post-colonial state and its late colonial predecessor was the historical exception to a *laissez-faire* attitude towards peripheral territories also shared by early colonial administrators.

169 Boone (2003). Similarly to this chapter, Boone argues that "some zones were governed intensely, through tight, top-down power, while others were left to their own devices, granted extensive autonomy, or simply neglected and not incorporated into the national space" (2003: 8). Yet she claims that it is "local-level configurations of power and interest" rather than "exogenous determinants" that primarily account for the differing political topographies of the state (2003: 10). This may be so with the agrarian states she discusses but not in the context that concerns me here. There is no doubt that the specificity and nuances of spatial relations in the oil state are also locally embedded: domestic and local politics are not meaningless in attitudes towards different regions. (An example is the way villages where the president was born get put on the national map, regardless of whether they are in the anonymous hinterland, as in the cases of Sassou Nguesso or Bongo.) But ultimately, oil seems to be the overriding explanatory factor for the comparative neglect or importance of a given region. In an attempt to rediscover rural politics one must not overstate the impact of the local on the national state; Boone forgets that some local actors matter more than others and that some matter not at all. In the contemporary politics of the Gulf of Guinea oil states, at least, there is scarcely a

neither the means nor the will to administer the totality of "their" territory in the interventionist manner attempted by most only decades ago.
According to Jeffrey Herbst, the broadcasting of power over sparsely
settled lands with only limited available means is the defining question
African rulers have faced for several centuries, and he argues that they
have come to remarkably similar conclusions across space and time.[170]
Herbst finds that in pre-colonial Africa, control of territory was always
secondary to the control of people and that:

> cost calculations led leaders to formally control only a political core that might
> be a small percentage of the territory over which they had at least some claim,
> because the cost of extending formal authority was very high.[171]

Arguably, a similar process is again in motion in response to the decay
of state "technologies of reach",[172] the worldwide de-linking of power
and territory, and the concurrent humbling of the state brought about by
globalisation.[173] Important features of this pattern are, however, unique
to the political trajectories of states in the Gulf of Guinea. They include,
firstly, the long-term structuring role of the private sector in the governance of the strategic areas and, secondly, the irrelevance of people for
the calculus of state power. The paradoxical modality through which this
approach to territory is enacted is twofold, best thought out as a continuum. At one end, there is a process of territorialisation of power around
"useful spaces".[174] These are restricted areas such as the capital—the ter

---

"battle for the control of the countryside" (R. Rathbone, *Nkrumah and the Chiefs* (Oxford:
James Currey, 2000), p. 161, quoted in Boone, 2003: 8).

170 Herbst, (2000:55)..

171 Herbst (2000: 55). Mbembe (2002: 70) also writes that in the pre-colonial era,
territory was neither "the exclusive underpinning of political communities" nor "the sole
mark of sovereignty", and that "space was represented and used in many ways".

172 This is the expression used by Kopytoff (1987: 29) to describe the "available material and administrative technology of political control"; its absence is said to limit "the
political penetration that the center could achieve both in geographical extent and, locally, in depth".

173 See, for instance, Badie (1995).

174 Nigeria is somewhat different here owing to its federal system (all other states in the
region are centrally governed, with limited or non-existent regional autonomy). Minority allocation and representation pressures have led to the creation of 36 federal states
that encompass the totality of the territory, and pressures from aggrieved groups for

minus for oil revenues and the site for domestic reallocation—other major cities, elite home towns, oil-producing enclaves and infrastructures such as ports, oil terminals and offshore rigs. On the other end, there is the virtual abandonment of large portions of the country, the "useless spaces". (These are of course not fixed or rigidly defined: there are plenty of "interstitial" spaces, occasionally occupied or neglected.) The grip over the useful spaces is consolidated through the use of state and/or private violence and the discharge of some state functions onto oil companies that ensure the *mise en valeur* of the enclaves. These three processes—discharge of state functions, deployment of violence to allow accumulation, and neglect of useless spaces—are not occasional, but rather permanent features in the contemporary governance of the oil state.[175]

Useful and useless spaces

Foremost amongst the useful spaces are the oil-producing enclaves. The reference to enclaves[176] throughout this book has two complementary meanings. Firstly, it pertains to the enclave nature of the oil economy, which is outwardly oriented, does not provide for mass employment and has few if any linkages with other sectors of the domestic economy. Secondly, and more importantly here, it describes the tightly circumscribed areas where oil is extracted and where companies and the state ensure levels of "stateness" that are absent in the remainder of the territory.

---

further segmentation and the corresponding "piece of the national cake" continue. This may appear like a "paradoxical territorialisation" in which the state is enduring and even extending its grip over the national space, if only on the basis of the cupidity of local elites. In reality, this merely transposes onto the level of each state, and increasingly, each local government (of which there are 768), the violent struggle for resources that defaces the national politics of the other oil states in the Gulf of Guinea, as the recent conflicts in Bayelsa, Rivers and Delta States suggest. In this sense, grafted upon a national map of power nexuses and powerless peripheries one can find a federal state-specific patchwork of useless and useful territories. In one word, Nigeria multiplies by 36 (or 768) the conflict that in other countries of the region occurs at the level of the central government.

175 Cooper (2002: 63) sees the "highly disarticulated" territory of the Belgian Congo, with its "hodgepodge of zones of mineral and agricultural exploitation and zones of neglect" in a similar way.

176 See Nies (2003) for a discussion of the potentially destabilising role of enclaves in world politics.

These are amongst the few tracts of the "national" surface not to be effectively abandoned or even held by non-state competitors. This is the case with sections of the Doba and Logone regions in Chad, the Niger Delta in Nigeria, Angola's Cabinda enclave, etc. In addition, cities such as Pointe-Noire in Congo, Port-Gentil in Gabon and Port Harcourt in Nigeria, which are central to the oil business and logistics and where companies have a very high profile, share important characteristics with oil-producing enclaves.

In the oil-producing enclaves, companies have taken up primarily security-related tasks in response to lawlessness, warfare or state incapacity, as well as contributing substantially towards infrastructure maintenance or construction that is beyond the means of the state and is essential for their operations. On occasion, companies have also come to provide a measure of health, food security and education to small numbers of directly affected communities, as well as philanthropic assistance to a wider number on a more infrequent basis. There are many historical precedents for private sector enclave governance.[177] These include cases where no modern state system existed, as with Africa's pre-colonial coastal entrepots for the evacuation of goods and slaves.[178] They also include instances of weak statehood where companies take over vast regions and administer them in virtually independent ways. This was the case of mining concessions in the former Belgian Congo, the Zambian Copperbelt, eastern Angola, northern Mozambique and the Latin American mining and agricultural emporiums.[179] In some contexts, a company enclave could even mimic a state, with an educational system, armed forces and bureaucracy, although this is rarer in the context of oil companies because the number of their employees on the ground is frequently limited.

Company enclaves are the physical embodiment of a reality that is, first and foremost, a legal one that is explored in Chapter 5: the con-

---

177 See, in particular, references to company rule in section one of Chapter 4.
178 Examples include the many pre-colonial fortified enclaves in Africa's Atlantic Coast like Elmina in Ghana or the city of Luanda.
179 In his autobiography, Gabriel Garcia Marquez (2002: 24) refers to the territory held by the United Fruit Company, on which he was raised, as the "the hermetic realm of the Company".

struction or furtherance of a regime of exception to accommodate forms of external investment amidst state decay. In them, state and non-state forces collaborate in the production of authority and a modicum of order that allow resource extraction to take place. Present-day enclave settings have allowed tight collaborations between the state and companies rather than the primacy of the latter over the former. The performance by companies of many essential state tasks is in fact the precondition for the mutually beneficial exploitation of oil.

Information about oil-producing enclaves and what happens in them is fitful, as access can be limited or non-existent.[180] The case of Chevron's Malongo compound in Cabinda and the company's role in the province more generally is illustrative.[181] Malongo is a sprawling American-style suburb for company employees built in the 1960s, complete with an 18-hole golf course and protected by a surrounding minefield and electrified fence, its employees airlifted around the territory and barely engaging with the local economy. Outsiders' access remains restricted, and more so after the 2002 polemic over the alleged use by the government of company premises to interrogate suspects in its bid to win the war against separatists.[182] Malongo is clearly a dent in Chevron's image but is also essential to keep its labour force content and protected in a hostile context. Oil companies, and particular American ones, are keen on enclave policies such as the importation of food through company caterers rather than buying it locally and the seclusion of their personnel away from locals even where there is no real risk. But companies are not keen on accumulating responsibilities that increase their visibility and the expectations of locals. The trend of company involvement in enclave governance

---

180 For instance, one of the flashpoints of the Niger Delta 1990s controversies were the security arrangements for the protection of infrastructure and company personnel between companies and state forces. Despite international pressure, most companies did not reveal the relevant details.

181 See Suzanne Daley, "Malongo journal: Why this pampered paradise? It's the oil, stupid!" *New York Times*, 25 June 1998, and Daphne Eviatar, "Africa's Oil Tycoons", *The Nation*, 12 April 2004, for recent accounts of Malongo.

182 See Solomon Moore, "Angola strife threatens a key source of US oil", *Los Angeles Times*, 16 March 2003, and Adelino Gomes, "Luanda viola direitos humanos em Cabinda e petrolíferas são cúmplices", *Público*, 12 December 2002, for these unconfirmed allegations.

occurs on account of the absence of the state rather than because the company wants to accumulate state-like functions.[183] In practice, companies cannot avoid taking up these tasks.[184]

Beyond the sites of oil exploitation, the writ of the state and its partners is often reduced to the urban areas and their immediate hinterlands, where most of the population and government spending are concentrated. No meaningful economic activity of a legal kind takes place outside of these and the territory that is extraneous is perceived as "useless space". Though claimed by governments, it is in fact abandoned. In Congo, for instance, travel between the two major urban centres, Brazzaville and Pointe-Noire, remains possible only by air seven years after the end of hostilities. The train that links the two on an infrequent basis is a target for attack by bandits; there have been few if any attempts by the government to reassert its writ over the vast expanse of Congolese countryside. Useless spaces can be virtually depopulated, like most of non-urban Angola, Congo and Gabon, or merely neglected, as in populous Nigeria, and no legal investment by the state or the private sector ever takes place there. Because they would be useless, counter-claims to authority by non-state groups emerge less often here than in the oil-rich regions. Alternatively, fiefdoms develop that articulate localised non-state regimes of control, often with the tacit support of the state.[185] As Daniel Pécaut writes, this "de-territorialisation is the concrete expression of the disappearance of institutionality".[186] Confined to its enclave and the mentality that goes with it, the downsized state is not interested in "making society legible" and is, like James Scott's pre-modern state, "partially blind". It knows "little about its subjects, their wealth, their landholdings and their yields, their location, their very identity"[187] and no longer seeks to incorporate

---

183 In contrast, many companies in the colonial period happily took over these tasks.

184 See section two of Chapter 5. See also Onishi (1999) for an investigation of the manifold responsibilities that have accrued to Chevron in the Niger Delta area of Nigeria.

185 Achille Mbembe calls this "private indirect government" (2001: 24-52). See also Roitman (1998).

186 Pécaut (2002).

187 Scott (1998: 30). "State knowledge" across the region (say, survey knowledge) has not been decidedly advanced since independence, and in some cases, colonial era documentation still provides much of the basis for state awareness of territory and population.

them. The "useless spaces" and the people therein virtually disappear from the radar of national politicians.

The means of coercion

The NOC explored in the previous section is not the only institutional survivor in the post-boom era. In apparent contradiction with the deterioration of other pillars of the state, the coercive apparatus either is in frank expansion or has at least not been jeopardised by economic decline.[188] Elites at the helm of the oil state make use of vast resources to protect themselves against threats. Their position and the resources it gives access to are a recipe for structural conflict around the state, even while the existence of the state itself is scarcely at stake,, as rivals want to take it over rather than destroy it. This continuing vitality of the means of violence encompasses a profusion of private security entities as well as the oil state's military, police, intelligence services and paramilitary forces, although these present varying degrees of efficacy. As with external representation (discussed in Chapter 6), violence and the willingness to deploy it remain credible state arenas. Moreover, through inflated arms procurement, joint ventures with foreign security companies and predatory economic activity made possible by the threat of force, they remain important arenas for self-enrichment in their own right.

Like the state, violence is itself "enclaved". Although vast tracts of territory are accessible by state forces, their role outside the "useful spaces" amounts to a roving and destructive presence. The Nigerian federal army may stage a vengeful attack against the the Tivs in the Middle Belt[189] or repress disturbances in the Niger Delta, and the Forças Armadas de Angola (FAA) may roam the Lunda provinces to crack down on illegal diamond diggers, but neither can provide everyday military coverage. Congo's militias may have spent the late 1990s plundering the Pool and the Nibolek regions, but could not construct a political order on the basis

---

188 Commenting on Angola's multilevel crisis of the last two decades, Tony Hodges notes that it has "crippled all branches of the public administration apart from the armed forces and the police" (2001: 44).

189 Human Rights Watch (2002).

of violence alone. As Pécaut observes in the case of Colombia,[190] roving means of violence may inspire a sense of ubiquity on the population—a perpetual sentiment of terror—but do not qualify as institutionalised daily presence. It is rather in the areas where oil is extracted and in the urban milieu that the state's power is felt and a political order sketched.

In oil-producing regions, this is often performed as an end-oriented task pertaining to both the protection of the oil that sustains the state and the punishment and elimination of armed competitors and saboteurs. Bids at secession or enhanced autonomy and other challenges to the state are recurrent in the Gulf of Guinea, where artificial boundaries and divergent histories are added to the incentives of oil wealth. In the postcolonial years these have included Cabinda, the Biafra attempt to leave the Nigerian federation, the more recent Niger Delta disruptions (discussed in Chapter 5), Bioko islanders' desire to part with Equatorial Guinea, and the fragmentary politics of Southern Chad and Anglophone Cameroon. All of these conflicts have their basis in wider and often older antagonisms but have been propelled by the perception of exclusion from the oil boom. Understandably, formal challenges that seek to question the state's fictions of control or legitimacy in oil-producing areas are brutally curbed because they endanger its very existence, and a permanent state of exception, premised on large numbers of stationed troops, is established in them.

Cabinda, the Angolan-controlled oil enclave, is an example of this.[191] Awakened by a sense of local identity[192] but decidedly encouraged by the abundance of oil, Cabindan intellectuals and politicians started to articulate their vision of a separate state during the anti-colonial war. Through the generous funding of President Mobutu of Zaire and France's Elf Aquitaine, a number of secessionist groups emerged in 1974-75 to argue with the departing Portuguese authorities for a separate deal.[193]

---

190 Pécaut (2002).

191 I thank Alex Vines for an interesting discussion on the subject of Cabinda.

192 Cabinda's claim to separateness is based on the 1885 Treaty with the Portuguese, which deemed it an entity in its own right, and the fact that it was locally administered until the late 1950s, by which time it was merged with the "overseas province" of Angola. See Martin (1976).

193 On Cabindan nationalist formations, see Mabeko-Tali (2001).

This was in vain, for Cabinda was firmly thrown into the hands of the new Angolan state without a shred of its autonomy preserved. Luanda certainly valued this, as it provided it with the lifeline to counter the decay of Angola's non-oil economy. The next two decades saw a complex interplay of interests preserving the status quo, with heavy military presence protecting oil investment while a diversity of actors, including the Angolan government and French and American oil firms, paid off the offshoots and splinter groups of Cabindan nationalism to keep peace.[194] More recently, and because of its military success against UNITA, the Angolan government has sought to play the military card.[195] The territory now contains about 30,000 soldiers, or an estimated 10 per cent of Cabinda's population, and human rights violations in the context of the anti-insurgency campaign have mounted.[196] The case of Cabinda holds two important conclusions for my purposes here. The first is that the power of the state to kill has not been undermined by its lack of power or willingness to do just about everything else, and that this power is more or less consistently deployed to protect and maximise resource extraction. The second is that state violence cannot be reduced to this simple enabling role.

This is made clearer by looking at the other meaningful arena for the state's display of its coercive power, the state capital and, to a minor degree, other urban centres. The public administration may no longer work at most levels and important activities, such as customs, garbage collection or the security of expatriate compounds, may have been farmed out to private entities. Yet the deployment of the means of violence is not only evident but goes well beyond the curbing or discouragement of dissent, and fundamentally relates to the self-affirmation of the state. The subject of state violence in the Gulf of Guinea is much too broad to be fully engaged here and the paucity of historical and ethnographic work should caution against generalisations. An analysis of state violence should allow for a large number of variables, including the history of each

---

194 These are currently FLEC-FAC, FLEC-Renovada and FLEC-Original (locally known as FLEC without an adjective).

195 Justin Pierce, "Cabinda: a Province Still at War", *Mail & Guardian,* 18 October 2002.

196 Human Rights Watch (2004).

state and particular political conjunctures as well as differing treatments on the basis of class, ethnicity and social context. The predominance of non-democratic regimes before the 1990s, for instance, often translated into forms of close surveillance; and in the context of Marxist-Leninist regimes, Congolese and Angolan cities were subjected to meticulous and repressive social control by a plethora of enforcement agencies absent elsewhere.[197]

Yet three key traits can be highlighted. The first trait of state violence is the often-operatic character that betrays anxiety about its sustainability as well as the exhibitionist desire to affirm itself: the erosion of infrastructural power leads the state to invest in the exercise of despotic power as a statement of its continuing strength.[198] This is typical of weak statehood, but the resources available to the oil state allow it to confound reality with displays of force that poorer states cannot afford to enact.[199]

A second trait is that this coercive power is overwhelmingly meant to secure the state from internal, as opposed to external, challenges.[200] In key urban battles of the last decade in Luanda[201] and Brazzaville, and in repression of opposition mobilisation such as Opération Villes Mortes in Cameroon, targets are disgruntled segments of the population as much

---

197 Hodges (2001: 50) refers to Angola's "efficient security services, developed with East German Assistance".

198 As Christine Messiant points out, the budget for the 2000 Angolan Air Force festival was $4 million, part of a string of festivities designed to underline "the prestige [of the regime], the normality of the situation and the power [*puissance*] of the state". See Messiant (2000: 17).

199 In her excellent analysis of state violence in postcolonial Luanda, Messiant refers to the "order of power" established after 1977 "independently from the [civil] war" which included the army, the Cuban military presence, several police forces including DISA, the secret police, neighbourhood informers and war veteran militias (Messiant: n/d).

200 This brings to mind the comment by a fictional character that the army is "for civilian use" (Naipaul, 1971: 105). The intervention of Angola's armed forces in Congo, the DRC, Zambia and Namibia is an exception, but even here combat operations tended to have UNITA rebels as a target. Angola has since rolled back its foreign engagements, although it is likely that it will go on using its military might as a regional lever. See Turner (2002) for Angola's role in the Congo war.

201 Maier (1996) contains a good account of the All Saints' Day 1992 massacres in Luanda.

as rebels. In addition to the conventional forces, this is accomplished through an impressive number of relatively well-paid and well-armed organisations such as Angola's Serviço de Informação (the secret police) and Polícia de Intervenção Rápida (the paramilitaries known as 'ninjas'), Congo's Réserve ministérielle, and Equatorial Guinea's Moroccan protectors of the presidential family. Their role is to fend off not only challenges from known enemies but also intra-elite challenges. Because of their power and resources, they are both a tool and a potential threat for the presidency, which promotes and undermines them almost simultaneously, so that outfits of this type multiply over time.[202]

Thirdly, there is the privatisation of violence, with widespread and long-term instances of dispersion of coercive responsibilities. Whereas the process of erosion of the state monopoly of violence is a global one,[203] in the Gulf of Guinea it is hard to imagine another context where it has been taken to such an extreme, and where privatisation has become essential for the smooth running of things. This occurs at several levels and involves state-sanctioned outfits such as foreign security and military companies, local security companies, and more ambiguous ones (useful, but less predictable) such as private militias or warlord formations. It also involves the overt utilisation of state armed forces for essentially private ends. These coercive organisations are not mutually exclusive, and may in fact overlap, act in coordination, or challenge each other. I will now briefly examine them in turn.

Coercive organisations

Although similar outfits were operative for years, the activities of private military firms (PMFs) have grown enormously since the end of the Cold

---

202 President Sassou, for instance, simultaneously beefed up the military and police forces of Congo and created three military intelligence services, whose main task is "the surveillance of each other". See "Sécurité, sécurité", *La Lettre du Continent* 362, 19 October 2000. Something similar occurred in Gabon with the creation of a "National Council of Security" whose control escapes all previously existing security outfits. See "Gabon: Guerre dans les services", *La Lettre du Continent* 342, 9 December 1999.

203 I thank Peter Singer for a number of insightful comments on this subject (conversation with the author, Berlin, 2 September 2004).

War to constitute a business worth an estimated $100 billion a year.[204] Africa has witnessed some of their highest profile actions and PMFs are indispensable to the domestic security of many of the states in the region. PMFs are very different from the old-style mercenaries that were an integral part of warfare in the continent for the past decades. They can be highly professional, corporatised organisations based in the West, Israel or South Africa, although there are differences in size and skills as huge as in other market-oriented sectors. PMFs are often staffed by former South African or Eastern Bloc soldiers as well as veterans from Western elite forces, and provide a diversity of services ranging from protection and logistics to the waging of war. These companies are prospering around the Gulf of Guinea as well as in other areas of the world where security is a scarce commodity. Clients include states, rebel groups and foreign business operators, especially in the mining and oil sectors. In a context of outright disruption PMFs play a crucial role in the protection of oil firm property and personnel, and may even allow operations to continue amidst inhospitable circumstances. They are also playing a more structural role as investment is increasingly thought out in terms of the "security umbrella" of PMF expertise and protection. As the situation deteriorates in places such as the Niger Delta—where MPRI, the giant US private military company, is currently training and restructuring the Nigerian Army— there is "an accelerating involvement of PMFs across the region".[205]

Direct military intervention by a PMF is a rare, if notorious, occurrence. The most well known is that of Executive Outcomes, a Pretoria-based company which intervened successfully in the 1992-94 Angolan

---

204 For PMFs see Singer (2003). I also found the contributions to the conference on "Private Military Companies in the Current Global Order", Rhodes House, Oxford University, 6 December 2004 helpful.

205 Comments by Patrick Smith, Editor of *Africa Confidential*, conference on "Private Military Companies in the Current Global Order", Rhodes House, Oxford University, 6 December 2004. He further argued that governments such as the US are "actively encouraging PMFs in the Gulf of Guinea", both in security provision and in the retraining of local armies. In addition to its Nigeria programme, MPRI is developing "an integrated team of defense, security, and Coast Guard experts to provide a detailed set of recommendations to the government of Equatorial Guinea concerning its defense, littoral and related environmental management requirements" (MPRI website, http://www.mpri.com/site/int_africa.html, accessed 30 September 2004).

civil war on the government's side.[206] Sonangol reportedly hired EO at the behest of Chevron, Elf, Petrofina and Texaco, initially to secure the important onshore Soyo facilities from a UNITA takeover. Its involvement ended up being much more extensive,[207] to the point that the company claimed to have won the war.[208] More frequent than such "cowboy tactics"[209] is the hiring of companies for routine provision of robust policing".[210] This was the case of another leading PMF active in the mid-1990s in Angola, DSL, which was contracted by a number of oil companies, Sonangol and De Beers to provide for security in onshore sites, offshore oil platforms and transport routes.[211] In Angola, where a 1997 law requires the hiring of Angolan security companies (doubtlessly for fear of the "institutionalisation" of PMFs), it often happens that PMFs create local subsidiaries in joint venture with personalities close to the presidency but continue to operate in essentially the same terms.[212] It seems clear that the everyday running of the oil business cannot take place without private security and that in extreme circumstances, it has both the means and the willingness to make the difference.[213] But their role goes further. Private security, in the form of companies or borrowed foreign troops, has become pivotal for the protection of power-holders themselves, as well as state institutions.

---

206 On EO in Angola, see O'Brien (2000) and Vines (2000)

207 According to O'Brien (2000: 52), EO had an estimated 1,400 employees active in Angola for the duration of the conflict.

208 This led to the government giving oil and mining concessions to companies close to EO.

209 This was the pejorative remark made by an executive of an established PMF about the methods of Lt. Colonel Tim Spicer, a former SAS officer with extensive involvement in some of the more high-profile PMF actions of the last decade (conversation with the author, Oxford, 6 December 2004).

210 Interview with K.A. O'Brien, RAND Corporation, Cambridge, 10 May 2004.

211 O'Brien (2000: 53).

212 Even when the companies are local, the pool of clients is minimal and they are *de facto* oil company-controlled: for example, a Nigerian-owned outfit called "Spies" that was employed by Elf-Aquitaine to protect its installations merely disbanded after the men were replaced by another security provider. See Soares de Oliveira (2002: 10-14).

213 See "Soldiers for Sale", *Africa Confidential,* 19 July 1996 and Harding (1996).

In addition to these alternatives, power-holders have spawned a growing number of private militias, particularly in the last decade, or tolerated emergent militias for their own political purposes. These charismatic formations, which often attain the proportions and hardiness of a war party, are usually staffed by aggrieved youth, are accountable only to their leader, and can be seen as the ultimate consequence of the privatisation of the state: personal war machines to eliminate opponents and consolidate power. In cases such as Congo during the tenure of Pascal Lissouba, militias may be invested in to the detriment of (or in addition to) a national army that the presidency no longer fully relies on.[214] Likewise, when a rival such as current President Sassou takes over, he may decide to continue his own militia despite full control of the armed forces.[215] In Nigeria, ethnic militias of all kinds have sprung up in the last decade, and particularly since democratisation in 1999, in alliance with local politicians.[216] Their existence cuts both ways: in the Niger Delta, in particular, it presently seems that groups armed by PDP governors and other PDP officials and factions for electoral intimidation and assistance with oil smuggling in the years after 1999 have now taken on a life of their own.

While they are older creations, the armed forces of Gulf of Guinea states have gone through the same process of "privatisation" that has spawned private militias and led to the hiring of PMFs. In this faction-ridden environment, it is difficult to analyse them as unitary institutions. If there is one common thread, it is the armed forces' commitment to enhancing resource accumulation for their leaderships and, in most cases, the state's civilian leadership, although civilian-military relations are as fraught as in any other context of the developing world and the army is still perceived as an intermittent threat. Because of the aforementioned privatised alternatives and the general unreliability of quantitative infor-

---

214 For Congo's many militias, see Bazenguissa-Ganga (1996).

215 Not fully trusting his new army, Sassou Nguesso has kept sections of his militia from integrating into it as well as creating a new, more reliable militia, the Commandance des unités spéciales (Comus), run by the Minister of the Interior, his cousin.

216 For the Delta region, see section three of Chapter 5. See also Gore and Pratten (2003) and Harnischfeger (2003) for other phenomena of private violence with ambiguous links to "governors and influential politicians".

mation,[217] budget expenditure does not accurately illustrate the extent of state investment in the armed forces, except for countries that, for historical reasons, have developed clear-cut commitments in this area. Gabon, for instance, spends comparatively little on defence, yet the small presidential guard and armed forces are essential elements of President Bongo's hold on power.[218] The most important source of regime survival, however, is the presence of a 650-strong French marine battalion next to Bongo's presidential palace, certainly not an item one would find in the Gabonese budget.

The centrality of the armed forces to domestic politics in Nigeria and Angola is much more obvious, as it stems from the civil wars fought in both countries.[219] By the time both conflicts terminated—in 1970 and 2002, respectively—the armed forces had acquired a size and importance that precluded their dilution in the manner presupposed in post-conflict contexts. Part of the problem is logistical—how to demobilise an army without prospects for social and economic reinsertion, and without increasing banditry or the threat of further armed challenges. But the crux of the issue is that army commanders have become major political actors and want to keep that status.[220] Having grown from the puny size and minor part in politics moulded by the departing colonial power into a formidable force in the aftermath of the Biafra war in 1970, Nigeria's federal army could not fail to play a key role in the country's political and economic life. This position was consolidated by the long period of military rule that, with the exception of the four-year hiatus of the Sec-

217 A recent survey argues that, in addition to poor statistics, military expenditure in the region is difficult to assess because donor pressure for low military expenditure leads to the manipulation of figures. See Omitoogu (2003). For the available data see IISS (2003).

218 The Gabonese army is 3,200-strong, the air force 1,000-strong and the Presidential Guard that protects Omar Bongo 2,000-strong. Official defence expenditure was $105 million in 2003. See IISS (2003: 213, 327).

219 Note that the absence of a prominent role in postcolonial politics does not amount to permanent irrelevance in domestic politics, as the case of São Tomé's ragtag army shows. On two occasions in the last decade, the poorly paid and mostly shoeless soldiery has made its voice heard through putsches. See section two of Chapter 5 for a discussion of the background.

220 See Campbell (1978).

ond Republic, lasted until 1999. Though out of politics for the present time, the senior officers of this 68,000-men force are ubiquitous players in all manner of legal and illegal activities throughout Nigeria, including the theft of oil.[221]

This is apparent in Angola's case as well. The 110,000-strong FAA, buoyed by their success against UNITA, the continuing strife in Cabinda and the tough regional neighbourhood, as well as a culture that sees internal and external enemies galore, is not destined for major downsizing in the foreseeable future.[222] In a 2004 address, Finance Minister José Pedro de Morais confirmed that defence expenditure would not significantly decline in the next five years, and unconvincingly pointed to peacekeeping, postwar reconstruction and the new "duties of states in a globalized world" as the future missions of the Angolan armed forces.[223] Its real activities are very different. Having profited hugely from wartime scams such as the acquisition of over-priced weapons and their sale to the enemy, the military leadership is now busy carving for itself a predatory role in Angola's peace economy, especially in diamonds, land acquisition through expropriation and, of course, the creation of security companies. According to some analysts, this transformation of national militaries into "tools for private gain by the political elites" is likely to be furthered in coming years.[224]

Conclusion

This chapter has sought to tease out of the variegated domestic political experience of the Gulf of Guinea oil states a number of important shared traits. They include the following:

---

221 See section three of Chapter 5.

222 Hodges (2001: 63) labels it "the primary and arguably the only real manifestation of a 'strong state'".

223 José Pedro de Morais, address at the Royal Institute of International Affairs, London, 18 April 2004. "Defence" was reportedly claiming "over a third of government revenues" some 18 months after the end of the Angolan civil war. See "Holding the Cash", *Africa Confidential*, 21 November 2003.

224 Dietrich (2000: 1).

• A brief historical moment of capital-intensive state building character-ised by ambivalent attempts at creating a "modern" state and exercis-ing infrastructural power. This was to be achieved through oil revenue disbursements rather than the construction of fiscal sustainability or the promotion of labour-intensive activities in sectors such as agriculture. These attempts were undermined because state institutions were weak and poorly staffed, but factors such as the expectation of unrealistically high revenues, flawed policies, the urban and unproductive biases of the state, and the predatory logic of political actions (which I examine in the next chapter) were central.

• In response to the decline brought about by the post-boom crisis, a privatising trend that allowed the radical disengagement by the state from whole spheres previously seen as the state's own, sometimes by subcontracting tasks, and other times by abandoning them altogether. The result has been a "failed" state, as described in the previous chapter. Yet "privatisation" in the broad sense used here has allowed the much-diminished state to concentrate on the institutions and methods that are essential for the preservation of the oil relationship with the international system, the maintenance of a minimal grasp over the domestic sphere, and the "success" of power-holders. These include access to money (through oil-backed loans), access to expertise (through NOCs) and ac-cess to the means of violence (through both public and private sources). Characterised by a peculiar pattern of neglect—of useless spaces and of useless people—the privatisation trend also conforms with the stream-lined, post-boom goals of power holders more keen on self-enrichment than on unfeasible visions of a "modern industrial state".[225]

• Through boom and bust, the state's mission, ways of wielding power, institutional purposes and capabilities have changed considerably, but mismanagement, theft and authoritarianism remain at the centre. The predatory manner in which the oil state has been governed explains the essentially arbitrary character of its political life and has stark implica-tions for the likelihood of any improvement.

This picture is not cast in stone and individual state experiences do not mirror it on all accounts, though it is an accurate portrait of long-term trends across the region. As everywhere else, the politics of the

225 Arnold (1977: v).

Gulf of Guinea are constantly changing; failed and weak states, far from being static, are permanently in a condition of upheaval. Some of these developments may even appear to contradict the "privatisation" scenario put forth in this chapter. The post-2003 oil boom, for instance, has resulted in the return of some elements of big-state thinking, especially in the inescapable construction sector. Long-planned, gigantic ministerial buildings are rising up all over Abuja; Angola is seeing a frenzy of public investment in badly needed (and not so badly needed) infrastructure[226] and recently announced record amounts of non-oil taxation.[227] The craving for big capital expenditure, it seems, has not been eliminated from the aspirations of elites. The new boom, however, does not encompass a commitment towards either institution-building, real administrative coverage, or the welfare of the population. The conditions, incentives, and institutional capability for structural change in the way the state is perceived by dominant actors and state power is exercised are lacking—especially at a moment in time when very high oil prices again reward the status quo or allow it to endure through cosmetic changes. This is as true for the elites analysed in the next chapter as it is for the many international actors that directly or indirectly contribute to the functioning of the oil state.

One should therefore not be fooled by a new era of (transient) oil prosperity into thinking that attitudes towards the state are slowly changing away from the picture outlined here. In fact, one should not discount the possibility of furtherance and deepening of the privatising logic that may well lead to additional de-structuring of some of the final islands of "stateness", such as the NOC and the means of coercion, that still exist.[228] This outcome would signal the arrival of a hyper-privatised state-

---

226 The vision of a "New City of Luanda", reportedly meant for hundreds of thousands of people, is the prime regional example of the return of big-state thinking. While the authorities claim that "the idea is to create something like Dubai", the emphasis on malls, convention centres, health clubs and luxury condominiums seems to suggest that "big plans for the capital do not include the city's poor"; see "Oil boom fuels Dubai dreams in Angola's capital", *Reuters News*, 22 November 2006.

227 On non-oil taxation see "Alfândega arrecada USD 84 milhões", *Jornal de Angola*, 13 June 2006.

228 I thank Nick Shaxson for alerting me to the possibility of further privatisation in the Angolan case—specifically, of Sonangol subsidiaries—as part of a "retirement plan" for

like entity, asserting itself even more viciously through the exercise of despotic power only tempered by the limits of its means and the narrow scope of its goals. This would still be accepted in the international system as a "proper" sovereign state—all empirical prerequisites for oil state sovereignty long abandoned—but would in fact be no longer recognisable as such to those living under its aegis or to those looking in with a modicum of objectivity. Arguably, this is the direction of the trends presented here.

President José Eduardo dos Santos. Experience in the region, including that of the Ahidjo-Biya transition in Cameroon in the early 1980s, shows that no real political clout is preserved by an ex-president who no longer controls economic opportunities. This irrelevance extends to all honorary positions, like Assembly Speaker or President of the ruling party, because power-holders themselves have laboured hard to render these institutions meaningless. Therefore, one cannot exclude the possibility that, in some cases, the price of retirement may be the private ownership of important elements of the oil economy.

3

THE PETROLEUM ELITES

The ship of state is the only ship that leaks from the top.

Sir Humphrey Appleby, Yes, Minister

Human action [in Africa] is seen as stupid and mad, always proceeding from anything but rational calculation.

Achille Mbembe, On the Postcolony

What successive civilian and military regimes have decried as the moral failings of corruption and bad leadership has actually been the modus operandi of politics itself.

Andrew Apter, The Pan-African Nation

For many years, the study of African elites has not been a popular choice with academics. This can be accounted for in several ways, but a preference for "civil society" movements and what are seen as previously neglected elements of the continent's politics is perhaps the most important.[1] Yet it is obvious that in a context of poor institutionalisation, extreme centralisation and unaccountable decision-making, a "very great deal of politics [will be] concerned with a very small number of people".[2] With this in mind, this chapter investigates the present-day political behaviour of those in control

---

1 See Daloz (1999) for a healthy critique of the "politique par le bas" approach that still dominates African studies, especially in France. See also Bakary (1990), which contains a helpful bibliography. This book makes only oblique reference to popular politics, but this has to do with focus, not with underestimation of the political relevance of "subordinate" social groups.

2 Clapham (1985: 61).

of the institutions and money flows of the oil state. Its chief argument is that the outcomes described in this study derive as much from the choices of petroleum elites as from the more frequently mentioned, and doubtlessly very significant, structural constraints of oil dependence, weak postcolonial statehood, and the actions of external actors. As the protagonists and only domestic beneficiaries of the state, petroleum elites enable the oil relationship by signing contracts and providing a moderately reliable context for companies to work without fear of expropriation or destruction. In turn, they deal with oil proceedings as essentially private rents. This unquestionably rational course of action—pursued in ways that impact negatively on the lives of the majority—is premised on scarcely any political project beyond the medium-term illicit enrichment that oil monies afford. As any loosening of control would be detrimental to elite interests, and as current arrangements allow for fabulous private wealth, no incentive structures are in place to bring about benign change.

The chapter proceeds as follows. Section one discusses the diminutive dimensions and internal politics of Gulf of Guinea elites, their complex relationship with the oil state and the ample rewards made available to insiders. It then tackles the issue of leadership, investigating the extent to which the oil business has allowed for great concentration of resources and power in the hands of heads of state and their small coteries. Section two studies the particular uses of the languages of corruption and reform by elites, before inquiring about the forms and extent to which elites have engaged in capital accumulation, and whether their strategies are rational and sustainable in the long run. While much that is described and analysed here will be familiar to students of African politics, in and outside the Gulf of Guinea, I argue that the specificity of oil elites is twofold. Firstly, the resources available to them are of a wholly different order than whatever is available in other states of the continent, which permits the pursuit of lifestyles and policies that cannot be replicated elsewhere. Secondly, their predatory and short-termist ethos would in other contexts lead to their swift demise, but that on account of the continuing availability of oil rents, elite strategies are sustainable.

I. ELITES, PATRONAGE AND LEADERSHIP

For our purposes here, the elite consists of those who, whether in government or not, can influence government policy orientations—though not necessarily particular decisions—and directly or indirectly benefit from them.[3] It is made up of a governing segment (the decision-makers) and a non-governing segment (the merely privileged). Confusingly, a position in government may not amount to membership of the governing segment, which in turn can be held without direct involvement in government. The governing segment may be very small and consist of those with direct access to the president and influence over his decision-making process: typically, this involves family members, presidential advisers, a select number of ministers, army men, businessmen and foreign middlemen. The non-governing segment includes a much larger number of people who broadly support the governing segment and have access to state largesse in the form of financial disbursements and economic and social opportunities. This comprises all whose lifestyles have been carefully sheltered from the vagaries of state decay. The political power of the elite stems from the control of economic opportunities entailed by state control for, in the Gulf of Guinea, the possession of the means of administration is coterminous with the possession of the means of accumulation.[4]

In discussing poorly institutionalised political orders, there is a tendency to be imprecise with concepts that, however valuable and clearly delineated in theory, cannot be easily reconciled with political practice.[5]

---

3 This classification is based on Vilfredo Pareto's work; see Friedrich (1965: 259-67). I also found Bottomore (1993) and Aron (1960), among others, useful in approaching the study of elites. I decided, nonetheless, to adopt a definition that is more empirically relevant to my study than the ones these authors suggest. In particular, both authors' assumption of socially differentiated elites (bureaucratic, ecclesiastical, intellectual, governing, etc.) makes little sense in the more fluid Gulf of Guinea context. Furthermore, I think that reference to a privileged link to the state as an unwavering characteristic of Gulf of Guinea elites is vital.

4 Since there is no productive (as opposed to extractive) economic activity in the oil state, the classical Marxist assumption that the elite is that which controls the means of production is unhelpful. A similar point was made more than two decades ago by Sklar (1979). The author noted that because of the predominantly rentier character of the state, class relations were "determined by relations of power, not production" (1979: 537).

5 According to Karl, for whom "the analytical distinction between these three levels [state, regime, government] is important and should be specified". Her understanding

This is the case with assessing the links between elites and the oil state: where does one start and the other end? While I believe that "it is essential to conceive of states as administrative and coercive organizations that are *potentially* autonomous",[6] in light of the context it is hard to address the state as an entity that is *in fact* autonomous from the interests of elites. On the one hand, the state has a history of its own and is relatively free from both foreign and domestic non-elite pressures; it is a subject of analysis in its own right, not reducible to a set of political actors. On the other, the state has been effectively captured by elites, who use it more or less efficiently to pursue their goals. There is no bureaucracy empowered to accomplish objectives autonomous from those of the elite; the state never slights the interests of its masters.

This does not mean that the state is a unitary actor or that it is coterminous with the elite, for two reasons. Firstly, because the state is itself the site of intra-elite struggle for resources and the object of greed by counter-elites. The point is obscured in cases where elite "pacts" provide for stability, but in general the state should be thought of as an arena where different claimants to the "national cake" fight, wrangle and get together to divide the spoils. Secondly, because elites, rather than fostering a "*L'Etat c'est moi*" attitude that would burden them with both the prizes of the state (which they already enjoy) and its liabilities, instead manipulate a public-private divide that spares them from having to take

---

of the state resembles Weber's definition provided in Chapter 1. Regime is referred to as "the ensemble of patterns within the state determining forms and strategies of access to the process of decision-making, the actors who are admitted (or excluded) from such access, and the rules that determine how decisions may legitimately be made." Finally, the government consists of "the actors [...] who occupy dominant positions within the regime at any give moment in time". See Karl (1997: 14). I find much of this conceptual meticulousness unhelpful here. The assumptions that underpin such distinctions—for instance, the institutionalisation of political order and the concurrent separation between actors and institutions—may not make empirical sense; the "state" may only remotely resemble Weber's model. This is not to say that the distinction between actors and structures should be abandoned, for "only states without a past look like the simple instrument of their dominant class" (Lonsdale, 1981: 154).

6 Skocpol (1979: 14), my italics.

responsibility for the latter.[7] This means that the elite and the state are different things and that they are both valuable as discrete concepts. It also means that their boundaries can be elusive, and this "elusiveness must be taken seriously".[8]

Shrinking clientelism

In the boom days, the oil state was the pinnacle of a classical broad-based clientelistic system whereby the elite and the masses were linked by vertical networks of solidarity, with political acquiescence or support exchanged for material redistribution of resources.[9] Although the patron-client rapport is by definition unequal, massive public spending and subsidies, as well as illicit opportunities, kept a variety of (mostly urban) constituencies across the social spectrum happy. Conversely, the privatising trend described in the previous chapter led to a decline in broad-based clientelism and "the emergence of a regime unresponsive to social pressures".[10] This does not mean that control, repression and horizontal affinities have wholly replaced unequal exchange and vertical linkages, but that the latter are much more tenuous and exclude most of the population.[11]

The boom-era clientelistic logic—which retains the support of the vast majority of the population—still makes an occasional appearance in social groups close to a generous powerful kinsman or, more importantly,

---

7 See Vallée (1999b). This was the argument put forward in the discussion on debt in section two of Chapter 2.

8 I am adapting a point made in the broader context of state-society relations in Mitchell (1991: 78).

9 The literature is extensive; see especially Clapham (1982). On patron-state relations more generally, I found Gellner and Waterbury (1977) and Schmidt et al. (1977) useful.

10 Clapham (1982: 16).

11 This may point to a radical shift in perceptions of wealth at the regional level that cannot be investigated here. While political power in the Gulf of Guinea was never closely associated with control over territory, control over people has been a central feature for centuries in ensuring "wealth, power, and mutual security", according to Vansina (1990: 251-2). But in capital-intensive oil economies, control over the precious commodity is everything, and people are irrelevant. Their ever-expanding numbers mean that they are no longer scarce and their labour is not necessary.

during elections,[12] but it is an episodic appearance, not the structural fact of yesteryear. The boundaries of solidarity have shrunk to a geographically and/or socially confined minority. There are partial exceptions to this in all states. Nigeria's federal system ensures that state-level elites access their share of oil rents, even if it does not trickle down from there. Gabon, in particular, has struggled to maintain the policy of ethnic and regional division of spoils (what locals derisively call "geopolitics"),[13] although even there, human indicators show that this barely reached the majority at the best of times. In Angola, the privatisation of small businesses benefited tens of thousands of people, mainly in the MPLA heartland.[14] But the general trend is that of an ever-diminishing number of beneficiaries of state largesse.[15] The elite still seeks to mobilise "those rituals and discourse that customarily symbolize legitimate authority", like kinship, inheritance, nationalism and "shared substance",[16] but in a context where support is dependent on concrete handouts and services, this is often formulaic and counts for little.[17] Elite legitimacy in the oil state is primarily external.

---

12 In recent years, the holding of elections in Nigeria, Cameroon and Gabon coincided with the one-off disbursement by the government of goods and services that were previously widely available but are no longer taken for granted. Such wild spending normally means that any fiscal restraint there may be is abandoned and austerity agreements with the IFIs are violated. It is assumed that the Angolan ruling party will do the same in the forthcoming elections.

13 See Auge (2003).

14 Hodges (2001: 53-4).

15 Writing on Nigeria, Andrew Apter notes that with the abandonment of a broad-based patronage system, "unilateral powers of plunder and patronage were [...] accorded to a restricted circle, less expensive to maintain than multiple lines of misappropriation and more loyal to the center" (Apter, 1999: 291).

16 Shore and Nugent (2002: 14).

17 The forms by which the elite seeks to acquire legitimacy in the absence of clientelistic distribution go beyond the purposes of this chapter and should be included in a broader discussion of the popular politics of the oil state. Legitimacy claims by elites, which may vary over time, are intrinsic to the trajectory of each state. President Bongo points to the record of social peace and ethnic sharing of spoils; the MPLA highlights its nationalist (no longer revolutionary) credentials in defending the motherland from foreign invasion and championing the only national, non-sectarian project for Angola. President Obiang

For those inside the elite circle the rewards are substantial. Elite members can access foreign exchange with ease and, in most cases for the greater part of the postcolonial period, benefit from a monetary policy guaranteeing overvalued currencies. They can access cheap loans that do not have to be repaid,[18] lucrative import licenses and favourable state contracts. They can also gain from the economic activities created by state weakness, such as the importing of fuel[19] or grain for the home market. Outsiders are barred from participating in many investment opportunities unless they partner with members of the elite, which makes the latter beneficiaries of all significant economic activity. Elites can access privatised state property at nominal sums. They have access to good-quality education, which includes overseas scholarships, and access to first-rate health care. Whilst there are no real property rights in the oil state—all one's possessions can be confiscated if one falls foul of the President—the property of elite members is more secure than anyone else's, and their houses are safer because of police or private protection. More importantly, elite members can take their resources outside the country to place them safely in bank accounts or invest them in real estate in Western cities.

Who constitutes the elites of Gulf of Guinea oil states? There is no one historical trajectory. It is to the individual past of each state that one should look for an understanding of the socio-cultural prerequisites for elite status and self-identity as well as why certain groups and individuals ascended to the summit of the postcolonial state, while others did not.[20]

---

of Equatorial Guinea points to his termination of his uncle's murderous presidency, his "godly status" and his delivery of the oil bonanza.

18 See Reno (1998: 193ff) for banking sector reform in Nigeria in the Babangida era.

19 In view of the evidence of extensive involvement by Nigerian military officers and politicians in the smuggling of oil, it seems that some insiders are profiting twice from the country's fuel predicament. They steal crude oil for sale abroad and they are contracted by the state to import refined products to cope with the shortfall that they help create.

20 The literature dealing specifically with elites and the politics of oil is minimal and there are no comparative studies. Although analysts with experience of Gulf of Guinea politics can present anecdotal portraits of the region's elites, one would need a series of contemporary country-specific monographs making use of biographical interviews and close, daily contact to answer these questions fully. This is difficult to effect, however, as such spheres are by definition hard to penetrate and pose ethical questions related to "be-

In most cases, postcolonial elites originated in the leading non-white social groups of the colonial period that staffed the lower and, sometimes,
middle ranks of the colonial state and inherited it at independence.[21] In
Angola, the end of the war in 2002 consolidated the domination of this
group, which has historically supported the MPLA, over opposing counter-elites.[22] In Nigeria, the unresolved tension between dominant Northern elites and the more entrepreneurial and better educated Southern
elites explains much of the post-independence turmoil. Even where the
contest for power has been decidedly won by one segment, the elite does
not form an unvarying entity. Its heterogeneity is visible even in small
societies where huge divergences fester despite the fact that virtually all
members of the elite are personally acquainted—which may be the case
of all states under review here, with the exception of Nigeria.[23] What
they have in common is control of material and political resources in the
oil state, and the willingness to maintain both.

The emergence of new social forces and interests and the permanent
recreation of political alliances mean that elites are open to occasional
additions to their membership. Recruitment can take place through a
number of avenues, whether regional, tribal, religious,[24] political (especially in the case of anti-colonial militancy), or merely personal: though
one can find a core to each elite, they tend to be omnivorous as to their

---

ing accepted" and "allowed in". See Shore and Nugent (2002) for a number of interesting
essays that discuss the methods and ethics of research on elites; see also Cohen (1979) for
an old but excellent ethnography of an African elite which holds important lessons for
work on the Gulf of Guinea.

21 Exceptions include Chad, which is presently ruled by Northern warlords who had
been excluded by the rule of the mostly Southern elites until 1979 and have since excluded them in turn. The diverse patterns of elite formation must be understood historically
and on a case-study basis and go beyond the scope of this chapter.

22 On Angola's competing elites, see Christine Messiant's important work, and particularly Messiant (1994).

23 Think of the fractious, though bloodless, politics of São Tomé and Príncipe, where
family ties unite most members of the elite. This is studied in Seibert (1999).

24 This includes membership of established faiths, but also membership of secret societies such as the cults that predominate in Nigerian universities and the freemasonry.

intake.[25] Co-optation, a feasible practice in the cash-rich oil state, is amply exercised to defuse threats by potentially destabilising individuals or groups. Cameroon's and Gabon's once-vibrant oppositions have been quietly brought into the fold over the last decade.[26] Nigerian and Congolese politics, perhaps the more instability-prone in the region, constantly present the image of politicians from previous, seemingly incompatible, regimes resurfacing as part of a new power arrangement.[27] The 2002 MPLA-UNITA settlement, whereby top cadres of the former Angolan rebel movement were provided with the high status and the means to live comfortably under MPLA rule, is another case in point.[28] In the aftermath of the 2003 attempted coup d'état in São Tomé, politicians not only declared an amnesty for the conspirators but also guaranteed that they would be included in the sharing of the oil revenues.[29] The capacity of the state to buy off rivals or, in some cases, create "opposition" parties that support the regime goes a long way to explain the resilience of most

---

25 This explains why there are so many prominent Bakongo and Ovimbundu in the Creole- and Mbundu-dominated MPLA; why an important Yoruba segment has always collaborated with the Northern "Kaduna Mafia"; why the predominantly coastal elites of Gabon have included notables from all over the country, etc. The exclusionary family- and clan-dominated politics of Equatorial Guinea and Chad are exceptions here.

26 In Gabon, this culminated in the 2002 inclusion in government of Father Paul M'ba Abessole, the once-feared leader of the opposition and former mayor of Libreville. President Bongo calls the inclusion of 29 of the country's 35 parties in the presidential majority his "convivial democracy". See "Bongo forever", *Africa Confidential,* 21 January 2005.

27 President Sassou Nguesso of Congo took initial care to sideline men such as Lissouba and Kolélas, whose popularity is a constant threat, but everyone else, including pre-civil war senior political figures, has been welcomed back to Brazzaville. The chastened return of the ageing Kolélas in late 2005 (made possible by his acceptance of Sassou and abandonment of competing political ambitions) means that most pre-war public figures are now back in the country. See Bazenguissa-Ganga (1996a) for a study that pays attention to the durability of political actors in Congolese politics.

28 Craig Timberg, "Between old allies, a profound divide: former Angola rebels see leaders prosper", *Washington Post,* 16 September 2004. UNITA's leeway has decreased greatly. It no longer controls the smuggling of diamonds and cannot use domestically the resources it holds abroad. Ironically, UNITA "now depends almost exclusively on the state budget". See "Election, what election?", *Africa Confidential,* 2 December 2004.

29 See section two of Chapter 5.

power-holders, including resilience in the context of (limited) democratic challenge. Yet the balance is frequently precarious.

Tensions

Although experience is varied, one should not assume that the existence
of resources galore is enough to make for a "smoothly operating whole".[30]
For what elites in the Gulf of Guinea do not countenance is what Pareto
called "circulation", i.e. their replacement. Previous experiences with
voluntary surrender, including Cameroon's Ahidjo-Biya transition in the
early 1980s and the Congolese elections that ousted Sassou in 1992, had
such bad consequences for access to oil rents by the losing side that a
zero-sum approach is widespread. Counter-elites who wish to take over
the state apparatus or non-governing segments of the elite who want to
preside over the spoils rather than merely benefit from their distribution
are therefore the subject of unrelenting persecution when bribery and
cooptation fail. The sources of discontent in the oil state are never-ending. Firstly, there are popular dissatisfaction and loss of legitimacy bred
by the lack of broad-based distribution of oil wealth, leading in some
cases to rebellion by excluded groups such as "youth",[31] though there
have been no regime changes in the Gulf of Guinea on the basis of popular politics.[32] Secondly, and closely linked to the previous point, there
are challenges by permanently marginalised counter-elites, composed
of those with a claim to elite status who are excluded from accessing
the spoils. Counter-elite bids often ride on top of popular discontent at
(especially ethnic) exclusion from the oil bonanza. Perhaps the obvious
examples here are the UNITA bid for power in Angola and the attempted
secession of Biafra in Nigeria.

---

30 Clapham (1982: 33).

31 "Youth" is used here in its Nigerian sense, i.e. men between early adulthood and
about 40 years old who are hard hit by the economic crisis, left outside clientelistic networks, and are unable to attain the financial independence and property that would allow
them access to full adulthood, marriage, etc. Oil revenues, because they flow to a state
apparatus controlled by senior male politicians, considerably heighten inter-generational
tensions.

32 The closest thing to a successful popular rising, the Congolese "*trois glorieuses*" of
1963, took place well before oil was an important source of revenue.

Thirdly, and more importantly for our subject here, there is plenty of strife that is internal to the elite. The elite is not a "class" with a constant notion of where "its" interest lies: impermanence, instability and conflict define it as much as collaboration. There is something of the Byzantine court about intra-elite politics in the Gulf of Guinea.[33] Direct access to resources is not to be taken for granted, for changes in the governing segment may encompass a change in the "constituency of beneficiaries", as happened in Cameroon when Paul Biya proceeded to re-centre patronage on his Beti constituency.[34] And in other contexts, individual ambitions and the very large amounts of money at stake have led to a deadly struggle for spoils. Equatorial Guinea's attempted coup of early 2002, which resulted in a high-profile and much-criticised trial of nearly 200 regime insiders, was essentially a family quarrel for succession.

Other rivalries, though phrased in ethnic terms, are also about competing interests of men who share similar backgrounds. Though portrayed as a North-South conflict that is part of a tradition of civil unrest dating back to 1959, the Congolese strife of the last decade is inseparable from the ambitions of Bernard Kolélas, Pascal Lissouba and Denis Sassou Nguesso, all seasoned Brazzaville politicos.[35] In cases such as Angola, the overwhelming threat of a counter-elite—in the form of the Ovimbundu leadership of UNITA—may have pacified internal elite (i.e. MPLA) politics, especially if coupled with a studious fulfilment of the material needs of all its members, including those of critical ones.[36] But elsewhere, and even in Angola where the UNITA threat has dissolved, the assumption of a tight-knit, quietly thieving elite is erroneous. The contradictions and

---

33 As an illustration see the wonderful portrait of Gabon's inner circle and its many rivalries in Vallée (1999a: 73-94).

34 Jua (1993: 132).

35 Clark (2002: 186).

36 For this purpose, President José Eduardo dos Santos is careful to give out so-called "Christmas bonuses", some as high as $30,000, to "virtually all members of the political establishment" (Hodges, 2001: 56) and to tolerate the business interests of critical voices within the MPLA. It is noteworthy that when the opposition weekly *O Angolense* published the list of twenty richest personalities in Angola, Lopo do Nascimento, the historical figure of the MPLA said to represent the party against the more venal presidency, was ranked second just after Eduardo dos Santos.

disparate appetites ensure that the challenge is continuous, although incumbents are in a good position to cope with it.

The imperial presidency

Overseeing the workings of the oil state is the President. While it is generally inadvisable to attribute too much significance to individual actors in the study of political processes, presidents across the Gulf of Guinea are undeniably powerful. Differently from rulers of the ideal-type "privatised state", who are often portrayed as the mere representatives of a broader shadow constituency,[37] petro-presidents have real individual authority. The powers afforded by this extraordinary concentration of resources are such that even systems that start as more or less collegiate oligarchies, such as General Babangida's regime, soon develop into essentially "personal rulerships".[38] Within the context of the institutional legacies and political patterns they inherit, presidents shape the way things are done. While leadership styles are very different, from the cult-like status of Obiang[39] to the designer-clothes sophistication of dos Santos and Sassou, a great deal of the domestic and international politics of the oil state is to be understood by the study of its leader.

In some cases, the leader's unquestionable domination was always there, as with President Bongo, despite the single party theatrical until 1990, and President Obiang of Equatorial Guinea. In others, it was laboriously built up in office on the strength of oil money and further centralisation, and, more importantly, through the piecemeal de-institutionalisation of decision-making and the eclipse of previous organs, such

---

37 Bayart, Ellis and Hibou (1999: 40-1).

38 For this transition see Amuwo (1995). Obasanjo also came to power in 1999 with the active support of Northern notables but quickly proceeded to forcefully retire more than a hundred senior army officers (mostly of Northern extraction), curb the ambitions of this previously dominant segment of the elite, and tightly assert personal control.

39 A vignette is illustrative here: when two French journalists covered the arrival of President Obiang at Malabo's airport in 2003, they recorded the chants of the singers and dancers (originally in Spanish): "Ô bon président Obiang, Ô merci pour le pétrole, Ô bon président Obiang, merci pour la liberté!". See Serge Enderlin and Serge Michel, "La Guinée équatoriale de Monsieur Obiang, ses palmiers et ses pétrodollars", Le Figaro, 17 July 2003.

as the single party or the armed forces, that may have played key roles until then. President Babangida moved the oversight of the Central Bank from the Ministry of Finance to the Presidency,[40] while General Abacha took control of the oil sector away from NNPC to his office. The paradigmatic case is President José Eduardo dos Santos, who arrived at the helm in 1979 as the unassuming party consensus choice following the death of the charismatic Agostinho Neto. By the late 1980s, he had become Angola's all-powerful oil czar, sidestepping the MPLA in decision-making, sidelining most of the bureaucracy, and nominating his minions to all sensitive posts. The case of President Sassou Nguesso is similar. In contrast with his reign in the 1980s, which had also revolved around single party structures and the army, Sassou's rule since the 1997 takeover is "structured around concentric circles that are largely autonomous, which never meet, and which answer to no one other than [the President himself]".[41] No one's role is taken for granted and positions are allocated on a temporary basis.

All strands of the state and the economy are ultimately tied to the "imperial presidency", around which grows a "parallel state [with] a lock on national finances, managing payments from oil-backed loans, arms deals and other lucrative off-budget expenses".[42] A recent US Senate investigation of corrupt practices related to Equatorial Guinea, for instance, noted that "some E.G. officials and their families had come to dominate certain sectors of the [...] economy and [...] had become virtual economic gatekeepers for foreign companies wishing to do business in the country".[43] Throughout the region, foreign investors must pay those

---

40 According to the (politically motivated) investigations against President Babangida during General Abacha's subsequent rule, this was what made possible the "dedicated and other special offshore accounts" run by the presidency, wherefrom an estimated $12 billion allegedly disappeared.

41 François Soudan, "Sassou a-t-il changé?" *Jeune Afrique/L'Intelligent* 2136, 18-24 December 2001.

42 "Angola: Holding the Cash", *Africa Confidential,* 21 November 2003.

43 See United States Senate (2004: 99). The report goes on to say that "the E.G. President controls several E.G. businesses which virtually monopolize the E.G. construction, supermarket, and hotel industries and generate significant revenues in other areas as well. The E.G. President's son apparently dominates the timber industry and also has significant revenues in other economic sectors. The E.G. President and his wife also appear

who are close to the presidency or involve them as partners if they want their business ventures to succeed. Trading contracts for the crude oil that is owned by the state are also handed out by the presidency. Oil accounts are managed personally or, in one case, even held under the President's name.[44] Essential supplies for the state are acquired above market prices from politically connected companies and middlemen.[45] Even in the much more confusing and multifaceted politics of Nigeria, in which Bongo-style politics is improbable, power-holders have exercised extraordinary influence over resources and opportunities. General Abacha, for instance, centralised all oil trading decisions at Aso Rock (the presidential palace), to the great benefit of a handful of Lebanese businessmen and selected army officers.[46] President Obasanjo, in turn, remained his own Oil Minister for six years in office (he recently nominated a trusted oil adviser, Edmund Daokuru, as a nominal "Minister of State for Oil" but continues to micro-manage the sector) and allows little real independence to institutions such as NNPC.

Those who hold important posts are very close to the President, who personally controls all crucial nominations but may allow "barons" to develop fiefdoms in exchange for political and military support. Over the

---

to control significant parcels of E.G. land which they have lease or sold to some foreign corporations."

44 See the discussion on Equatorial Guinea's oil monies and the Riggs Bank scandal in section two of Chapter 5. In the late 1970s and early 1980s, President Ahidjo of Cameroon also kept the country's oil revenues in foreign bank accounts at his discretion.

45 An example is the Angolagate scandal, concerning an estimated $633 million in arms were acquired at inflated prices that allowed the skimming of considerable sums for Angolan politicians and French intermediaries, including President Mitterrand's son, Jean-Christophe. See Fabrice Lhomme, "Sous L'Angolagate, la corruption ordinaire", *Le Monde,* 24 January 2001, Fabrice Lhomme, "Affaire Mitterrand: les secrets d'Angolagate", *Le Monde,* 13 January 2001, and Fabrice Lhomme, "Un témoignage éclaire les dessous des ventes d'armes à l'Angola", *Le Monde,* 23 April 2001, Stephen Smith and Antoine Glaser, "Les hommes de l'Angolagate", *Le Monde,* 13 January 2001.

46 The Lebanese traders included the Chagouri brothers and Ely Calil, who was allegedly the key financier of the 2004 attempted putsch in Equatorial Guinea that is briefly discussed in Chapter 5. For a profile of Calil, whose fortune was made in Nigerian oil trading deals in the Abacha and Babangida days, see "Ely Calil: Smelly, the missing essence in a coup plot", *The Times,* 5 December 2004.

years, a small number of personalities has played the "musical game of chairs" and occupied the relevant posts in defence, finance, the national bank and the national oil company. More importantly, there is great continuity as to the advisers of the president. The fifteen or so personalities that have revolved around the Futungo de Belas—the presidential palace complex that is the byword for Eduardo dos Santos' coterie—for the past decade and a half have a stranglehold over the economic life of Angola.[47] Characters such as former Oil Minister Rilwanu Lukman have been recurrent in Nigerian decision-making circles over more than thirty years. The tight nature of the inner circle has been, if anything, deepening over the years, for everywhere one finds a growing tendency towards the empowerment of the President's close relations. In addition to the familiar faces of regime barons, the inner circle increasingly features more of the president's own family. President Bongo's daughter Pascaline is vice-president of Elf Gabon; his son and putative successor, Ali, is minister of Defence; his son-in-law, Paul Toungui, is minister of Finance.[48] In Congo, Sassou has placed his uncle, his cousin and several nephews, as well as his fellow Mbochi tribesmen, in sensitive positions.[49] In Equatorial Guinea, the oil sector is under the effective command of one of the President's sons, the forestry sector under another's, and much of the cabinet hails from his mainland clan.[50]

But even if their rulers would wish them so, African regimes are not dynastic and no clear mechanism exists to guarantee peaceful transitions

---

47 "Futungo 'boys' are key to Angola power", *Africa Analysis*, 17 April 1998. See also the section on Angola in Africa Confidential, *Who's Who in Southern Africa* (Oxford: Blackwell, 1998).

48 See the useful dossier "Gabon: les 50 hommes et femmes qui *comptent*", *Jeune Afrique-l'Intelligent* 2191, 5-11 January 2003. I thank François Gaulme for a fascinating discussion of Gabonese politics (Paris, 20 April 2003).

49 See "Nepotists' nirvana", *Africa Confidential*, 30 April 2004, for a useful guide to President Sassou's trusted advisers. In addition, his son Denis ran the SNPC oil marketing operation in London and SNPC itself was run until 2005 by Bruno Itoua, a close associate and family friend whom Sasson Nguesso calls his "nephew" and who meets the President several times a week. This led a Congolese journalist to describe the oil sector in the following words: "father gives oil to cousin, cousin gives oil to son, son sells, son rings up father—it is a family venture" (author interview, Brazzaville, 7 November 2005).

50 See section two of Chapter 5.

that do not lead to the upsetting of elite lives. Presidents, in turn, prefer not to nominate a successor, in case it hastens their ousting, although would-be dauphins prop up constantly. The succession to Bongo, now in his 40th year in office, has been the talk of Libreville for two decades. José Eduardo dos Santos, after playing the retirement card in 2003,[51] has confirmed that he will be running for reelection in 2008. Both Bongo and Obiang seem keen on having their children succeed them, but this is very unpopular in both instances.[52] While it is questionable that any of the mooted presidential succession picks will come to pass, it is clear that incumbents want to guarantee that the levers of the state will remain close at hand that their fortunes and statuses will not be attacked by successors. In turn, the fear that smooth, unproblematic successions are very unlikely boosts their resolve to stay in power.

II. KLEPTOCRACY AND REFORM

The study of firm-state relations in the Gulf of Guinea in Chapters 4 and 5 shows that practices often labelled "corrupt" are in fact the normal way of proceeding in the oil state. This has been so with virtually all power-holders in the last decades. Those states in the region that are assessed by Transparency International invariably rank amongst the ten most corrupt in the world,[53] and capital flight is a significant problem in all. Because "foreign direct investment is discouraged by high corruption levels",[54] plunder and poor governance could be expected to impact negatively on the willingness of investors to brave the Gulf of Guinea and keep generating the resources that allow the prosperity of elites. But owing

---

51 Nick Shaxson, "Angolan president says he will step down", *Financial Times*, 24 August 2001.

52 This has not prevented Bongo and Obiang from pushing their sons for succession. Fear of Teodorin Obiang's takeover may be at the root of a number of internal elite conspiracies, as discussed in section two of Chapter 5.

53 The Transparency International 2006 Corruption Perception Index (CPI) ranked 163 countries in terms of the degree to which corruption is perceived to exist among politicians and public officials. In ascending order of perceived corruption, Chad was ranked 156th, Equatorial Guinea 151st, Angola, Congo-Brazzaville and Nigeria 142nd, Cameroon 138th.

54 Rose-Ackerman (1999: 2-3).

to the nature of the oil business, which is unruffled by such risks, actions that would in other contexts undermine the state are here paradoxically sustainable as they do not imperil the revenue stream. That does not mean that the oil state is bulletproof, but that elite strategies of gross mismanagement and theft that have led to state implosion elsewhere can be sustained for much longer.

Following Susan Rose-Ackerman's typology, I see the oil state as a kleptocracy where corruption is organised at the top of government by a ruler or set of rulers facing "a large number of unorganized potential bribe payers".[55] Two sets of consequences proceed from this. The first pertains to the internal politics of corruption. Instead of "competitive corruption", where each official, minister or notable is pursuing separate paths to personal accumulation, often with outside partners, to the detriment of the interests of his nominal peers, there is "monopolistic corruption", where patronage systems are under the control of a strong leader.[56] Even in Nigeria, where at any one time a myriad actors are pursuing divergent, conflicting and apparently decentralised agendas, oil sector decisions are highly concentrated. (This comment is made in the context of government relations with the oil sector. Once oil monies enter Nigerian politics, the struggle for the spoils is chaotic and certainly "competitive", not "monopolistic".) The second consequnce refers to the relationship with bribers, which are mostly foreign companies related to the oil business, including services firms. Powers of negotiation are much increased when major decisions are in the hands of one or a few "gatekeepers" and not diffused across a vast and poorly paid bureaucracy. This is heightened by the existence of other investors keen to enter the market, and the fact

---

55 Rose-Ackerman (1999: 114-15).

56 Charap and Harm (1999). While the paper tends to highlight differences in leadership, I find that this dichotomy is strongly related to the nature of the particular source of wealth. Some commodities, such as alluvial diamonds, tend to detract from the need for a collective approach, whereas others, such as petroleum revenues, necessitate it. Congo during the Lissouba years (1992-97) provides a good example of this. Differently from the oil rent, which was dealt with in a centralised manner, Congo's role in the international economy of diamond smuggling was the subject of freelancing strategies by many politicians, allegedly including President Lissouba himself. See Misser and Vallée (1997) for the best study of diamond politics in Africa, including chapters on Congo and Angola.

that some of the investment may already have been sunk.[57] Added to this is the growing business sophistication of elites (as evidenced by the brief analysis of NOC strategies in Chapter 2). The language of high finance and increased familiarity with the technical and managerial complexity of modern business culture are leading to the modernisation and sophistication of plunder. According to Global Witness, mechanisms of misappropriation have diversified to include offshore money laundering and overpriced military procurement as well as the contraction of oil-backed loans.[58] Even allowing for the substantial weaknesses of the oil state, this shows elites to be much more powerful and resourceful than the mere tools of foreign capitalism they are normally assumed to be.

Accumulation

In the 1970s, some observers saw in the elites of oil states the makings of an *haute bourgeoisie*: their capital was certainly being gathered in an unsightly bout of "primitive accumulation"[59] but would, it was thought, ultimately allow the emergence of modern economies unthinkable in capital-scarce African states. Reality could not possibly be more different, even at the time. Oil monies do not form the seed capital of a non-oil or post-oil productive economy. Elites are focused on "rent seeking, not economic activity"[60] and live off easily accessed state resources, having neither the capacity nor the inclination for freestanding production. As a Nigerian analyst put it, "rigor, discipline, investment spirit and other Weberian capitalist ethics [are] conspicuous by their absence",[61] at least domestically. Research shows that there are multiple paths of accumula-

---

57 This is particularly clear in the research that I conducted in regard to Angola, which consistently showed the presidency to be able to block entrants into the domestic market, favour firms, award contracts, and keep control over all economic opportunities in the country.

58 Global Witness (2004: 36).

59 Marx defined the "primitive accumulation of resources and labor" as capitalism's initiatory process that comes into being "dripping from head to foot from every pore, with blood and dirt" through the "expropriation of the mass of the people by a few usurpers" (quoted in Freund, 1998: 98).

60 Rose-Ackerman (1999: 131).

61 Amuwo (1995: 7).

tion within each country[62] other than oil, and that elites are fast in profiting from them. These include legal businesses (as a rule, short-term
investments) and insider (and strictly legal) access to foreign currency
at unrealistic exchange rates.[63] They also include appropriation of public
monies through the holding of high office and illegal activities from the
"black market" to drugs,[64] fraud and, particularly in the Nigerian case,
oil theft.[65]

But alternative sources of income cannot compete with oil rents and
are in fact largely dependent on them. Elite domestic business interests
are concentrated in the limited linkages to the domestic economy created
by oil, such as services and imported goods. Furthermore, misappro-
priated oil revenues are not reinvested in the domestic economy. The
substantial share that is taken abroad is either stashed in private bank
accounts or reinvested in the profitable markets of the world economy,[66]
as Table 7 shows. Needless to say, western banks (not only in the obvious tax havens but also in the world's most reputable financial centers)

---

62 See Geschiere and Konings (1993), which concludes that "informal" and "formal" (i.e.
state) paths of accumulation are actually "straddled" by political and economic actors.

63 Shaxson (2007: 9-10) retells how Angola's politically connected individuals in the
mid-1990s could exchange their worthless kwanzas (worth 20 times less in the parallel
market) for hard currency at the official rate.

64 Illegal strategies in the oil state are often complementary to state-centred rent seeking rather than alternatives to it. See, for instance, United States Department of Justice
(2001) for the dimensions of Nigerian (allegedly official) involvement in the drugs trade.
See also Bayart, Ellis and Hibou (1999) which includes discussions of criminal activities
by Nigeria's and Equatorial Guinea's elites.

65 Estimates of oil theft in Nigeria put it at around 10-15 per cent of daily production,
i.e. more than the total daily production of oil states such as Chad or Congo-Brazzaville.
Most of this consists of "large scale theft [...] involving maritime tanker transport" and
sophisticated alliances of influential politicians, army and navy officers, oil company employees, local militias and international crime syndicates. See Davis and Kemedi (2006)
for an invaluable study of the phenomenon containing a wealth of information as well as
analysis.

66 As Manuel Castells argues (1998: 91), "capital flows from African countries to personal accounts and profitable international investment "provide evidence of a substantial
private accumulation that is not reinvested in the country where the wealth is generated".

benefit immensely from these transfers, and their governments have accordingly been mostly uncooperative in tackling this perverse system.[67]

	Period Covered	Real Capital Flight ($ millions)	% of GDP	Net External Assets ($ millions)
Angola	1985-96	17032.5	267.8	9179.9
Cameroon	1970-96	13099.4	185.6	7364.4
Congo-Brazzaville	1971-96	459.2	49.6	-3986.6
Gabon	1978-96	2988.7	87. 0	717.7
Nigeria	1970-96	86761.9	367. 3	98254.4

Table 7. Indicators of capital flight ($ millions 1996)[68]

A great deal of the revenue that stays in the domestic economy is simply lost through mismanagement and the white elephant projects briefly discussed in Chapter 2. The most important destination for oil revenue that is not sent abroad is, nevertheless, consumption. Very little academic research has been conducted on the character of money spending across the Gulf of Guinea, with the partial exception of Nigeria's 1970s extrava-

---

67 The key work on this subject is Shaxson (2007). Another noted critic of capital flight and tax evasion has called attention to the fact that such looting can only take place with the "connivance of an extensive pinstripe infrastructure of banks, lawyers and accountants who provided the means for tens of billions of dollars to be shifted offshore": see Christensen (2005) for an eye-opening article on the subject. The Economist suggests that 40 per cent of Africa's privately held wealth, which often consists of siphoned-off public monies, is held offshore (Guest 2004) and a similar percentage for Nigeria has been mentioned. Anecdotal evidence shows that the favoured place for investment is property markets in Europe and North America, not financial markets, although the fortunes of some heads of state have been placed in the hands of investment managers who have invested them wisely. The case of President dos Santos' private wealth that was managed by Swiss private asset managers has been unearthed in recent investigations. See e.g. Sylvain Besson, "La fortune du président angolais José Eduardo dos Santos était gérée depuis la Suisse", Le Temps, 30 August 2003.

68 Source: Boyce and Ndikumana (2001)

ganza,[69] but the anecdotal evidence strongly indicates that consumption is the principal terminus for the oil rents that are not taken abroad.[70] The cities of the Gulf of Guinea are littered with imported luxury goods, expensive cars, overpriced restaurants, designer clothes, extravagant palaces.[71] The oil bonanza has allowed Gulf of Guinea elites, which are inserted in "global networks of wealth, power, information and communication",[72] to satiate their desire for foreign material possessions as well as their whims. During a November 2005 trip to Congo, exemplary instances of this were brought to my attention. There was the exorbitant Marseilles/Pointe-Noire yacht race,[73] which was entirely financed by the Congolese government and included the partial refurbishment of Pointe-Noire's upscale coastal strip and a lavish reception ceremony during which President Sassou gave the French winner a cheque for $50 million CFA. There was the mausoleum-like monument to Pierre Savorgnan de Brazza in downtown Brazzaville—a project very dear to the President—that now holds the remains of the French explorer. There was also the $40 million airport being built in Sassou's home village in the scarcely populated northern region.[74]

---

69 See Apter (2005) for a study of the "cultural extravaganza" that was FESTAC, the enormously expensive 1977 World Black and African Festival of Art and Culture held in Lagos. See also Barber (1982) and Watts (1994).

70 The "social and cultural processes that underlie needs, generate demand, and are satisfied in consumption", which are immensely relevant for the politics of the Gulf of Guinea, cannot be discussed here at greater length. For a brief discussion of this see Carrier (1996).

71 Frequently quoted is the recurrent presence of Cameroon, Congo and Gabon in the top five of world importers of champagne per head. In 1984, for instance, Congo was the world's fourth largest per capita consumer of champagne; Gabon was the first. See Howard Schissel, "Top of the Pops", *West Africa,* 3 June 1985.

72 Castells (1998: 92).

73 Local wits dubbed this "*la course des piroguiers*".

74 This last project is perceived by many Congolese as part of a wider "battle plan" in the event of a resumption of hostilities. It would allegedly allow Sassou to benefit from "foreign reinforcements" (conversations with the author, Brazzaville, 5-12 November 2005).

Several non-mutually exclusive explanations for this pattern of wastage can be suggested. The first is that the elite utilises the display of wealth and the rituals of power it affords—military parades of new jet fighters, uniforms and weapons, for instance—to affirm its status and the authority of the apparatus it dominates. The need to project an impression of potency is made more pressing by the empirical decline of the state. The second is that resources are "dilapidated" in the financing of patron-client networks.[75] In view of the shrinking of patronage relations to a small minority, it seems unlikely that this would consume the enormous resources funnelled into classical clientelistic systems.

The third explanation is that resources are spent in the accumulation of social capital necessary for the reproduction of elites.[76] The messiness of political competition and regime transitions in Africa should not lead one to underestimate the capacity of elites to reproduce across time.[77] This is apparent in the manner elites have sailed the rough postcolonial seas through periodical stunts of rhetorical reinvention such as the successful transition from dictatorship to some form of electoral process, from *dirigisme* to cutthroat capitalism[78] and, more recently, from opaqueness to "transparency".[79] To equate elite survival with the uninterrupted flow of oil monies would therefore be simplistic.[80] However, their current

---

75 Bayart, Mbembe and Toulabor (1992: 243) quoting Sara Berry.

76 Veille (1984: 25).

77 Bayart (1991: 220).

78 The case of Angola's "oil nomenklatura" is a good example. The elite initially feared reform because it thought it could not possibly replicate its hold of the state apparatus in the new, unfamiliar circumstances of a privatised economy. Soon enough, though, it found that the latter provided not only the opportunity for a self-imposed "recycling" but an even better context in which to prosper as well. See Ennes Ferreira (1995).

79 Witness the enthusiasm of Angola's Deputy-Prime Minister Jaime when he stated that "in obedience with the imperative of transparency in the management of the public good, the government and the national assembly, on the one hand, and national and international public opinion, on the other, will be informed of the use of public monies in general, and of oil monies in particular". See "O governo não embarcará em euforia despesista", *Semanário Angolense*, 23 October 2004.

80 Some authors see African elites as remarkably resilient. According to Bayart, Ellis and Hibou (1999: 115), "the social groups which benefited from collaboration with the European occupier, from nationalist mobilization, from the accession to independence,

status is premised on it. Without oil revenues, which are quickly spent and need continuous servicing, elites could not possibly subsist in their present form and lifestyles.

This is made clear when one looks at the incipient "entrepreneurial classes" in the region, which according to classical economic theory should be pushing for an open, competitive, efficient management. To start with, it is questionable whether we can take their entrepreneurial status at face value. Nigeria is an exemplary case. Differently from other states with a tradition of heavy-handed colonial bureaucracy and little economic freedom, Nigeria has a long-standing class of indigenous businessmen. But their achievement of economic success did not derive from "any intense mode of accumulation arising from advanced productive technologies, entrepreneurial innovation, or the organizational discipline of a dynamic, cohesive, capitalist class".[81] Rather, it came about because of Nigeria's oil endowment and the Federal State's nurturing of their fortunes that started with the first Enterprises Promotion Decree in 1972.[82] This mandated joint ventures with local businessmen for many foreign-owned operations, which meant access to windfall earnings as "sleeping partners" without having to go through the task of managing complicated businesses.[83] Interviews with entrepreneurs have shown that consumption and real estate acquisition were the first and second uses of the proceeds from joint-venture operations.[84] Three decades later, this is precisely what the Angolan elite is seeking to do with its imposition of joint ventures on most foreign businesses operating in the country. In short, local entrepreneurs are predominantly parasitical and their non-

---

from nationalization of the economy and then its liberalization, are today well placed to benefit from the eventual criminalization of the continent". My argument is not that they would necessarily "circulate" in the absence of oil, but that the specific lifestyles and manners in which they hold on to power would be significantly upset.

81 Lubeck and Watts (1994: 205).

82 See discussion in section three of Chapter 4 and also Akinsanya (1993).

83 Bienen (1988: 238) claims this was the reason why businessmen were not keen on outright expropriation: it was easier to simply let technical partners move on with the business.

84 Biersteker (1987: 148). Ironically, Nigerian entrepreneurs have exhibited competitive and innovative behaviours that are only indirectly connected with oil rents, but this is almost exclusively in informal or illegal sectors.

oil activities are oil sector-dependent. As Peter Lewis writes, there is no constituency for "productive capitalism" amongst private sector elites who are part of the same rentier economy as their public sector counterparts.[85] This is because they could not survive as such out there, away from the umbrella of oil rents.[86]

Attitudes towards reform

In the profoundly corrupt oil state, the language of anti-corruption and reform is paradoxically important, for at least four reasons, all of which involve hypocrisy on the part of power-holders. The first is to assuage the international community by making the right noises. This is particularly important in states such as Gabon, Congo and Cameroon, which because of indebtedness or dwindling oil resources are seeking a good rapport with donors and IFIs. The second reason is to occasionally show the peo-

---

85 Lewis (1994: 438). There are exceptions such as Conoil, a Nigerian independent oil producer with interests in Nigeria and the Nigerian-Saotomean JDZ, described as "the closest thing Nigeria has to an integrated oil company outside the state sector". Yet even Conoil owes much of its assets to political connections with the Babangida presidency and, more importantly, the last Minister of Oil under General Abacha, who "persuaded Elf to relinquish" valuable acreage and sell it to Conoil. See Jonathan Bearman, "Nigeria's Conoil doesn't Miss a Trick", *African Energy*, April 2004. In turn, the spectre of these political links has resulted in the failure of Conoil's bid for a good stake in the JDZ. President Obasanjo, who exercises great influence in JDZ bidding rounds, decided to favour Energy Equity Resources, a company without expertise created for the licensing round but owned by Aluko Dangote, a financial backer of Obasanjo and Vice-President Abubakar before both fell out with each other. The notion of a politically aloof entrepreneurial career is unfathomable in Nigeria. See "New Eldorado: Obasanjo Steps in to Take over JDZ Licensing", *African Energy* 74, May 2004; for a profile of Alinko Dangote, one of Nigeria's leading businessmen, see Stephen Ubanna and Dominik Umosen, "The True Nigerian Entrepreneur", *Tell* 39, 30 September 2002.

86 In a volume of essays on the "reinvention of capitalism" outside the Western industrialised states, Bayart notes that "plenty of phenomena that seem to contradict the expansion of capitalism can in time [be] propitious to the accommodation of the capitalist mode of production and thought in societies where it was previously alien" (Bayart, 1994: 42). Pursuing the same line of reasoning, Bayart and his coauthors have also hypothesised that crime can have "productive consequences" in Africa's recreation "of itself" (Bayart, Ellis and Hibou, 1999: 116). To my mind, this is certainly not taking place with oil revenues in the Gulf of Guinea.

ple that "the President listens". This goes a long way towards providing the public with some illusion of responsiveness on the part of the authorities. It was particularly at show in the mid- to late 1990s in Angola, when the president publicly sacked numerous officials and brought in "reformers" to clean up the mess, or more recently in Chad, when the president suspended three ministers for having signed oil contracts which he himself had approved.[87]

The third reason is to get rid of enemies and critics. Undoubtedly Nigeria's General Abacha provided the best example of this when he started out his period of extraordinary plunder in office with a very popular "war on corruption" reminiscent of General Buhari's effort in 1983-85.[88] This included an investigation of what had happened to the $12 billion Gulf War windfall that effectively sidelined Babangida-era politicians not on good terms with the new regime.[89] In Congo, President Sassou did not hesitate to have his predecessor Pascal Lissouba convicted on corruption charges to ensure the latter would not return to the country.

The fourth reason to engage in "reform" is that, in extreme circumstances of absolute mismanagement, of which there are many, it may just be the only thing left to do. The system occasionally needs revamping or enhanced sophistication, and reform-lite shorn of its elite-unfriendly implications is not unwelcome in such circumstances. This happened in the Angolan case already mentioned, when economic disruption was not only in the public's eye, but in the state coffers as well. It may again be

---

87 Andrew England, "President tells oil companies to leave Chad", *Financial Times* 28 August 2006.

88 The search for culprits for parastatal inefficiency and bank and tax fraud led to the arrest of numerous members of the elite, while other known thieves were given ministerial posts. As William Reno notes, "the end to tax evasion", which focused exclusively on the regime's critics, was initially welcomed by reform-minded outsiders, in a perfect example of how IFI-inspired "reforms" can be used as a resource to cope with internal rivalries. See Reno (1998: 204-5).

89 The investigation resulted in the Okigbo report that subsequently fell into obscurity and was never the basis for prosecution of those involved. In a recent television interview, President Obasanjo claimed not to know its contents. One of Nigeria's main newsmagazines lamented that "the Okigbo report seems to be as dead as Sani Abacha". See Dare Babarinsa, "Fall of the Mighty", *Tell*, 18 April 2005.

happening at the present time for different reasons, as the presidency adapts its grip over the economy to a peacetime gear.

None of the states under analysis has a clear, consistent assemblage of independent-minded technocrats allowed to do their thing such as Indonesia's Berkeley Mafia or Chile's Chicago Boys.[90] But the rhetoric of the technocrat is significant, and especially since the 1980s, when Western-induced reforms were forced upon the region, "reform" is normally heralded by the nomination of one such team. The reform drive is always presented as a moment of rebirth, and may coincide with the mostly deceptive but nonetheless strong-worded dismissal of the previous lot of presidential appointees. (They will then languish for a few years in cozy ambassadorial posts or other such out-of-the-way sinecures, and return unannounced at some future stage.) The president uses the occasion to acknowledge policy failure and, by creating scapegoats, distance himself from the results, while making vague promises that are often interpreted by the IFIs to mean that the next team will be more open to their recommendations. Following the August 2002 elections in Congo, Sassou dropped 80 per cent of his "old" cabinet and brought in a "a new cabinet of young technocrats from the private sector", which had the public hoping that the President "is serious about shaking up corruption, nepotism and mismanagement".[91] In the aftermath of Cameroon's 1997 elections (which he won handsomely), Paul Biya enacted an 80 per cent renewal of parliament and 45 per cent renewal in ministerial posts.[92] After years of hostile relations with the IMF and the World Bank, President Eduardo dos Santos nominated a "star economic team"[93] headed by a former IMF

---

90 Nigeria had her moment under the "super-permanent secretaries" (i.e. senior civil servants with political clout) during General Gowon's rule (1966-75), but that was hardly a primer for efficient and uncorrupt administration. Differently from other areas of the developing world where the "Washington consensus" was partly bought into by segments of local elites, especially those trained in US universities, sub-Saharan Africa saw little if any genuine conversion to supply-side economics. See the thought-provoking Santiso (2005) for an essay that explores Latin American dealings with successive "paradigms" before turning towards, the author argues, a "political economy of the possible".

91 "Old Guard, New Guard", *Africa Confidential,* 30 August 2002.

92 Eboko (1999: 99-ff).

93 Interview with Angolan economist, Lisbon, 25 September 2003.

official to shore up its credibility and improve the chances of a post-war donors' conference.

Reformist technocrats tend to fall into two camps. The first is that of the faux technocrat. He has the impressive credentials, the international CV and the ease with reform language, but is in no way an outsider to the elite game. While presented in the guise of renovation, he has held previous appointments and is close to the President. His credibility will be deployed to improve relations with the international community, push through basic reforms that do not imperil the elite, and improve macroeconomic management—but under the imperative of furthering, and never undermining, the elite's grasp of the oil state. High profile examples include the aforementioned Angolan reformist team[94] and Casimir Oye Mba, nominated Prime Minister of Gabon at a time of great pressure in the early 1990s,[95] but the current consensus around "good governance" and engagement with moderate reform efforts will tend to promote more such figures into prominent positions.

The second camp is that of the true reformer, who may be pressed upon the president as the only way to deal with debtors and external criticism or be part of a genuine if partial progressive agenda. This seemed to be the case with President Obasanjo's recruitment of an impeccable team of technocrats to clean up some of Nigeria's decades-long mess. These included Ngozi Okonjo-Iweala, a Harvard and MIT-trained former vice-president of the World Bank, as Finance Minister,[96] Obi Ezekwesili as Minister of Solid Minerals,[97] Charles Soludo as Governor of the Central Bank of Nigeria and Nuhu Ribadu as chairman on the Economic and Financial Crimes Commission (EFCC). Amongst their high profile targets

---

94 The two central figures here are José Pedro Morais, Finance Minister, and Aguinaldo Jaime, Deputy-Prime Minister. President dos Santos nominated them but kept all the high-profile rent-seekers in the cabinet (Angola Country Report, EIU, February 2003, p. 12).

95 Vallée (1999b: 73-8).

96 See Simon Robinson, "The Corruption Cop", *Time,* 11 October 2004 for a profile of the Nigerian Finance Minister.

97 Ezekwesili was formerly Special Assistant to the President and a founding member of Transparency International, and leads Nigeria's fight against corruption. She is also responsible for the Nigerian Extractive Industries Transparency Initiative.

were the "419" scam industry,[98] the opaque banking sector, the estimated $3 billion siphoned off by Abacha to the safety of Western banks[99] and the debt burden. Their ostensible goal was the restructuring of important elements of the Nigerian economy. Many of the measures they have enacted—such as the online publication of how revenues are shared at the federal, state and local levels[100] and the submission of the names of 14 state governors to the British authorities over money laundering charges[101]—were unthinkable only a few years ago.[102]

The actual results of the reform process fall well short of the hype. The reform team seems to understand that it cannot take on the most prominent Nigerians and its actions are of necessity "highly selective".[103] Many, including former President Babangida, are exempt from the anti-corruption zeal. Some of the gravest issues such as the combating of oil theft by well-connected crime syndicates are notoriously neglected. In turn, opponents and expendable allies of the President have been subjected to a high degree of surveillance and repression. In addition to the widespread election fraud of 2003,[104] this seems to be part of a pattern whereby reformist, democratic language is used but "democratic principles [are thrown] away on a daily basis".[105] This double standard has led analysts

---

98 This is the Nigerian name for fraudulent schemes by confidence men (419 is the relevant section of the Nigerian criminal code which outlaws fraud).

99 See "Swiss government to release N66.5b Abacha loot", *The Guardian* (Lagos), 22 September 2004, "Late Nigerian dictator looted nearly $500 Million, Swiss say", *New York Times,* 19 August 2004; and Haig Simonian, "Swiss court orders repayments of Abacha funds to Nigeria", *Financial Times,* 17 February 2005. Official investigations have also shown that nearly two dozen banks in Britain hold accounts linked to Abacha family members.

100 www.fmf.gov.ng (accessed 23 June 2004 and 14 May 2005; the information seemed to be incomplete and out of date).

101 In addition to the dismissal and arrest of Plateau State Governor Joshua Dariye and Bayelsa State Governor Diepreye Alamieyeseigha, other cases against major figures on anti-corruption grounds include those of former Senate President Adolphus Wabara and Tafa Balogun, the former head of the police.

102 See Evans, Hull and Davis (2005) for a good discussion of the pervasive of money laundering activities in Nigeria and by Nigerian politicians abroad.

103 Peel (2005: 8).

104 See footnote 167 of Chapter 6 for a discussion of this.

105 Peel (2005: 8).

to note that dealing with enemies through resort to the anti-corruption agenda is a time-honoured practice in Nigerian politics rather than something new. It is often not clear if Obasanjo himself, who was not paying Finance Minister Okonjo-Iweala's salary,[106] was always supportive of her agenda and that of her colleagues.[107] Shortly after having hired Okonjo-Iweala, Obasanjo sought to have the planning and budget departments removed from the Finance Ministry to the Presidency. While this did not take place and the minister was allowed to go on and negotiate a substantial debt write-off for her country, it highlighted the extent to which the reform process was subsidiary to President Obasanjo's broader political goals and could on occasion be nullified by it. Once the usefulness of Mrs Okonjo-Iweala had been spent and she became an unwelcome, thrifty presence on the eve of a necessarily big-spending election year, she was eased out of government. In Nigeria and elsewhere in the region, true reformers have limited scope for action and their political fortunes, in the absence of a strong body of supporters, can change overnight. The reason for this is that the reforms they propose are highly detrimental for elite vested interests. Elites stand to lose from any relaxation of their economic and political control and rightly see stringent approaches to reform as threats to themselves.

This means that while there are individuals supported by small constituencies who work towards improving the management of the oil state, the political and economic costs of reform are too great for elites to even

---

106 This is close to $240,000 per year, as opposed to the average ministerial salary of $6,000 per year, and is being paid by the UN as part of a programme to repatriate top Nigerian expatriates. According to Paul Vallely, "The woman who has the power to change Africa", *The Independent*, 27 June 2006, the minister gave up this salary in the face of harsh criticism in 2005.

107 Countless examples of the ambivalence of Obasanjo's reformism could be mentioned. The Nigerian press was particularly incensed at the 2003 granting of marginal oil fields to 31 Nigerian companies mostly headed by "former dignitaries of NNPC" very close to Obasanjo. See "Oil Barons Back in Business", *Africa Energy Intelligence,* 6 March 2003. This contrasted with the pledge for transparency made by President Obasanjo in 1999 when, shortly after being elected, he revoked similar contracts awarded to insiders in the last days of the Abacha/Abubakar dictatorship. Perhaps an even better example is the Nigerian policy towards the São Tomé and Príncipe oil sector, discussed in section two of Chapter 5.

consider them seriously.[108] Reform has thus become, as Bayart puts it in
a different context, a "pidgin language" between local rulers and outsiders, perpetually on the table, ever in motion, yet lacking political will
for truthful implementation.[109] The point is not new, having been made
by many scholars in the broader Sub-Saharan African context,[110] but it
deserves to be underlined since the lesson remains unlearnt. However
severe the "crisis" discussed in previous chapters, opaque approaches to
governance remain preferable to transparent alternatives.[111] Proper accounting, for instance, might give access to loans from the lenders of
last resort and the regularisation of relations with other creditors. But it
would also substantially diminish the leeway of local rulers, and, bearing
in mind the precariousness of political life in the region, many prefer
things as they are. These are the political conditions under which, Reno
points out, "the logic of the political network overrides considerations

---

108 Commenting on this, Callaghy expressed scepticism about the positive impact of
"technocratic enclaves", noting that "bureaucratic rescue personnel in little rubber boats
are not going to have a major impact on the patrimonial sea of African political economy".
See Callaghy (1987: 107).

109 Bayart (2000: 226).

110 Clapham writes that most reform measures strike "directly at the basis for economic and hence political control maintained by African governing elites" who cannot,
therefore, "be expected to acquiesce readily in their own marginalisation". See Clapham
(1996b: 812).

111 Reform is in turn likely if the costs of non-reform are too great. When an audit of
SNH became a key demand for re-engagement with the IFIs and qualification for a PRGF
and HIPC status, President Biya accepted that it was in his best interest to ease the un-
questioned presidential monopoly over what now amounts to a fast disappearing source
of revenue. The resultant highly critical audit of SNH, conducted by British consultants
Gaffney & Cline in 2000, is one of the best analyses of NOC workings in the Gulf of
Guinea. Cameroon's shrinking oil reserves, together with the declining importance of
oil revenue as a source of patronage, have created the context for significant efforts at
reform—paradoxically, the chances for success are closely intertwined with the growing
irrelevance of the sector under reform. Similarly, the state of public finances in Congo
over the past years has been conducive to some form of reform. There is, however, no
guarantee that the course will be stayed if oil revenue streams increase considerably.

of efficiency".[112] The upshot of this is that reform is initially postponed[113] for as long as possible and is then pursued in a dishonest way. When reforms are enacted, either they have a self-serving purpose that makes them palatable for elites, or there are loopholes aplenty to counteract their stringency.

Sustainability

The perennial character of the crisis of the oil state and its enduring indebtedness and political instability have led commentators to question the wisdom or indeed the rationality of the political actors at the helm of the state. Surely it would be in the interests of elites to build a sturdier state edifice, guarantee that at least some of the resources are used to buy the social peace that can only benefit them, access credit at normal interest rates and not those of oil-backed loans, etc. This is to misunderstand a number of things about petroleum elites. One is the tenuous nature of elites' hold on power and their fear that the proper institutionalisation of the state might detract from their hold over resources. Another is the limited extent to which elites, if they wanted to, could pursue their goals in a different manner: many simply do not have the competence, the human resources, and the technology of administration to run a full-fledged modern state. Yet another is the fact that elites may feel their goals are being realised in the present state of affairs, and that no meaningful adjustment is warranted. One should be careful not to attribute to Gulf of Guinea elites goals such as state-building in manners pursued elsewhere, provision of public goods and broad-based prosperity, which they do not share with their critics. They may simply not have a normative vision for their polity beyond the wellbeing of their friends, family and narrowly defined community.[114]

---

112 Reno (1998).

113 The title of an article on the resumption of talks between the IMF and the Angolan government read, in an apt display of the predictability of outcomes, "Angola Dances IMF Minuet—again", *Africa Analysis,* 19 March 1999.

114 In this, they present a remarkable difference from countries in the Middle East (and most other oil exporters in the developing world) where great amounts of oil money *have* been spent trying to buy off the population, however inefficiently. As mentioned in Chapter 2, in the Gulf of Guinea this happened only to a moderate extent in the 1970s

Elite political choices may therefore be seen as selfish but not irrational. As Chabal and Daloz write, the current arrangements are "most satisfactory, at least from the microsociological perspective of the individuals and communities they serve".[115] The neglect of efficiency and production of tragic outcomes for the majority of the population have nothing to do with irrationality; that need not be the subject of lengthy debate. In the case of Russia, for example, it is commonly agreed that the Soviet state was, in the early 1990s, methodically imploded for the benefit of an oligarchic minority: the detrimental effects for just about everyone else's wellbeing are never marshalled as evidence to disprove the successfully self-serving plot.[116] Only in Sub-Saharan Africa must the analyst contend with the idea that political action is senseless.

It cannot be denied, however, that the timeframe within which this rationality is pursued is distinctly short-term and that the perceptions of time by the various actors of the oil game are very different. While oil firms may think in decades, local elites will be only intermittently concerned with whether or not certain strategies may be so damaging as to diminish their own possibilities of future enrichment. As analysts of Nigerian politics know, those who hold public office there often act on the basis of extraordinarily short-term concerns.[117] Even in contexts such as Angola or Gabon, where elites tend to frame their statist ambitions and expectations of survival in a more forward-looking manner, this is frequently tempered by urgent considerations regarding political challenges, the risk of inadvertently triggering default with creditors, or

---

boom years, and seems not to be taking place in a meaningful way at the present time. Why the spending patterns of African oil states should diverge so much from the norm and why, in relative terms, they give so little attention to trying to buy off legitimacy are questions beyond the scope of this book.

115 Chabal and Daloz (1999: 15).

116 There is a good body of literature on Russia's 1990s brand of public sector reform/ privatisation. Among many, see Gustafson (2002) and Hoffman (2002).

117 An example of this is the way in which NNPC's operational capital was plundered to the detriment of its outstanding obligations to joint venture partners in 1996 and 1997: this jeopardised operations more than any oil rig attack by disaffected youth. See "The Oil Hostage", *Africa Confidential,* 29 August 1997.

plain greed.[118] Moreover, everywhere across the Gulf of Guinea elites find it hard to sustain cooperation for a prolonged amount of time, and endure in a condition of "turbulent equilibrium" that is akin to that of Mafia formations rather than secure ruling classes.[119]

There is nonetheless a considerable difference between oil states' elites and those of other states in Sub-Saharan Africa, where "sub-optimal" rent-seeking leads "leaders to steal so much from the state that [it] begins to crumble and, along with it, the opportunities for future [rents]".[120] The future rents of the oil state are not dependent on a state capacity that can be nurtured or eroded, but exist by virtue of the willingness of foreign oil firms to brave difficult environments in search of petroleum. Subject to the flow of oil rents, the deportment, lifestyles and decisions of petro-elites are sustainable.

Conclusion

This chapter has argued that petroleum elites have absolute domestic control over oil profits and that this is exercised in a self-serving way, leading to the enrichment of a few and the deepening impoverishment of the many. Stating this is neither to exaggerate the relevance of elites nor to demonise them. Richard Joseph's claim that Nigerian masses believe their votes "should yield immediate dividends to their communities and themselves in the apportioning of the 'national cake'"[121] is probably true for the populations of other Gulf of Guinea states as well. There is

---

118 This may explain the behaviour of the MPLA towards the Angolan state. On the one hand, the MPLA has a developed nationalist mythology, a discourse that purports to have state-building ambitions for Angola, and a "visceral belief in its legitimacy to lead" for which it "has drawn up plans until 2025" (Vines *et al*, 2005: 21). On the other hand it has plundered the state and exported its hard currency as if there were no tomorrow.

119 See Gambetta (1988: 158-75). Although there is no space here to pursue the analogy between the leaderships of weak or failed states, on the one hand, and crime syndicates, on the other, this is certainly an interesting avenue of inquiry.

120 Herbst (2000: 133).

121 Joseph (1987: 3). Joseph further notes that each political act "is not just an act of individual greed or ambition but concurrently the satisfaction of the short-term objectives of a sub-set of the general population".

certainly a moral universe of clientelism and particularistic attachments that is shared by elites and the rest of the population, although attempts at collapsing the divide between the two entirely avoid the fact that the latter are mostly excluded from access to the benefits of oil wealth.[122] Occasional efforts at soothing popular discontent may take place sporadically but nothing on the scale of the inefficient, but doubtlessly massive, attempts at buying off the population in the Middle East or Venezuela. The point remains that Gulf of Guinea elites are first and foremost politically sustained by their relationship with the international system, not by a putative domestic legitimacy, in the same way that they depend on external rents, not on internal tax gathering.

The deep-seated structural constraints of African oil economies are insufficient to account for the predicament of Gulf of Guinea states. There is no doubt that even if petro-elites were benevolent, substantial problems would remain in what are otherwise backward, non-competitive and unappealing economies. But, as things stand, predatory rule by elites is a key explanatory factor for decay.[123] In addition to generic bad handling that brought about state ruin, specific political decisions in the context of economic decline, such as the rollback of public goods provision instead of a curtailment of ostentatious consumption and expenses of representation, account for the vertiginous regional drop in human indicators. The patrimonial mode of state management, with its attendant lack of bureaucratic rationality and calculability, has greatly undermined the state,[124] whose decay stems from the contradiction between the economic logic of proper state management and the political logic

---

122 This is the case of Patrick Chabal, Jean-Pascal Daloz and also of *Politique Africaine*-related authors such as Bayart. They argue that there is no "contradiction between the values and attitudes of Africans and the ways in which they are governed" (Clapham, 1994: 433-4). Their hypothesis has the merit of exposing the weaknesses of class-based analysis and of revealing a shared language of politics. Nonetheless, it seems at odds with the fact that the material basis for bringing together the "*haut*" and the "*bas*" in a clientelistic embrace has eroded with the privatisation of the state and the decline of clientelism. Bazenguissa-Ganga (1996a: 11) adopts the opposite approach, defending the existence of two categories of actors, "rulers" and "ruled", which are inserted in "different modes of social existence".

123 Lewis (1994: 438).

124 Watts and Lubeck (1983: 107ff).

of empowered actors. Most conceivable reforms to this management style would undermine the leakage potential of the oil economy, and are therefore unwelcome.

Can such strategies last for long? When discussing the "adaptability and resourcefulness" of ruling classes, Albert Hirschman cautions against underestimating them because the idea that (and he goes on to quote Lenin) "there is absolutely no way out of a crisis [for them] is a mistake. There is no such thing as an absolutely hopeless situation."[125] As in other moments in history, Gulf of Guinea elites remain the invaluable gatekeepers, the "tricksters who [profit] from their function as political and cultural intermediaries [with the outside] for economic gain".[126] Although dependent on resources procured in the international economy, elites have remarkable internal autonomy from both domestic non-elite constituencies and from foreigners keen on striking a good long-term partnership with those who alone permit outsiders' access. Threats to their status come not from popular politics or external impingement but from intra-elite rivalries. However effective from the viewpoint of private misappropriation, the appalling political order they have constructed is devoid of a vision for the state and blind to the notion of a national community. Gulf of Guinea elites are not state-building elites. And in much the same manner, challenges to elite domination are profoundly non-ideological, consisting of a naked struggle for spoils by like-minded political actors.

---

125 Hirschman (1981: x).
126 Bayart, Ellis and Hibou (1999: 43).

4

OIL COMPANIES AND THE STATE IN
THE GULF OF GUINEA: A HISTORY

Especially in Africa, and especially in the natural resource business, multinational firms and local political elites have worked together, in each other's interests rather than in the national interest

Bill Emmott, *20:21 Vision*

Like it or not, oil companies become entangled in the politics of countries where they are the main source of income

"African Energy", *Africa Confidential*

Karl Polanyi once referred to the "metaphysical extraterritoriality" of the bankers who, in his view, held together the economic order of the gold standard.[1] This expression perfectly captures the size, importance, sheer economic strength and, most of all, elusiveness of multinational oil companies that are amongst the world's leading non-state entities and are, indeed, wealthier than many developing countries. Consequently, the state-multinational firm rapport is often misconstrued. Views of the development-bearing corporation are pitched against equally unsophisticated portraits of foreign investment as the Trojan horse of imperialism. The lesson that partnerships between host-state and firm are based "on a functional exchange of benefits and opportunities"[2] remains unheeded by much writing on the subject. The aforementioned bias towards "West-

---

1 Polanyi (1957: 10).
2 Strange (1992: 6).

West issues"[3] permeating research on the politics of the international economy has led to a decline in interest in unequal commercial transactions between developed and developing countries. That which remains is more often than not informed by dependency-type dichotomies that misrepresent the impact of foreign investment and severely neglect the role of local actors and institutions in shaping the links between companies and the domestic realm.

Here I seek to partly remedy such readings by highlighting the mutual constraints and opportunities that have characterised the oil relationship in the Gulf of Guinea across time. My aim is to chart a history of oil investment in the Gulf of Guinea from the early twentieth century to the early 1990s, mostly concentrating on the subsidiaries of five multinational firms that are currently responsible for roughly 90 per cent of oil output across the region. These are Total (previously TotalfinaElf, a conglomerate bringing together two French companies, Total and Elf-Aquitaine, and Belgium's Petrofina), Royal Dutch/Shell and Chevron, ExxonMobil and AGIP, the upstream arm of Italy's former national oil company, ENI. The focus is on companies' relationships with host states and their approaches to local politics.

This chapter makes four claims. The first concerns the long-term nature of oil investment across the Gulf of Guinea and its origins in colonial sponsorship. The second is related to the manner in which oil firms are constitutive of local politics by virtue of their presence and their disproportionate contribution to state revenues. Thirdly, far from the stereotype of heavy-handed tactics, companies have for the most part been politically tame and have prioritised an intimate rapport with incumbents. This, together with the evident host-state dependence on foreign technology, ensured their relatively unscathed emergence from countless local mishaps. Fourthly, state absence does not mean the absence of the state. What is often portrayed as the retreat or failure of the state and the absorption of its functions by the private sector is rather a dynamic mode of political action being pursued by state actors for their own benefit. This process of "discharge" (already outlined in Chapter 2)

---

3 Strange (1994) is one of the best discussions of their centrality to IPE research. Risse-Kappen (1995) is an example of an influential collection that mostly shirks North-South issues.

is a leitmotif of state relations with large-scale extractive industries in the region, dating back to the onset of colonial exploitation. It has decidedly concurred to place the private sector in a pivotal position vis-à-vis the local state and society, and to permanently shape their interaction with the global economy. But, and this last claim is of the essence, it has also positioned the state in the role of articulate enabler and partner that no economically deterministic analysis will grasp.

The chapter advances in a chronological manner, with some thematic detours to underline common issues. Section one studies the efforts by oil companies to research and develop oil resources across the region until the end of the colonial period. It uses evidence from other extractive industries active then to flesh out key elements of state-firm relations. It seeks to show the close links between the colonial state and investment in the oil sector and the importance of incumbency for business success in years to come. Section two focuses on oil development from around 1960 until the eve of the 1973-74 oil crisis, examining the establishment of a postcolonial *modus vivendi* between the newly independent states and oil companies and the manner in which the descent into political turmoil did not hinder oil sector investment. Section three discusses the brief post-1973 era of economic nationalism that affected all states in the Gulf of Guinea and the way in which state-oil firm relations were spared and even deepened when other investors suffered hostility or expropriation. The section ends with the demise of the Cold War and the simultaneous technological discoveries that would fundamentally intensify the insertion of the Gulf of Guinea in the international economy, a sweeping change that is the focus of Chapter 5.

Before proceeding, I want to add some comments on the methodology employed in this analysis. Even while neglecting less prominent companies for the sake of focus, the scope of inquiry remains broad. This is compounded by substantial gaps in our knowledge of oil company activity and, in some cases, of the company itself, beyond more or less agiographic firm-sponsored histories.[4] As Mira Wilkins points out,[5] the

---

4 Hast (1991) is a useful directory of company histories that, while analytically limited, provides the best available short descriptions. See also Neff and Williford (1995) and the useful Horsnell (1999).

5 Introduction to Hast (1991: x).

history of Western business abroad has been mostly overlooked, even in cases where operations of parent companies received a measure of scrutiny. This is certainly the case with the scarce bibliography on the present subject.[6] References to it lie strewn across footnotes and parenthetical remarks but have rarely benefited from focused inquiry. Petroleum-related activity in Africa has simply not been the subject of sustained interest by oil experts, with the exception of Nigeria. Institutional memories are notoriously selective, and Africa seems to play a very slender role in those of the major oil companies. Such neglect owes to its peripheral character in the pre-1990s political geography of petroleum. With the exception of Elf, and to a much lesser extent Shell, none of the majors has traditionally derived a significant proportion of profits from the region, which is also why an Africa angle is absent from most company histories. Neither has oil ranked high for analysts of regional politics. Again with the exception of Nigeria, its importance, and that of the private sector more generally, is either underplayed or merely stated without careful study.[7] There are reasons that partly account for this. Firstly, as mentioned in the Introduction, it is very difficult to conduct the necessary field research in many of the relevant locations. Secondly, companies are uninterested in collaborating with investigations into the murkier side of their operations and

---

6 Company histories tend to concentrate on corporate development, the rapport with the home country, the firm's role in particular moments of international crisis (e.g. the Suez occupation, the 1973 boycott) and its internal politics. Correspondingly, they mostly underplay investment in peripheral locations and the linkages between "direct foreign investment and the colonial experience". See Wilkins (1977: 594). Notable exceptions include Bamberg's close study of BP's operations in Iran, although one could argue that, in view of their centrality to the company's overall activity, it could hardly have been otherwise. See Bamberg (2000).

7 In the Angolan case, this is particularly galling: there is no article-length treatment of Chevron operations in Cabinda despite its centrality for Angolan state survival. This should be the pivotal research clue for Angolanists, but it is not. Most importantly, few attempts have been made to scrutinise the operations of particular companies outside the usual host-state/firm bilateral equation. Fine accounts of Shell's presence in Nigeria, for instance, utterly ignore its role in Gabon; and Elf's activities in Angola have never benefited from the attention they garnered in Congo-Brazzaville.

fear jeopardising their otherwise hassle-free rapport with host states.[8] Thirdly, oil multinationals, mirroring the opaque accountancy of NOCs, rarely publish separate annual reports for local subsidiaries.

For these reasons an exhaustive, encyclopedic study of the subject is presently unattainable. But the lack of knowledge about oil firm operations in the region is such that a singular company focus, while welcome as an important contribution in its own right, would not of itself unravel the broader political processes that have accompanied oil investment in the Gulf of Guinea.[9] Firm-specific research by company historians has allowed partial access to archives but results have often been anodyne and mostly presuppose a pro-business approach.[10] It is not surprising that few, if any, company historians ever come up with unflattering portraits of their employers.[11] Independent investigations have been more critical but their data are frequently incomplete and unsystematic. In the absence of breakthroughs such as the opening up of Elf's secrets,[12] authoritative knowledge of individual company operations in remote locations and across a substantial timeframe remains beyond our reach.[13] A more feasible alternative to both the exhaustive approach and the micro-studies

---

8 In his survey of imperial business in Africa (Hopkins, 1976a and Hopkins, 1976b), A. G. Hopkins expressed optimism that companies could be persuaded to open their archives, especially those regarding their earliest activities. The potential impact of revelations concerning the late colonial period seems quite limited indeed, but such materials remain mostly unavailable.

9 This assertion is made in the context of the current need for an overview of corporate activity in the Gulf of Guinea. In the medium term, the opposite applies: only a series of company-specific studies by dedicated business historians can ameliorate the seriousness of this analytical gap. Whether meaningful access to company records can be achieved is a different matter.

10 A case in point is Howarth (1997) perhaps the most uncritical and celebratory history among a number of company-sponsored commemorative editions.

11 Unless the object of derision is an internally agreed one, like a vilified CEO (e.g. Henri Deterding, the Royal Dutch head who ended up as a Nazi sympathiser) or a string of bad management decisions: in short, happenings accepted as negative that have safely made it into the company's lore.

12 See section three of Chapter 5.

13 As noted by Hertner and Jones (1986: 1) the fact that most companies publish only aggregate data instead of separate data for each subsidiary means that even a "quantitative" analysis of individual operations will remain essentially speculative.

is that of a general historical overview of corporate operations. While not resolving all questions, it furnishes an unprecedented long view of corporate practices in the region, from which country- and company-specific inquests can proceed.

I. THE PRIVATE SECTOR AND THE COLONIAL STATE

Central Africa's late nineteenth- and early twentieth-century engagement with the private sector was neither auspicious nor particularly reward-ing.[14] It led to the premature deaths of a considerable number of Africans and the stripping away of important assets in an unsustainable manner. After about four decades of rogue capitalism that yielded both fantastic gains and a sizeable number of investment failures, the 1920s saw the institutionalisation of the more successful operators and an ameliora-tion of their methods of extraction. Some widely disseminated instances of predatory activity notwithstanding,[15] the early history of business in Africa remains under-researched to the extent that a leading authority on the subject classified perceptions of it "as having the status of beliefs rather than knowledge".[16] The relevance of this handicap goes beyond the issues that concern me here and may well constitute a serious omission in our understanding of the colonial experience itself.

Faced with such a scarce body of evidence and analysis, one must beware of careless generalisations about the consequences of colonial non-state investment across sub-Saharan Africa. What is known points to a number of different outcomes according to the specific setting, the

---

14 There is a longer history of private sector engagement with sub-Saharan Africa that cannot be broached here. See, for example, Austen (1987).

15 Joseph Conrad's turn of the century works—the short story *An Outpost of Progress* and the novella *Heart of Darkness*—and the string of anti-Congo Free State publications such as the 1904 Casement Report jointly form a grievous indictment of King Leopold's murder-ous private domain. Ascherson (2001) and Hochschild (1999) are widely read histories of the plunder of Congo. Coquery-Vidrovitch (1972) is an impressive and unsurpassed history of concessionaires in what was then French Equatorial Africa (present-day Chad, Central African Republic, Congo-Brazzaville and Gabon). The labour conditions preva-lent across the region until the late 1930s were memorably denounced in André Gide's diaries of his journey to the region (Gide, 1927).

16 Hopkins (1976b: 267).

sector concerned, the timing of involvement and the size and goals of different firms.[17] Colonial companies varied tremendously in terms of profitability, structure, method of employment[18] and, as one moves away from the first decades, the degree to which they can be said to have negatively impacted on local economies and societies. But they did share a key feature. Their establishment was underpinned by a framework created, maintained and sometimes upset by the colonial state, whether directly as with state-sponsored firms, or indirectly, as with investment by private interests.

Early state-firm partnerships

A look at the shoddy or absent administrative arrangements of most colonies during the early decades of occupation is informative here, because lack of institutionalisation in fact disguised economies shaped by the subcontracting of state functions by the state itself. In turn, this was a pragmatic attempt at fulfilling necessary tasks that surpassed the financial resources available to the weak colonial state and, crucially, the political will it could muster at the metropolitan level. Colonial firms, Kahler notes, "relied upon the colonial state not only for the creation of markets",[19] but also for measures of state support that served to maintain their profitability in a difficult environment, such as tariffs and subsidies. This made "the development of a market economy in Africa [...] a po-

---

17 The reference work for private investment in Africa remains Coquery-Vidrovitch (1983). The essays contained in this collaborative effort by about seventy authors are of very uneven quality (that on Elf Aquitaine is particularly poor), but as an ensemble they remain matchless, especially in regard to the colonial era. In her introduction, the editor underlines the need for the "accumulation of a series of monographs on [individual] companies" without which one is circumscribed to "an excessively distanced vision" of their activities (1983: 8). Hopkins (1976a) and Hopkins (1976b), is a survey on colonial business. It looks at around fifty studies "which can be regarded primarily as contributions to the business history of expatriates firms operating in colonial Africa", concluding that "the number of company histories which meets professional standards is very small indeed" (1976b: 267).

18 Hopkins (1976b).

19 Kahler (1981: 385).

litical act"[20] that confirmed Polanyi's perception of market economies as dependent on "continuous, centrally organized, and controlled interventionism".[21] This was so regardless of the colonial state's quasi-invisibility on the ground. It is noteworthy that, in the Gulf of Guinea, the arrival of modern business coincides with the establishment of the colonial state and that, therefore, their development proceeded in tandem.

Such was the case with the concessionary companies that took over the day-to-day running of much of AEF in 1899.[22] After protracted debates on how best to exploit this vast and impoverished territory, it was decided that forty companies would be given territorial monopolies of resource extraction, mostly of wild rubber and ivory. Coquery-Vidrovitch[23] recognises the influence of a strong entrepreneurial lobby eager to privatise exploitation, but ultimately stresses the metropolis' desire to do empire on the cheap. This is evident throughout. In the law that opened room for concessionaires, for instance, there was explicit recommendation that no sovereign functions should be delegated to the companies,[24] but in practice this was ignored. The same law stated that militia forces should not be put under the direct command of the companies, while adding that militiamen's salaries were to be maintained by the companies themselves.[25] Provisions were also made for state regulation,

---

20 Kahler (1981: 384).

21 Polanyi (1957: 157).

22 As Roland Pourtier notes, the consequences of company actions in AEF are sufficiently differentiated to merit separate analyses, as in the work of Coquery-Vidrovitch. The portrait presented here is nonetheless congruent with the record of the majority of operators. See Pourtier (1989: 136, vol. 2).

23 Coquery-Vidrovitch (1972: 25).

24 Coquery-Vidrovitch (1972: 52). Pourtier (1989: 81-7, vol. 2) points to 1899 administrative references to "a protection, or sovereignty, tax" to be collected by the companies for the colonial state. Only in the 1920s did companies' tax collection abuses lead to devolution of fiscal responsibilities back to the state.

25 Coquery-Vidrovitch (1972: 105) describes about three decades during which companies made extensive use of violence with the tacit complicity of the colonial state. This took three forms: 1) pressure over the labour force through intermediary chieftains; 2) in their absence, direct and uncontrollable application of force; 3) applied force with the unstated support of the local administration.

but the means were never put in place.[26] In reality, concessionaires soon developed into what became known in Portuguese Africa as "majestic companies", in an expression of both abnormal size and quasi-sovereign deportment in the backlands they oversaw.[27]

The parallel with present-day or even contemporaneous oil firms should be qualified. Concessionary companies were undercapitalised and understaffed to a dramatic extent; they were technologically backward and mostly failed as business ventures. They certainly failed to provide the colonial state with the fiscal returns initially expected;[28] oil, on the other hand, has disappointed few, despite price instability. Furthermore, concessionaire methods were labour-intensive to a degree that their demographic consequences are still measurable today, whereas the oil sector is a capital-intensive one with very limited, and skilled, labour needs. The point to bear in mind is not so much the actions of companies as the continuity of rationale of the colonial and postcolonial state: concessions are seen as delegation of *mise en valeur*, in terms of both economic exploitation and the construction of vital infrastructures. Through them, effective powers accrue to the private sector, some of which concern the governance of whole sections of the nominally "state-administered" territory. This was not lost on state purists such as AEF Governor-General Antonetti, who flatly declared that he had never seen anything like this "abandonment of the obligations of the state, in regard to both the defence of the natives and that of its functions and public interests".[29] But

---

26 The history of British concessionaires in Central-Northern Mozambique is of comparative relevance here. See especially Newitt (1995) and Vail and White (1980).

27 The historical experience of what one scholar calls "the power bloc" of the colonial state, mining companies and the Catholic Church in Belgian Congo, while unique, holds important comparative lessons as a partnership "[...] which profoundly changed the course of history in Central Africa". See Vellut (1983: 127).

28 As noted in Gide (1927: 104).

29 My translation, quoted in Coquery-Vidrovitch (1972: 508). Austen and Headrick quote a 1919 report by the Lieutenant-Governor of Gabon to the same effect: "Across this vast region, the company has enjoyed an absolute commercial, industrial and agricultural monopoly, almost the right to administer high and low justice, *since we have ceded to it* our administrative posts, since our administrators defer to it, since it has its own armed men and it polices the region itself" (Austen and Headrick, 1983: 39, my italics).

neither did he overlook the fact that this strategy was pursued as state policy rather than being the result of pressures by non-state actors.

The disappointing performance of company rule and a more assertive and dedicated metropolitan look on things imperial meant that, from the late 1920s, the colonial state would have a more visible role. Indeed, only in the late twentieth century would one encounter widespread instances in which state institutions are equally tenuous and their functions so thoroughly appropriated by the private sector. Alternatively, the late colonial period and the first decades of independence were characterised by attempts at conforming to more conventional state formats. The experience of extensive delegation of sovereign functions would, however, remain an important element in the historicity of these societies, and certainly a credible future policy option in response to economic decline and state dereliction.

The partnership between state authorities and private interests does not mean that they saw their interests as coterminous, for they certainly did not, as complaints by officials such as Antonetti show. This is a widely disseminated myth that historians have laboured to undermine[30] by showing that a degree of tension has always existed between mineral extractive companies and public authorities. Any approach such as that which orients this chapter, stressing the shared goals of the two, must not lose sight of the recurrent divisions. This said, colonial governments consistently accorded competitive advantages to national firms in products believed to have a strategic value, chief among which was petroleum.[31] The obvious example here is that of Anglo-Persian, in which the British government came to take equity interest;[32] others include Shell-BP in Nigeria and several French state-owned oil firms in AEF. Disagreements

---

30 One of the most important recent contributions on colonial history cautions against "assuming coherence" in the agendas of the private sector and the colonial state (Cooper and Stoler, 1997: 19). Hopkins (1976b: 272) comments on the "sometimes [...] strong and occasionally hostile views" expressed by colonial officials on the subject of expatriate firms in Africa. Jones (1981: 78) notes that a measure of hostility towards oil companies was recurrent in government circles throughout the colonial period.

31 Jones (1996: 77).

32 See Ferrier (1990), Bamberg (1994) and Bamberg (2000) for the history of British Petroleum and the rise and demise of its Iranian concession.

there may have been but—especially but not exclusively in the realm of mineral extraction—the policy convergence is undeniable.

Oil companies

Multinational firms are firms of a particular nationality with partially or wholly owned subsidiaries within two or more national economies that expand overseas primarily through foreign direct investment.[33] Mineral extractive firms, it has been pointed out, differ from others as their investment location is dictated by geological factors, a fact that in turn weighs heavily upon "the technological and organizational parameters within which the enterprise functions".[34] Oil firms have historically consisted of large bureaucratic organisations involved in capital-intensive and very risky forms of investment on account of both the hazardous nature of exploration and "uncertainties regarding cost and completion time [and] subsequent performance".[35] More often than not, the leading oil players have been of Anglo-American extraction, despite continental European competition that is almost as old.[36] Indeed, of the so-called "Seven Sisters"[37] that dominated oil production for most of the twentieth century and remain key players in the business, only Royal Dutch/Shell, with its composite British-Dutch character, qualifies as a partial exception.[38] The five American companies, which have recently re-consolidated into the large "super-majors" Chevron and ExxonMobil, emanated from the

---

33 Gilpin (2000: 164).

34 Schmitz (1995: xi).

35 Jones (1996: 76).

36 Grayson (1981) explores the history of European state-owned energy companies, the so-called "national champions".

37 These were British Petroleum, Royal Dutch/Shell, Exxon, Chevron, Texaco, Mobil and Gulf, which still accounted for three quarters of non-US and non-Communist world production and 60 per cent of refining capacity by the late 1960s. See Penrose (1968: 88). In addition to Penrose study (particularly pages 87-149) see Sampson (1975) for a study of the pivotal role of the seven companies.

38 For the early history of Royal Dutch/Shell see Gerretson (1953-57) and Forbes and O'Beirne (1957) as well as Howarth, *A Century in Oil*.

1911 breakup of the Standard Oil monopoly[39] that had dominated the business until then. All are fully integrated petroleum companies combining extensive production capacity with transport, refining and marketing facilities, although the balance between these elements has varied considerably within each company and in different regional settings. As the century progressed, the search for oil by the oil multinationals took on worldwide dimensions in response to new oil-dependent industries and the displacement of coal as key source of energy, and as previously known reserves proved insufficient for the period leap in world petroleum needs.

Some colonial firms profited from inordinately high returns on investment as a consequence of state-supported "monopsonistic control over local resources".[40] This was the same with the oil sector, where the impact of preferential access was delayed but eventually bore fruit. In the Gulf of Guinea, major oil finds took place only towards the end of the colonial era in the late 1950s, even though exploration had been a major concern for decades. But by then, companies benefited from incumbency and extremely favourable terms that would transpire into the post-independence setting. The oil sector was the subject of tailor-made policies that favoured metropolitan companies and substantially reduced access by foreign entrepreneurs.[41] Furthermore, the oil sector, like mining, was one of the beneficiaries of the late colonial spurt of infrastructural improvement, most of which had the interests of investors in mind. A major implication of this is that oil firm investment can only be understood from the viewpoint of its origins in the "greenhouse" of colonial state nurturing.[42]

Our inquiry needs to start here because early investment by a large company has almost invariably translated into sustained domination of oil

---

39 For the history of pre-1911 Standard Oil see Hidy and Hidy (1955). For the history of Exxon, the main successor company, see Sweet and Knowlton (1956), Larson *et al* (1971) and Wall (1988).

40 Hopkins (1976b: 287).

41 Jones (1996: 77).

42 Kahler (1981: 388). US investors were, at least in the British Empire, not subjected to any type of discrimination as regards access and often benefited from this enabling framework as well. See Jones, *Evolution of International Business*, p. 288.

production until the present time. Moreover, subsequent fiscal regimes and national oil sector structures developed at least in part in interaction with these foundational firms. According to Geoffrey Jones, first mover advantages were mainly the product of concession agreements, which typically granted companies ample rights over a given area for about five decades.[43] This gave first investors both remarkable access and the guarantee that later entrants would not jeopardize their grip over resources. Such arrangements proved unimpressive for newly independent states. Yet "the lack of bargaining skills and the technical know-how on the government side, and the control over technology, capital and markets, on the other"[44] meant that private operators would remain the empowered partners for more than a decade. Renegotiation on more favourable grounds was piecemeal and did not jeopardize company standing, as I shall now show.

Production starts

Oil exploration in Nigeria was first conducted in 1907,[45] the earliest exploration in the whole region. Intermittent attempts with mostly disappointing results took place in the following decades. The harsh conditions in which this long period of research took place are underlined in contemporary accounts.[46] In 1937 BP and Shell formed a partnership and obtained an exploration license covering the totality of Nigerian territory.[47] No meaningful exploration was conducted during the war years, but the partnership was reactivated soon afterwards. In 1949, Shell-BP reduced its license to a 60,000 square miles area in southern Nigeria. In 1951, an equal ownership joint venture between the two companies was established. BP's interest remained "financial rather than operational [for] it was Shell who provided the management and technical advice".[48] BP's

---

43 Jones (1996: 76).

44 Ibid.

45 Frynas (2000).

46 See Shell-BP, *The Shell-BP Story* (Port Harcourt: Shell-BP, 1965) for the oil-man-as-brave-explorer-of-darkest-Africa. See also "Nigeria: Growing Source of World Oil", *International Petroleum Review*, July 1963.

47 This account draws mainly on Bamberg (1994: 109-13).

48 Quoted in Bamberg (1994: 109-13).

secondary role and its eventual expulsion by the Nigerian government in 1979 underline the extent to which Shell—Nigeria's key producer up to the present time—was and is the pivotal private sector actor in the local economy and politics. Oil was first struck in commercial amounts at Oloibiri in January 1956, with Nigeria's first crude export dating from February 1958. Meanwhile, Shell-BP relinquished some of its acreage in the region, thus permitting the entrance of other companies, starting with Mobil in 1959. By the time of independence the following year, oil output was quietly rising but remained a modest source of income for the state.[49]

The search for oil in Gabon,[50] and indeed in what was then French Equatorial Africa, was started in 1947 by a French State subsidiary, with the first oil shipment to France dating from a decade later.[51] Although these were modest finds (Gabon's output remained limited until a major Shell find in 1967), they were enough to attract the attention of both Mobil and Shell, which entered into joint venture arrangements with the French state subsidiary the following year. Shell's and Elf's production grew steadily in the first years following independence in 1960[52] but by then Gabon still possessed a fairly diversified, if fundamentally extractive, economic landscape. Independence brought no changes to the investment framework other than the alteration of the French subsidiary's name[53] and the nominal granting of 0.575 per cent of the company's capital to each of the four new states of the former AEF.[54]

---

49 Production was then at a mere 17,000 bpd. See Bevan, Collier and Gunning (1999: 26).

50 See Yates (1996: 55ff) for the first decades of oil exploration in Gabon. Pourtier (1989: 194-5, vol. 2) makes reference to superficial "geological prospecting" taking place as early as 1931, while Bouquerel (1967: 186-8), mentions the presence of American geologists as early as 1926. As in Nigeria, the strenuous conditions of research in Gabon and the technical difficulties of operating there are underlined in period accounts.

51 Gardinier (1999).

52 Mobil's success rate was low, and the company left Gabon in 1963. Despite the myriad investors, the only significant producers in Gabon until 1983 were Shell and Elf. See Pourtier (1989: 197, vol. 2).

53 From SPAEF (Societé des Pétroles d'Afrique Équatoriale Française) to SPAFE (Société des Pétroles de l'Afrique Équatoriale).

54 Pourtier (1989: 195, vol. 2).

Angola's experience was equally incremental. Oil had been known and used by local inhabitants of Northwest Angola as a source of energy since at least the 16th century.[55] This certified presence led to insistent drilling in the beginning of the century by the Belgian-owned Companhia de Pesquisa Mineral de Angola and, subsequently, by the American Sinclair Oil Corporation. Both companies made numerous discoveries but these were deemed non-commercial and interest in Angola's oil reserves only re-materialised in the early 1950s. Petrofina's first commercial find in 1955 at Benfica in the outskirts of Luanda soon led to awareness of the country's potential. This was followed by the 1958 discovery of the Campo de Tobias field and the company's decision to build a large refinery in Luanda in the same year.[56] The late 1950s also witnessed the entry into Angola of a company that would come to play the key private-sector role in the Angolan economy: the Cabinda Gulf Oil Company (Cabgoc), a subsidiary of Gulf Oil later purchased by Chevron. The company acquired from the Portuguese state the right to prospect and exploit in the Cabinda enclave for the duration of five decades, with the possibility of renewal for a further twenty years. It did not have to comply with "certain requirements usually asked by the Portuguese authorities such as having the main office in Portuguese territory, having a majority of Portuguese nationals on the board, or transferring 20 percent of the capital stock to the colony of Angola. [...] The company was free to decide the level of production as well as the storage and sale of whatever it found in its concession zone".[57]

---

55 Caley (1997: 75).

56 The company had an important presence in Angolan fuel distribution since the 1920s. See Dumoulin (1997) for Petrofina's presence in the country. See also Clarence-Smith (1983) for an overview of Belgian investment in Angola, which was allowed even at a time of stringent economic nationalism because of Belgium's perceived harmlessness and as a bulwark against economic domination by Anglo-French capital.

57 See Dos Santos (1983: 107) and also "Les sociétés multinationales en Angola", EurAfrica, February 1975. The last decade of colonial rule witnessed a more flexible framework for external investment in Angola but this deal was unprecedented in 1957, when great anxiety regarding foreign ownership was still the rule. Clarence-Smith (1983) outlines the many strategies used by Belgian investors and the Portuguese government to give a more Portuguese face to the mainly Belgian DIAMANG.

As Fred Cooper noted, "independence" is often an unhelpful marker from which to assess the political trajectories of the continent.[58] There was more often than not a great deal of continuity in the productive structure of the former colonies. This is particularly so with the oil sector, a nascent and, for once, relatively discreet one in local politics. While smaller business interests in the Gulf of Guinea feared independence and were visibly associated with the more reactionary elements of the European presence, the oil sector did not perceive majority rule as a hindrance to success[59] and kept aloof from the most divisive issues of the day. Five decades had passed from the beginning of exploration across the region to the first important oil finds. This unusually long-drawn process was frequently the subject of pessimism: Pearson quotes an early 1950s assessment flatly claiming that "Africa has no oil".[60] By 1960, several countries in the region were experiencing intense and optimistic prospecting for commercially viable oilfields. Very recently, some of these bids had proven successful. At the end of the day, however, oil remained a secondary source of revenue in the mostly agricultural economic landscape of the region. Beyond a small circle of technical and entrepreneurial expertise, the perception of wealth to come was absent from the concern and imagination of politicians and societies across the Gulf of Guinea.

II. POST-INDEPENDENCE TURMOIL

A careful look at oil production in the Gulf of Guinea circa 1970 will show the reader that seniority matters: as George Frynas notes, the companies that currently dominate the Nigerian upstream were all already active in the country then.[61] The same applies to Gulf's presence in Angola, and Elf's in Congo-Brazzaville, Gabon and Cameroon. Despite the fact that only Nigeria had by then achieved anything close to the status of a large

---

58 Rather than taking it for granted, Cooper asks what "difference the end of empire meant, as well as what kinds of processes continued even as governments changed hands". See Cooper (2002: 15).

59 Austen (1987: 211).

60 Pearson (1970: 13).

61 These are Shell, Mobil, Chevron, Elf, AGIP, and Texaco (Frynas, 2000: 11).

producer, heightened interest by additional oil companies contributed to a growing understanding of the region's potential. The rapport with host states, however, remained mostly within the format of the late colonial period. I will now explore the events and political trends that caused this to change.

Companies and politics

One of the key concerns of the bibliography on state-firm relations in the postcolonial era is the extent of firm involvement in local politics. In the Gulf of Guinea, it is undeniable that all companies have on occasion taken clear and consequential political decisions; but few would portray themselves as having done so. For the most part, their involvement has not taken the interventionist, unpopular character of multinationals' meddling elsewhere. From the militant perspectives of the United Fruit Company (sponsor of the overthrow of Guatemala's democratically-elected government in 1954), the Union Minière (financier of the Katanga secession in Congo) or ITT (backer of General Pinochet's 1973 coup), oil investors in the Gulf of Guinea have pursued unproblematic policies that are favourable to incumbent regimes. Elf-Aquitaine, which occasionally aided in the disposal of unhelpful leaders, is often pointed out as a different story. But even in this case the company's muscled actions are inseparable from the agendas of important domestic actors and, at any rate, the exception to a mostly friendly exchange with local rulers. On the basis of this rather discreet record, oil companies in the Gulf of Guinea have claimed to lie outside the realm of politics.

Their taking solace in an apolitical stance should not, however, obscure the fact that their impact was and remains momentous. The decision, in the context of civil war or widespread unrest, to support the powers that be carries as much weight as that of sponsoring anti-government groups, although the latter is of course much more subject to international opprobrium. And even when few policies of great host-state impact can be pinpointed, it is clear that the structural importance of companies for state budgets yields a quiet influence that cannot be overstated. This means that, nuances and corporate styles aside, oil firms have intervened politically and in a thoroughly partisan manner. This includes anything from consistent lobbying on foreign policy directions in the developed world in favour of host states to the payment of taxes to an embattled sovereign. Elf Aquitaine provides the best example of political involvement across the spectrum for the better part of the postcolonial years. A look at oil politics in the Gulf of Guinea cannot be separated from this company—in regard to its actions, of course, but also in terms of what Elf was as a political actor.

A noteworthy feature of the early postcolonial setting was the presence of Western European "national champions" (i.e. state-owned firms).[62] Owing to preferential access in France's case, one would expect a measure of resilience by a state firm of the former colonial power. But Elf was a technologically vigorous operator pursuing aggressive exploration strategies rather than a non-competitive hanger-on. Elf was created in 1965 as a tool of national energy policy by General de Gaulle.[63] Its initial success was underpinned by the soon-to-be-expropriated Algerian oil and natural gas wealth: in just a few years, Elf grew under the leadership of Pierre Guillaumat into France's first industrial group. The choice of Guillaumat, former Defence Minister and Gaullist insider, illustrates the multiple goals bestowed upon the new company, which soon became one of the pillars of the "Paris-African complex" of relations between France and its former colonies.[64] In states outside the French *chasse guardée*[65] in Africa such as Angola, Elf took up the role of *de facto* representative of the French state. In spite of great diversification that would make Elf the world's eighth largest oil company by the 1980s, its production epicentre would be the Gulf of Guinea,[66] which by the mid-1990s still accounted for 60 per cent of the company's output and 70 per cent of its proven reserves. As insider André Tarallo noted, "the future of the company was being played in the Gulf of Guinea [...] finding petroleum [in those years]

---

62 See Grayson (1981).

63 The other French major, Total, was only 35 per cent state-owned and was generally seen as politically unreliable, at least to the extent envisioned by de Gaulle. See Dominique Gallois, "Les destins croisés d'Elf et de Total", *Le Monde,* 9 October 1999.

64 Chafer (2002: 343-63).

65 "Private hunting ground", the expression used to define the area of French influence in Africa, i.e., most of the former French colonies and later on, other Francophone states such as Rwanda, Burundi and Congo/ Zaire.

66 Sub-Saharan Africa's importance for Elf further increased with the forceful end to Elf's Algerian involvement in 1971.

was a matter of life or death".[67] Elf would indeed come to secure the necessary business opportunities, but via relationships and operational procedures that later came to haunt the company's reputation.

The other European state company with a key role in the regional oil industry was AGIP, the upstream subsidiary of Italy's ENI.[68] In the aftermath of the Second World War, Enrico Mattei had been entrusted with the management of the then-ineffectual Italian NOC. He proceeded to upgrade it into one of Europe's most significant industrial conglomerates—and one that subsequently became an important actor in the Gulf of Guinea's petroleum sector. National champions came to behave in the international arena in remarkably similar ways to their private counterparts.[69] They were certainly no less agile in securing commercial opportunities and, as the case of newcomer AGIP shows, their strategies could be exceedingly creative. Further to this, both companies,[70] and Elf in particular, soon developed highly personalised and occasionally illegal agendas that are not necessarily coterminous with those of the home state.[71] But these were strands that intersected with, rather than displaced, the fact of state ownership. They were pursued in tandem with the unambiguously nation-centric goals of securing sources of energy and furthering national prestige, which therefore cannot be underestimated

---

67 "J'ai eu à gérer l'"indivision africaine" du groupe Elf", *Le Monde,* 24-25 October 1999.

68 The Ente Nazionale Idrocarburi (ENI) was founded in 1951 as the state-owned holding company of AGIP and thirty-five other companies with the goal of ensuring that "Italy had its own international petroleum supply, independent of the 'Anglo-Saxon' companies" (Yergin, 1991: 503). Tightly run by Enrico Mattei up to his untimely death in 1962, ENI burst into the international oil scene in the 1950s with a string of oil deals that were uniquely favourable to the host state (Penrose, 1968: 142-4). See Frankel (1966) for the momentous history of the Mattei years.

69 Moran (1992: 11) and (1981).

70 AGIP remained the fiefdom of the socialist minority till the end of the First Republic in the early 1990s.

71 See Médard (1997) for an analysis of Western corruption in Africa with good discussions on the peculiarities of the French and Italian roles. On Elf as a nexus of political, personal and even criminal networks, see, among others, Glaser and Smith (1992) and Glaser and Smith (1997); Verschave (1999); Verschave (2001); Pean (1983); and the seminal ARTE documentary, "Elf: une Afrique sous influence" (2000).

in the broader picture of the national champions' corporate cultures and *modus operandi*. Nor can one forget that such strategies were pursued jointly with local power-holders, not in defiance of them.

This is clearly brought out in the case of Gabon, the centre of Elf's regional policy. Sharing the stage with a mostly French-staffed Shell subsidiary, Elf was the preeminent producer and undisputed foreign actor in the country's politics and economy.[72] In 1967, Shell's Gamba concession came on stream and with it began a series of further licenses to foreign companies and commercial oil finds. That same year, Gulf Oil moved into the Gabonese onshore, as did AGIP a year later. From then on, petroleum extraction would permanently dwarf all other economic activity in the country. By the early 1970s, Gabon had materialised into a sparsely populated "oil emirate" of immense wealth.[73] It was also Elf's jewel in the crown, ensuring 54 per cent of the company's oil production within ten years. Despite President Bongo's spurious calls for "Gabonisation"[74] and occasional but not very credible threats of "Libyan-style" takeover, firm-state relations were amongst the best in the region. A substantial change did occur in the aftermath of Gabon's astutely scheduled affiliation to OPEC in November 1973—just in time to reap the benefits of the oil

---

72 This included a presence that went far beyond oil production. According to Auge (2003: 97-102) the Elf-Gabon business empire of the early 1980s included the Société des Transports Maritimes (79 per cent), Société Gabonaise de Pêche Industrielle (100 per cent), Société Gabonaise de Céramique (100 per cent), Société Agricole de Port-Gentil (75 per cent), COGER (59 per cent), Société Sucrière du Haut-Ogooué (39 per cent), Société Gabonaise de Cellulose (27 per cent), Société Gabonaise de Forages (60 per cent), Gabonaise de Peintures et Laques (32 per cent), Gabonaise Industrielle de Construction (52 per cent), Centre International de Recherche Médicale de Franceville (100 per cent), Société de Ciments du Gabon (70 per cent), Société de Transports Maritimes (79 per cent).

73 In the aftermath of the Teheran and Tripoli accords, the signing of an Elf-Gabon Oil Convention in April 1971 had already improved the modest fiscal take of the state. See Daverat (1977: 41).

74 "Bongo warns oil firms", *West Africa,* 29 October 1973. As late as 1989, and the situation is fundamentally similar today, Pourtier (1989: 199, vol. 2) noted that "the near totality" of technical and managerial expertise in the extractive industries was ensured by expatriates, who constituted an estimated 10 per cent of Gabon's labour force. Nigeria is the exception here, as the labour forces of oil majors operating there are around 90 per cent Nigerian, but even there, Nigerians are underrepresented at the technical level.

embargo. Gabon's share of profits, along with the framework for foreign direct investment in country, was overhauled in 1974. A mandatory 25 per cent of state ownership in Elf was implemented, thereby increasing the state's revenue and making a fourth of the production available to it. While Gabonese personalities, including members of President Bongo's family, would take up central roles in the new joint ventures, they remained basically foreign-run. This resulted in a much better deal for the state, but also underlined that Gabon would not pursue full ownership as a strategy for the oil sector.[75]

Early threats

The late 1960s also brought about the first serious political trial for oil companies in what was to become a particularly heated corner of global politics. Nigeria's Biafra war of 1967-70 was the litmus test of investor resolve in the region; but while it led to occasional doubts on the part of companies as to whether to stay or leave, it also showed that investment remained profitable even amidst a great deal of disruption. The oil sector had grown exponentially in the first five years of Nigerian statehood as several other companies joined the Shell-BP consortium in the Delta region.[76] The immediate political impact of this trend was moot and it was to be several years until "rumors of the potential importance of petroleum began to enter into political discussions".[77] By 1964, however, the increased revenue stream and the perception of an unfair regional allocation formula had become a potent engine in domestic politics.

While not the root of the attempted secession of Nigeria's Eastern Region in 1967, the promise of oil wealth or, from the federal viewpoint, the prospect of losing it, certainly exacerbated the ensuing conflict. The Nigerian civil war impacted on the private sector in terms of both output and the political choices it had to confront as the investment turf was hotly fought for. On account of France's pivotal role in supplying their

---

75 Pourtier (1989: 195ff, vol. 2).

76 Shell-BP also entered an agreement (on a 50/50 basis) with the Nigerian government in regard to the building of a refinery at Port Harcourt designed to meet Nigeria's refined products needs. The refinery was inaugurated in 1965. See "Nigeria's Oil Revolution Begins", *West Africa,* 27 February 1965.

77 Pierson (1970: 137).

war effort, the secessionists had a close relationship with Elf, and unconfirmed reports suggest that attempts were made to cede exclusive rights
over petroleum production to French business concerns.[78] Interestingly,
this support did not have long-term consequences for bilateral relations
and French investors were still welcomed once the Federal Government
had won the war.

There were other, more visible, short-term consequences for oil
companies, in the form of all-out disruption of the Niger Delta. This
led to a decline of around 90 per cent in Shell-BP's onshore production.
Yet Gulf's (admittedly less important) offshore operations escalated by
a staggering 300 per cent during the same period.[79] This is not to underestimate the extent to which oil production was affected in the first
two years of the conflict,[80] although the Federal Government's eventual
reoccupation of most of the Delta would mean that it controlled the oil-
producing region by late 1968. Alternatively, this showed that, firstly,
offshore production is politically and literally insulated from onshore
strife and, secondly, that companies would not revise their commitment
towards the extraction of Nigerian oil even in the context of such an apparently destructive event as a drawn-out civil war. The lesson that the
oil sector could and would weather out just about anything was not lost
on anyone. As Khan remarks, the war was the litmus test for the oil sector in Nigeria: despite the many political troubles they have faced since,
"foreign investors continue to stay".[81]

Meanwhile, Gulf Oil in Angola was facing two threats. The first was
the prospect of sabotage by anti-colonial guerrillas hidden in the lush for-
ests of Cabinda, and the second the international backlash over collabora-

---

78 A much-propagated rumour of the period was that of a deal between the Rothschild
Bank, former employer of Georges Pompidou, and the rebels concerning the supply of
arms in exchange for mining rights. See Wauthier (1995: 132-3, 185-9).

79 Khan (1994: 9).

80 Yergin (1991: 556) notes that the onset of the Nigerian civil war removed some
500,000 bpd from the world market. This was one of the most important of such inter-
ruptions prior to the oil embargo of 1973-74, and happened simultaneously with widespread disruption of oil routes in the Middle East. By November 1968, Shell-BP was
producing 200,000 bpd out of war-damaged areas in comparison with a pre-war figure of
500,000 bpd. See *Africa Report*, February 1969, p. 39.

81 Khan (1994: 5).

tion with a Portuguese government then fighting a vicious multi-front war against three competing liberation movements.[82] The high profile of Gulf's Angolan operation came precisely at the moment when it found an unusually rich oilfield, which remains on stream today. Accordingly, Gulf dug its heels in and decided to stay for better or worse. By 1970, oil was colonial Angola's fourth export; by 1973, it had become her first. The contribution of Gulf towards Portugal's war effort was both unambiguous and substantial.

Of the two potential hazards, that of attacks against oil operations never materialised. This was primarily because Gulf was safely ensconced in coastal pockets or offshore, but also due to the colonial government's provision of a military *cordon sanitaire* around urban areas of the Cabinda enclave—much as the Angolan armed forces provide today and Cuban troops did until 1991.[83] This pragmatic alliance between private capital and state security forces was to be frequently repeated across the Gulf of Guinea, most recently in the Niger Delta troubles of the 1990s. The second hazard—that of an international backlash against the company—could not be avoided. In 1973, a UN special commission reported that "the growing oil output in Angola [illustrates] the effort of the Portuguese government to accelerate the process of colonial exploitation of the territory" and underlined Gulf Oil's pivotal role in it.[84]

Gulf's reaction was extremely resourceful: indeed, while its methods have been re-elaborated and somewhat refined, the premises that have guided corporate defence of investment in murky locations were rehearsed by Gulf at this early stage. Gulf's penchant for deflecting crit-

---

82 In an interview with the author in May 1998, former Governor-General of Angola Colonel Camilo Rebocho Vaz emphasised the cordial and hassle-free rapport between the Portuguese government and Gulf Oil.

83 In addition to this, Gulf operations in Cabinda were secured by the company's own "armed guards" (*Africa Contemporary Record,* 1970-71: B 582). The existence of private armies in Portuguese Africa was an important long-term feature of the local economy—the obvious example in Angola is that of DIAMANG, which possessed a militia between 1917 and 1974—and one always pursued in close articulation and harmony with the colonial ("state") authorities.

84 United Nations Report on Angola Petroleum Exploration, General Assembly Report A/9023 (Part III), October 1973, p.12 (United Nations Archives, New York), quoted in Freire Antunes (1992: 213). See also Wright (1998: 54-5).

ics was to be particularly valuable in the 1980s, when elements in the Reagan administration sought to pressure the company to relinquish its role as economic lifeline of yet another questionable government—this time that of the Marxist rulers of post-independence Angola. On both occasions, the company did not budge. The statement by Gulf Oil Vice-President E. E. Walker in response to criticism of his firm's involvement in Angola mirrors, in its flawless modernity, present-day reactions to external criticism of state-firm partnerships in the Gulf of Guinea:

> We can prove that the royalties paid to the [...] government of Angola are being used towards the improvement of life conditions of the people [...] as Gulf's money reaches the government, the military budget remains constant, whereas health and social assistance budgets have risen exponentially. [85]

As with Gabon, Congo-Brazzaville's search for oil was an overwhelmingly French affair, only more so. It was also a remarkably modest one until late in the 1960s, despite a promising find at Pointe-Indienne in 1957. Elf Congo, the local subsidiary, was created in 1970, "the same year of the official launch of Congo's 'Marxist-Leninist' experiment",[86] and put under the steady hand of André Tarallo, the Elf executive who would figure prominently in the scandals that engulfed the company's reputation three decades later. By the 1990s, Elf still accounted for three-quarters of the country's oil production, and the degree of profitability was its highest in the region.[87] In 1969, AGIP also acquired an offshore concession. Immediately afterwards, AGIP and Elf agreed on exchanging a share of each other's Congolese subsidiaries,[88] thus creating two joint venture companies with an effective monopoly of Congo's upstream. Under this arrangement, the two companies controlled Congo's oil, with no meaningful entry by other companies for more than a

---

85 House of Representatives, *The Complex of American-Portuguese Relations,* p. 563, quoted in Freire Antunes (1992: 214).

86 Clark (2002: 177).

87 Daverat (1977 : 45). Incidentally, it was none other than then-minister Pascal Lissouba, future president and arch-critic of Elf, who signed the 1969 agreement for Elf's offshore exploration (*Africa Report,* January 1969, p. 24).

88 The holdings were 35 per cent AGIP/65 per cent Elf and 65 per cent AGIP/35 per cent Elf.

decale.[89] Whatever the iconoclastic character of AGIP may have been in the Mattei years—the company had put paid to the 50/50 industry standard in its 1957 deal with Iran and become the largest buyer of Soviet oil—its involvement in Congo did not make waves. By the time of the first major oil find in 1972, Elf and AGIP were in near-perfect conditions for maximum return on investment.

Similarly to post-independence Angola, the leftward turn of domestic politics did not have the drastic impact in the oil sector one would expect. In fact, it was the combination of increased production and the 1973 hike in world prices that allowed the pursuit of Congo's incipient socialism in the form of costly policies regarding infrastructure, selective nationalisation and the setting up of agricultural cooperatives.[90] This high-spending strategy was complicated by overly optimistic projections by the foreign operators, which overstated Congo's recoverable reserves in the early 1970s. Together with the concurrent expansion of state-owned enterprises and bureaucracy and Congo-Brazzaville's political instability, this meant that the country found itself under a heavy debt burden much earlier than its neighbours did.[91]

As in many other instances, it was Elf that recurrently ensured that budgetary commitments were kept, even while other oil exporting states were enjoying the honeymoon of high prices;[92] as shown in the next chapter, this assistance had high costs. In order to regulate and compete with AGIP and Elf, and because it was the modish thing to do for an oil exporter, a NOC, Hydro Congo, was created in 1974.[93] Hydro Congo would come to be a dismal failure. It did not replace the foreign subsidiaries in research and production, nor did it attract long-term interest

---

89 Prospecting permits were granted to American, Canadian, Swiss and Brazilian oil companies with a view to increasing the number of investors in the country's upstream. See Thompson and Adloff (1994: 237). The impact of this was limited until quite recently, and it certainly did not affect Elf-Agip domination.

90 Clark (1997: 65-9).

91 Vallée (1988: 15-ff) dates the debt crisis from 1969, "in spite of its oil wealth or because of it".

92 Vallée (1988: 17) mentions the $100 million loan the company made available to the Congolese government in 1978 so that it could pay for modernisation work on the Congo-Ocean railway.

93 Thompson and Adloff (1994: 234).

from other investors; it did very badly in negotiations; and in contrast from the "cash-cow" character of other NOCs across the region, it became an unbearable financial and political weight upon the state itself. As the 1980s price slump approached, the position of the Congolese state was eroded in a silent but inexorable way.

III. NATIONAL ASSERTIVENESS AND THE BOOM YEARS

The early 1970s brought a head-on clash between the private sector and the state in most of the developing world. In the midst of outright expropriation or the raising of operational costs, rampant economic nationalism and accusations of imperialist exploitation, the highly symbolic oil sector found itself in the thick of political mayhem.[94] In some of the cases under review, the transition from colonialism to independence for foreign private investment had been smooth and orderly, with very superficial reconfigurations hiding a basically unchanged framework of external ownership. This was more so in the former French territories, while Nigeria embarked on a policy of indigenisation across the economy premised on increased amount of local participation in employment and ownership, but mostly falling short of nationalisation. Other governments, such as that of independent Angola, resorted to mass expropriation and the expulsion of European settlers. All states saw an increase in both public ownership and share of profits as a result of the momentous changes brought about in the period.

Yet even in states where great systemic change was enacted with the demise of colonialism or subsequent radical politics, the oil sector was left largely undisturbed. (The only post-independence ruler who would probably have not been pragmatic enough to refrain from interfering— Equatorial Guinea's Macias Nguema—had no oil investors to harass at the time.) Where the host-state's bargain was eventually renegotiated to the benefit of the state, as in Nigeria and Gabon in the mid-1970s, this never amounted to "creeping" expropriation—i.e. the step by step introduction of "crippling taxes, discriminatory licensing or taxation regula-

---

94 The militant 1962 United Nations General Assembly Resolution 1803 on "Permanent Sovereignty over National Resources" had created the context for later actions.

tions, or 'unfair' competition from state corporations".[95] The levying of more onerous terms notwithstanding, investment in the oil sector across the region remained attractive throughout.[96]

Despite previous bouts of nationalisation of foreign property,[97] it was in the aftermath of the Suez Crisis of 1956 that its potential value as a policy instrument became apparent.[98] The 1960s and 1970s, with emphasis on the period between 1970 and 1975, witnessed thousands of expropriations before the trend lost credibility in the following decade. Although their pursuit was widespread, the vast majority of such policies were enacted by a handful of regimes normally described as "mass expropriators".[99] Both Angola and Congo belong to the category, with Nigeria during the Gowon period a "selective expropriator".[100] Whichever the general drift of these national economies, the label certainly does not apply to their handling of the oil sector.

The fact that the latter emerged intact from a decade of economic nationalism may be attributed to two factors. The first is the tactical understanding by foreign firms that this was a time to accept, while guaranteeing a sizeable profit margin, a measure of change in the relationship with the host state. Faced with straightforward ejection elsewhere, it was not difficult for companies to make room for a degree of state empowerment

---

95 Williams (1975: 263).

96 In a noteworthy contribution to the small literature on the history of energy in Africa, Kairn Klieman provides an in-depth study of the Congolese (Kinshasa) downstream "petroleum wars" of the 1960s. Klieman portrays this as an instance, a decade before the dates normally flagged out by analysts, of "demands for economic sovereignty and control over petroleum markets". When they came to the African upstream, such demands were to prove far less intruding, and companies in the 1970s were much less intransigent. See Klieman (2006: 2).

97 These include Russia in the aftermath of the 1917 Bolshevik Revolution; Mexico in 1938; Bolivia in 1937 and 1952, and Iran in 1951. Nationalisation is here taken to mean the expropriation, with or without compensation, of private property for public use. See Williams (1975).

98 Williams (1975: 262).

99 See Kobrin (1984: 329-48) and Kennedy (1992: 67-91).

100 Kennedy (1992: 73) lists no less than 128 firms nationalised during the Neto presidency in Angola and 31 in Congo-Brazzaville throughout Ngouabi's rule (1968-77). Nigeria is credited with 24 expropriations over a seven-year period (1967-74).

that still left them in an advantageous position, at least in comparison with that of other oil provinces from which they were merely excluded.

The second factor was the centrality of oil profits for state budgets, which could be jeopardised almost immediately as a result of political meddling, and the understanding by host states that they had neither the expertise nor the integrated sector structures for replacing oil firms. It is noteworthy that this did not hinder expropriation efforts elsewhere, especially since political reasons rather than economic efficacy primarily account for nationalisation. Large-scale mining was expropriated in Zambia, Ghana and Zaire, for instance, in apparent oblivion regarding the lack of local capacity for running operations. But it seems that, in the context I am addressing here, there was simply no room for self-delusion: short of a psychopathic form of politics, Macias-style, it was impossible to deny a number of facts. These included the technological complexity of oil production, the hazardous nature of searching for petroleum as a capital-intensive and often fruitless endeavour and, finally, the already mentioned lack of local expertise, both technical and managerial. The Nigerian experience as the main regional innovator in the field of national control is illustrative here.

The limits of national assertiveness

Nigeria became a world-class oil exporter when all other states in the region were still modest producers or just ground for exploration. Therefore, a "break with the past came earlier" than elsewhere and national sector structures affirmed themselves in the decade and a half following independence.[101] Local empowerment in business and the civil service started in the late colonial years, and by the mid-1960s the number of nationals working in the oil industry was quite substantial.[102] But decisions concerning research and development—and, crucially, those pertaining to production limits—remained the preserve of oil companies. At independence, the sharing of profits between the oil companies and the

---

101 Bevan, Collier and Gunning (1999).
102 Of the 4,500 people directly employed by the oil industry, 2,000 were skilled Nigerians and 700 were expatriates. See S. R. Pearson and S. C. Pearson, "Oil Boom Reshapes Nigeria's future", *Africa Report* February (1971), p. 16.

Nigerian government took place on a 50/50 basis.[103] Soon afterwards, AGIP negotiated—in line with Mattei's policy of diversifying Italy's sources of crude—a separate agreement with the Nigerian government that gave the latter the option to buy 30 per cent of the company's local subsidiary if oil was struck. Despite this breakthrough, it was Shell-BP that "more or less determined both the production and the price levels of Nigerian crude oil" throughout the 1960s.[104] Although OPEC terms were introduced in 1967, only in the immediate aftermath of the civil war did the Nigerian government start to act on the basis of the enormous reserves it held, "gradually but continuously [altering] the terms of the relationship to its advantage".[105] Nigeria could afford this in the early 1970s as Africa's most populous state, a recent member of OPEC and up-and-coming producer on the way to achieving 2,000,000 bpd. The companies were tuned to the changing mood regarding fiscal terms and firm-state relations, and quietly accepted more onerous terms that did not detract from the ultimate profitability of their investment. According to Pearson, it was Shell-BP that "supplied the initiative for the Nigerian emulation of the Libyan changes by informing the Ministry of its desire to afford better terms to Nigeria".[106] These adaptable stances, together with Nigeria's need of the private sector, ensured that companies would navigate growing demands and remain at the helm of the country's oil production.

The first Enterprises Promotion Decree of February 1972 was a crucial step "on the path that would lead throughout the 1970s to one of

---

103 This followed the introduction of the Petroleum Profits Tax Ordinance of 1959.

104 Khan (1994: 15). By 1965, Mr Stanley Gray, General Manager of Shell-BP could still openly say, with a hint of a threat as to government price meddling, that "any increase in the per barrel cost of finding and exporting oil would tend to jeopardize the present programmes of each company whose targets are based on the assumption that existing conditions will continue". See "Nigeria's Oil Revolution Begins", *West Africa*, 27 February 1965.

105 The November 1969 Petroleum Law governing the granting of oil concessions had already made room for a substantial Nigerian government participation (from 30 per cent to 50 per cent) and for moderately increased financial arrangements. See Pierson and Pierson, "Oil Boom Reshapes Nigeria's future", p. 14-15.

106 Pearson (1970: 26).

the most comprehensive mandatory joint-venture programs in Africa".[107] The decree outlined the government's intention to divide the Nigerian economy into three sections: fully Nigerian, Nigerian equity participation of at least 40 per cent, and a third that would be unaffected by the decree. In regard to the oil sector, the preferred modality of action was forceful government participation in the capital of companies, pursued in tandem with the creation of national sector structures in the form of the Nigerian National Oil Corporation in 1971. NNOC took an 80 per cent stake in the Shell-BP joint venture (later labelled the Shell Petroleum Development Corporation) and 60 per cent in other companies, but did not interfere with the day-to-day running of operations, which remained in the hands of Shell.[108]

Expropriation was not considered an option, with the exception of BP's assets and the less politicised nationalisation of Esso's marketing structures.[109] The expropriation of the former had more to do with poor relations with the BP equity-holding UK government, particularly in the wake of President Murtala Muhammed's assassination in 1976,[110] than with objectionable actions by the company itself, despite references to petroleum supplies to South Africa and Rhodesia as precipitating cause for the break. The regionally unprecedented character of indigenisation policies in Nigeria notwithstanding, results were widely seen as disappointing,[111]

---

107 Biersteker (1987: 52).

108 In the face of serious financial problems, NNOC's successor NNPC had to partially divest from the joint venture, and now owns only 55 per cent.

109 Interestingly, although it nationalised Esso's distribution, the state was content with taking a 60 per cent share of the Shell Oil Marketing Company, in another example of the latter's favoured status.

110 Kirk-Greene and Rimmer (1980). James Bamberg (1994: 519) notes that, "on balance, it was not in the Company's interest to be closely associated with a declining power in an age of rising nationalism".

111 An interesting consequence of Nigerianisation policies that has not yet been researched is the extent to which ethnically based divergences have been brought into the now mostly locally staffed multinationals. In the interviews I conducted in Nigeria during September 2002, the issue was broached numerous times. Most interviewees, including those working in Shell, believed this was a defining element in the internal politics of the companies. A caveat was made in the case of Elf, where the ethnic theme was less evident to most of the interviewees.

at least in terms of ultimate control of the sector.[112] Even today, Nigeria's grasp of its oil economy is both heavy and ineffective, with ubiquitous national sector structures incapable of performing basic upstream, midstream and downstream tasks.

Marxist-Leninist pragmatism

The Nigerian bid for oil sector control, though a failed one, represents the most ambitious such attempt at the regional level. Host state impact on oil operations goes downhill from here, as the experience of Congo-Brazzaville shows. After the French-backed government was toppled in 1963—and especially following further radicalisation in 1968—the country's rulers pursued an equivocal path towards "socialism". Political opportunism and mostly affable relations with France, with the exception of tensions in the mid-1970s,[113] nonetheless guaranteed it would be *sui generis*. On the one hand, much foreign property was left undisturbed, and the investment outlook remained bright for well-networked foreigners. On the other hand, attempts at a planned economy and collectivised agriculture were pursued with great zeal until the late 1980s, becoming a local fixture in a bizarre combination of Marxist rhetoric and high champagne consumption. In spite or because of this, as already explained, the oil sector was never tampered with and the Elf/AGIP monopoly prospered.

This is not to say that relations were placid. As the national champion of the former colonial power, Elf was the subject of permanent criticism. It was also widely (and correctly) perceived as a meddler in Congo's notoriously fractious politics. Although evidence remains sketchy, it

---

112 Biersteker (1987). It is worth remarking that, by concentrating on poor results on this front, analysts fail to properly consider the relevance of a broad indigenisation of the labour force of foreign firms. This certainly does not mean a greater level of national control over the sector. It does, however, introduce a high degree of local technical and managerial expertise and leverage at the company level, as a comparison of oil investment in Nigeria (or less so, Angola) with that in Equatorial Guinea or Chad will show.

113 According to Patrice Yengo, this downturn in relations was caused by the state's sudden financial needs. These were brought about by the failure of President Ngouabi's ambitious three-year plan, which had been made possible by the first oil receipts in the early 1970s. See Yengo (1998: 481).

seems that Elf had an indirect role in sponsoring the 1977 ousting of President Ngouabi, which was conducted by means of his assassination. Ngouabi was about to force through changes in oil sector taxation to curb the budget deficit and keep up with his growing requirements and had threatened all-out expropriation if these were not accepted by Elf. The company's reaction was as confrontational as can be; it simply stopped paying taxes to the government and halted all new investment.[114] But the partnership was healed and furthered in the twelve years of the first presidency of Denis Sassou Nguesso.[115] Sassou Nguesso's takeover coincided with increased production from just 50,000 bpd in 1978 to 100,000 bpd in 1983, and the second oil shock of 1979-80, the windfalls from which were a crucial element in Sassou's construction of "socialist legitimacy" through means of state largesse.[116] As, in essence, the only meaningful providers of revenue for the country's budget, the two foreign companies were guaranteed a free hand. In addition both, and especially Elf, were involved in providing the government with loans. Their sympathetic contribution towards the extra-legal cash needs of Congolese politicians[117] and occasional help with funding emergencies, such as the government's incapacity to pay public sector salaries, also contributed to the acquiescence of local partners. A measure of tension did develop from the early 1980s, when calls for more investment by President Sassou Nguesso were rebuked by Elf, and especially from 1986, when the fall in oil prices led to complaints that Elf had, like a decade earlier, overestimated reserves.[118] But this was kept within the bounds of civility and does not compare with

---

114 Elf CEO Pierre Guillaumat's ultimatum was that "between the risk of bankruptcy and the risk of nationalization, we have chosen. Now it is up to you to choose." Quoted in "Congo: trading in a hostile world", *West Africa,* 22 August 1983, pp. 1936-7.

115 On the good relations with France during the Marxist years, Olivier Vallée notes that *"les hommes politiques français se satisfaisaient de la chaleur de l'accueil qui sourdait derrière la logomachie révolutionnaire et reconnaissaient l'élégance de la coupe des costumes Cardin du président Sassou"* (1999a: 120).

116 Clark (1997).

117 As documented in Verschave (2001). For references to Elf's "slush fund" for African heads of state see the excellent coverage by *Le Monde* and *Libération* from the mid-1990s, and section 3 of Chapter 5.

118 Thompson and Adloff (1994: 234-5).

the much more antagonistic approach to the company that was to characterise the transition to democracy in the early 1990s.

Through a combination of extreme pragmatism and commitment to the development of national expertise, Angola's post-independence elite managed not only to maintain a good rapport with the oil companies but also to develop local capacity and leeway far beyond those of the Congolese. This was exceptional, especially in the context of the broader economic policies of the ruling MPLA, which pursued an agenda of aggressive expropriation of settler property and sought to run the economy centrally. Coffee estates, cattle raising, diamond mining and the industrial and agricultural sectors were all subjected to Soviet-inspired reforms. The inherent inefficiency of planned economies, together with the lack of human resources and the mounting UNITA threat from the early 1980s, guaranteed that the Angolan economy would enter a state of long-term decline. The result was catastrophic to the extent that, by 1977, the attainment of "1973 output levels" had become the never-to-be-repeated chimera of civil servants.[119]

And yet this most radical of Africa's Cold War pawn did not expel the private sector from its oilfields. The idea certainly crossed future President Agostinho Neto's mind in the long years of exile and guerrilla warfare and was party policy until the eve of independence. As late as November 1974, the MPLA was denouncing the activities

of all oil companies which, directly or under any form of combination or association with Portugal and South African interests, pilfer the oil and gas of Angola.

---

[119] This debacle was not conducive to a measure of pragmatism in the running of the non-petroleum economy. By September 1979, the MPLA was still merrily expropriating the debris of the pre-independence economy. Companies nationalised in that month included Mabor, the national rubber producer; the Angolan Minerals Company; *Panificadora Santa Filomena* (the national bakery); *Empresa Licores de Angola*, a distillery; *Companhia de Cervejas do Sul de Angola,* a brewery; *Pedra de Agua Termal e Turística*, a mineral water and spa tourism company; *Sumos Naturais de Huila* and *Sociedade Refrigerantes*, two soft drinks companies, Cinangola, the national cinema laboratory, etc. (including hundreds of small family businesses in the services sector left behind by European settlers). The shrewd management of the oil sector during the same years appears even more astonishing from this viewpoint. See Roger Murray, "Angola: nationalization and the role of foreign capital", *African Business*, September 1979.

[…] with the inevitable independence of Angola all those companies which operate offshore or inland will be chased from our national territory and all their equipment and assets seized.[120]

But the collapse of the Angolan economy throughout 1975, and the brief stop in production towards the end of that year, made it clear that the new regime would neither be able to exploit the oil on its own nor survive without it. By March 1976, Gulf Oil had decided to pay an outstanding share of royalties to the MPLA,[121] thus recognising it as the legitimate government of the country despite Henry Kissinger's vehement opposition.[122] Simultaneously, the MPLA reassured Gulf that it would be business as usual and swiftly promoted the company's return to Cabinda. By May 1976, a high-level delegation of Algeria's Sonatrach arrived in Luanda with a view to helping the new government create its own NOC, Sonangol.[123] The Arthur D. Little consulting firm became an equally important source of apprenticeship in the first, fast-paced years of Sonangol's rise.[124]

---

120 MPLA statement quoted in "Cabinda and Angola", *Africa Confidential*, 22 November 1974.

121 A similar dilemma had appeared during the Biafra war. The decision to pay oil royalty payments to the Nigerian Federal Government instead of the secessionist movement (following the threat by the former to cancel existing contracts were it not to receive the monies) had an obvious impact on the course of the conflict.

122 The $100 million owed by Gulf to the transitional government had been provisionally put into an escrow account at the behest of the US government. See Dos Santos (1983: 107) and Wright (1998: 72-6). Towards the end of the year, the company would state that it was "operating under the same terms and at nearly the same production levels which prevailed before suspension". See Gulf Oil Corporation (1976: 8).

123 As explained in Chapter 2, Sonangol would mature into a fairly successful company within a decade. I thank a senior executive of Sonangol with experience dating back from the company's foundation for an important interview on its origins (author interview, Luanda, 23 January 2004).

124 Another lynchpin of the early years of Sonangol was the unprecedented 1976 agreement between the company and Petrogal, Portugal's NOC: Petrogal workers would stay on for two years with Sonangol to help it get on its feet and would subsequently be allowed to rejoin the Portuguese company at an equivalent position. Many accepted and most stayed more years, into the early 1990s. Although leadership was Angolan from the start, the intermediary staff was replaced only when qualified Angolans could take

The "Communist" period of the MPLA that would last until the end of the 1980s was not one of mere consolidation of oil investment. Although as yet showing little of the business savvy that it would subsequently show, the MPLA managed to bring in Elf Aquitaine, which started production in 1985, and a number of other companies, restore petroleum output to pre-war levels, and then raise it further. Relations with Elf were initially quite terrible owing to the widespread perception that it sought to bring in the *modus operandi* it had developed in France's former colonies. This was in no way ameliorated by the company's uncongenial first local director.[125] His timely replacement by Fernand Poimboeuf, who would remain at the head of Angola's Elf operations for more than a decade and develop extensive personal relations there, quickly normalised the rapport, and the company's steep learning curve of things Angolan further cemented the bond.[126] This of course, did not prevent Elf from cultivating separate links with UNITA, just in case.[127]

Angola's explicit attempts at diversification of investors as a tool of national control did not mean that it faced Gulf's operations with distrust. Indeed, the company was an ally of the Angolan government for the duration of eighteen years of US enmity, and a strong advocate for the recognition of the regime. As early as 1979, Gulf's impression of Angola's government as able "to understand the difference between a

---

the job, a transition more reminiscent of Ivory Coast or Gabon than of Marxist Angola. Thus both people and structures were kept from the late colonial period, an astonishing exception in view of the fate of the remainder of the Angolan economy. As a senior Sonangol official put it, "in no other area or sector was there such a degree of continuity; everywhere structures were destroyed" (author interview, Luanda, 24 January 2004).

125 Author interview with former Sonangol official, Luanda, 8 March 2002.

126 Author interview with Total senior official, Luanda, 6 March 2002.

127 Elf's meddling in Angola did not have as many consequences as in Congo-Brazzaville, but it was still mischievous. Apart from the 1974-75 support for Cabinda secessionists already referred to, there is evidence suggesting, that throughout a significant part of the Angolan civil war, Elf made payments to UNITA as well. In the run-up to the September 1992 elections, it is claimed that Elf turned $16-20 million to UNITA in exchange for "a leading role in the country" if the party won the ballot (quoted from the Elf Indictment in Global Witness, 2004: 38).

multinational and its home government" was being publicly conveyed.[128] The intermittent hostility of the Reagan administration did not change the thrust of the company's engagement with Angola,[129] which was kept after the acquisition of Gulf by Chevron in 1984.[130] This was so to the extent that an influential US magazine soon labelled the MPLA as "Chevron Socialists".[131] The issue of US policy towards Angola in the 1980s is far too broad to be engaged with here, but some points are worth making. Chester Crocker, formerly Assistant Secretary of State for Africa (1981-88), has accepted that Gulf/Chevron was occasionally pressured by the Administration. But he claims that its divestment was never seriously advocated,[132] both "because it would merely open a vacancy for competitors" (i.e. Elf) and, crucially, because the company had become "interpreter of Republican DC (*sic*) to the MPLA, and an important informal channel in its own right".[133] In short, though the enabling role of oil operators for the fortunes of the Angolan government was clear, their partnership was left undisturbed. The Angolan ruling elite reciprocated by allowing the oil sector, and itself, to prosper.

Francophone moderation

The "moderate" Francophone states were even more amenable to business pragmatism than the formally socialist Angola and Congo-Brazzaville. The development of the oil sector in Cameroon had been a much-delayed

---

128 R. Deutsch, "African Oil and US Foreign Policy", *Africa Report,* Sept.-Oct. 1979, p. 49.

129 In the aftermath of Ronald Reagan's election, Gulf was still commending the "business-like and non-ideological [MPLA posture] in its relationship with Gulf and other American and Western multinational companies". See Susan Gilpin, "Gulf Oil Ask for Sensible US Policies", *African Business,* June 1981.

130 As already noted, the history of Cabgoc is yet to be written. But the little we know of the methods of its holding company point to "years of illegal activity by Gulf on a large scale" (Horsnell, 2003: 23). See in particular McCloy (1976).

131 M. Massing, "Chevron socialists versus Reaganite guerrillas: upside down in Angola", *New Republic,* 3711, 3 March 1986.

132 Compare with the total ban on trade with Libya by President Reagan in January 1986, which led to the pulling out of American oil companies still in that country.

133 Author interview with Dr Chester Crocker, Georgetown University, Washington D.C., 28 July 2002.

affair, with a 1948 concession to a French state company yielding limited success for the following two decades. After intermittent drilling, there was major commercial find in the offshore Rio del Rey concession in 1972, and production started in 1977. The terms were not unfavourable to the state, especially bearing in mind the experience of Congo-Braz-zaville, but all details concerning the oil sector were kept hidden from public scrutiny.[134] The secrecy that has surrounded Cameroon's oil is fertile ground for conspiracy theories. A popular one is that the French discovered petroleum in commercial quantities in the 1950s but chose not to develop it until Cameroon was made "safe" from the UPC rebellion and OPEC dominance in the 1970s started to bite.[135] When most analysts determined that Cameroon would remain a modest producer in terms of both output and the duration of its production cycle, sceptics claimed that the country's reserves were "grossly underestimated" by companies unwilling to accept the onerous terms, by regional standards, imposed by Cameroon.[136] Ngu suggests that Elf was "the sole beneficiary" of oil production until 1980 "since the state had no formal oil policy concern-ing the sharing of profits", and reminds the reader of the "lamentation" of the Elf CEO regarding the creation of the NOC Société Nationale des Hydrocarbures.[137] Because of the pivotal role of oil earnings for the Cameroonian government's profligacy, no serious attempts were made to upset relations with Elf and Shell, the other main investor.

Gabon's business rapport with Elf was equally ambivalent but ulti-mately untroubled. Throughout the late 1970s Elf's dominance of Ga-bon's upstream became an intermittent source of local concern,[138] but alternatives were pursued fitfully. With the goal of diversifying invest-ment, the state sought to promote the role of Shell and that of newcomer Amoco. It also established the ill-starred Gabonese National Oil Com-

---

134 Ndzana (1987). Even in the late 1990s, it was practically impossible to interview the CEO of the Société Nationale des Hydrocarbures, Adolphe Moudiki.
135 See OCLD (1982). This is suggested despite the disappointing drilling results of non-French companies such as Mobil, which conducted offshore research for oil as early as 1965 (*Africa Report*, January 1968, p. 41).
136 Jua (1993: 134).
137 Ngu (1988: 141).
138 "Elf, oil and the dollar", *West Africa,* 30 August 1982.

pany (Petrogab) in 1979.[139] By the mid-1980s, however, Gabon was still yielding OPEC's poorest take per barrel of oil, and the country seemed no closer to possessing either technical expertise or regulatory capacity. The depth of the links binding the ruling elite with the oil companies, together with the close relationship between France and Gabon, nonetheless guaranteed that the relationship would not be jeopardised. The 1985 Shell discovery of the Rabi-Kounga giant oilfield meant that, just when a decline in production and oil prices seemed to threaten the viability of the state, it would have something to hang on to. Therefore no radical revision of contracts or upsetting of the firm-state rapport occurred. Short-lived disturbances underscored their partnership. During anti-regime protests in 1990, for instance, violent demonstrations led to the closing of company operations in Port-Gentil. In no time, French troops retook control of the country for President Bongo and oil production was restored.[140]

Towards the 1990s

This survey of the politics of oil in the Gulf of Guinea during the age of economic nationalism shows that, as a whole, the region saw little if any hostility towards foreign investors in the petroleum sector. Certainly, that age brought a sea change in perceptions of multinational company relations with the host-state, and the previously mute issue of host state control did come to the forefront. The nationalist rhetoric was sometimes deployed, occasionally by political actors with a serious inclination for such drastic measures but more often by fairly safe incumbents such as President Bongo of Gabon, agitating for a better deal from companies.

---

139 This was reportedly done on the advice of Occidental Petroleum's Armand Hammer, no stranger to oil majors' dominance. See footnote 110, Chapter 2, for more on Petrogab.

140 Most analysts have underplayed the extent to which an oil rapport frequently dismissed as "neo-colonial" in fact empowered President Bongo, as well as the pivotal role of Elf in allowing his very long grip over Gabon. The 1990 riots are illustrative here. Faced with anti-Bongo (and therefore, anti-French) riots in 1990, Elf had its personnel evacuated from the country: this was seen as the ultimate betrayal by President Bongo who publicly challenged the company to return or lose its preeminent role to "Gabon's new partners". The company swiftly returned.

(And companies could afford to accept such demands because previous terms were so unfavourable to the state that improving them would still leave investors in a very good position.)

But, all in all, no Mossadegh surfaced in the region and hostility towards oil multinationals was mostly not deployed as a political weapon. States were more interested in diversifying their pool of investors so as not to be dependent on one hegemonic firm than in expelling the ones already in. No attempts were made to achieve control over prices or the allocation of supply. And by the end of the period, oil companies remained at the centre of the region's economy. This is in contrast with their exclusion from most other major oil producing states in the developing world, including many OPEC members and others such as Mexico. It is also different from the treatment meted out to non-petroleum sectors of Gulf of Guinea economies, where outcomes were more mixed.

Companies themselves mostly kept a cordial and occasionally collaborative rapport with each other, with harmony the rule and rivalry the exception. This is not to say that there were no instances of intense competition until the 1990s or, for that matter, that there is no hint of inter-firm collaboration in today's race for offshore acreage—there are plenty. But for most of the period covered in this chapter, two types of situation led to remarkably non-competitive firm behaviour. These are, firstly, the unambiguous domination of one company over others (e.g. Chevron in Angola; Elf in the Francophone states), and secondly, the creation of joint ventures that ensured the lack of serious private sector tensions (e.g. Shell-BP in Nigeria, AGIP-Elf in Congo-Brazzaville).

The time for host-state demands on investors ran out in the early 1980s amidst worldwide recession, a fall in oil prices and the failure of the state-led development strategies that had informed earlier policy. The hostility towards FDI that had prevailed up to that time in many cases turned into a competitive pro-business rush to attract investors.[141] In the Gulf of Guinea's oil sector, the very tenuous nature of the nationalist bid for economic power during the 1970s meant that there was a

---

141 Coakley *et al* (2000: 1-15) points to strenuous efforts across the continent over the last decade with a view to revising mining and oil sector legislation to "end discrimination between local and foreign investors", diminish government involvement and guarantee lighter tax burdens.

great degree of continuity and that the following decade was not lived as a "return from the brink" by investors. In fact, it ushered in ever-closer relations between companies and cash-hungry local elites. By then, the state was fairly empowered in terms of negotiation skills and knowledge of the market but the transfer of technology and know-how had not been as successful, even in states such as Nigeria or Angola where many locals were trained by the oil firms.[142] The costs of the sector remained too high and no NOC could conceivably replace the majors, even in the comparatively simple onshore upstream sector. They needed foreign companies more than ever.

Chapters 1 and 2 explained the manner in which the prosperous oil years were followed by an age of political strife and socio-economic decline that continues today. By the 1990s, all states were crippled by unmanageable debts, overvalued currencies and rising inflation, and found it impossible or undesirable to keep up with public sector commitments. Everywhere the debris of high-spending policies (motorways, dams, public buildings) lay in disrepair, and the non-oil economy withered away.[143] The first two decades of independence had been marked by the seeming ubiquity and then slow demise of what James Scott termed the "high-modernist state".[144] But even amidst this, state retreat from most of the "national" territory (considered economically

---

142 Oil companies provided, and still provide, limited employment in terms of the broader needs of the economy. They have nonetheless been partly or wholly responsible for the training of whatever limited oil sector expertise is locally available to the oil state (i.e. not including the consultants and hired hands that are so important for oil state strategies). This is particularly clear in the case of hegemonic firms, such as Shell in Nigeria, which have trained many local experts. The assistance provided by NOCs from other developing countries may have been instrumental at a brief moment in time (e.g. Angola in the late 1970s) but was always insufficient. Of late, collaboration between some NOCs in the region has increased but it is too early to gauge its importance (see section two of Chapter 2). Employment by a major oil company remains the most attractive segment of the domestic labor market at all levels.

143 Callaghy (1995: 44) mentions that foreign company divestment was widespread throughout the 1980s: 43 of the 139 companies British companies in Africa at the beginning of the decade were no longer there by 1990. This also occurred in the oil states, and especially in Nigeria, but did not affect the oil sector at all.

144 Scott (1998).

useless), and sometimes even from oil producing areas (passed on to companies) had already started. In Angola, this took place abruptly, as a consequence of independence, foreign invasion and civil war. Elsewhere, this delegation and neglect were piecemeal, proceeding in tandem with the dereliction of the state and its lack of interest in full territorial coverage.

An important development that paralleled the decline of those institutions that had hitherto provided education, employment and health to some sections of the population was the rise of demands that companies should furnish them in substitution for the state. Such increasingly vocal requirements could elicit more popular support in areas of intense and decades-long onshore production that had seen little state investment or company philanthropy. In other parts of the developing world, any involvement by private companies in activities that were not narrowly commercial would have faced a barrage of nationalist sentiment and claims of interference. Yet in the Gulf of Guinea it was precisely their lack of involvement in activities that cannot be seen as falling within the ambit of a profit-oriented enterprise that created hostility towards companies. In the 1980s, these claims were still seen as manageable because oil multinationals were often confined to circumscribed areas with comparatively few inhabitants (whereas the colonial concession-aires discussed above had ruled over vast tracts of land), but they were mounting rapidly. As the following chapter shows, these demands have burst into the open throughout the 1990s, particularly in the Niger Delta, and will likely increase in the future.

An area where oil firms have had a history of autonomy that is resented by the population is that of security provision. As already mentioned, official and unofficial company security forces have invari-ably been part of large-scale investment in the region, and so has col-laboration with state security for the protection of company property, personnel and operations. The visibility and violence of such collabora-tions has increased in the last two decades with the threat or reality of political strife, with negative consequences for firm-local community relations.

Conclusion

In one of her last contributions, Susan Strange placed different forms of non-state authority along a continuum:[145] at one extreme, there are those entities that may contest and challenge state authorities, such as crime syndicates; on the other extreme are those whose activities reinforce and sustain the state. The role of oil multinationals in the Gulf of Guinea region is decidedly of the latter type. Far more successfully than the early colonial concessionaires that were meant to underpin state power but ended up undermining it, foreign investors in the oil sector have been the source of sustainability and prosperity for power-holders since independence. The activities of oil multinationals have proved compatible with all sorts of political systems and their centrality has increased with the demise of large parts of the administrative apparatus of oil states. Their continued presence on the ground is also a reason why potentially obscure failing states are the beneficiaries of international webs of complicity that further their standing and reputation on the world stage.

This chapter has sought to study the long history of engagement between states and oil companies in the region. Its guiding theme is that oil companies have acquired a role that far exceeds that of a conventional private sector operator. They provide governments with the bulk of tax receipts, are important international allies and are increasingly (and unwillingly) taking up tasks that were until recently expected to be performed by the state. In opposition to views that see any discharge of state functions as a zero-sum game harming state interests, I have shown that it can be pursued by the state itself as a pragmatic adaptation to unfavourable circumstances or as a solution to tasks it does not care to perform. Furthermore, such strategies can actually provide those in power with revenue streams superior to those they would capture via a decaying conventional state apparatus. And because most companies have a profit-oriented, apolitical approach, elites are able to harness the impressive power of the oil economy for their own political ends without the drawback of "company-creep".

Although Elf frequently participated in spectacular actions of political interference,[146] I have instead prioritised the structural impact of companies

---

145 Strange (1996: 92).
146 See section three of Chapter 5.

at the domestic level and the manner in which partnerships with local rulers have furthered the latter's power. The deleterious political role of the oil sector is thus related to the granting of political support and financing to regimes that are aloof from the demands and needs of their citizens. In this regard, it is the search for pragmatic, hassle-free, investment-advancing, profit-making firm-state partnerships, and not the aggressive behaviour many critics assume multinationals to take in the developing world, that is proving so detrimental for local broad-based prosperity, as Chapter 5 shows. The argument is thus explicitly set against dependency approaches that would grant scant leeway to the receiving end of the oil pact. This measure of local elite empowerment is of course subject to country-specific nuances that show great differences in host-state bargaining power, and is pursued within the framework of unambiguous technological and financial primacy by oil firms.

The Gulf of Guinea at the beginning of the 1990s presents the reader with a mixed picture, where all manner of speculation is possible owing to "the lack of information about the continent's energy potential".[147] While Africa's mining wealth is taken for granted, its oil deposits are perceived to be far more modest.[148] The region has a number of major producers (although just one that is world-class), some up-and-coming oil states, and many hopeful leads that adventurous investors may or may not wish to pursue. An analysis of the oil sector centred on this period shows that by then: a) it permanently shaped local politics and the interaction of these states with the outside world, and b) business entities that give a state such a vast share of its income have an obvious stake in local politics. But recalling the early 1990s, the Gulf of Guinea oil sector of the period is still, in the larger scheme of things, a bit-player, except for Elf that has most of its proven reserves there. Operations in Nigeria are important for Shell; Chevron values its Angolan assets and knows them to be worth more.

---

147 See "Finding oil in troubled waters", *Africa Confidential,* 3 June 1994. The article notes that "some 80 percent of Angola" has not been "properly assessed".

148 Writing in 1987, Ralph Austen mentioned the tendency of "publicists on the right and left to occasionally point to the diamond and metal resources of Central and Southern Africa as a "Persian Gulf of Minerals" vital to the industrialized portions of the world" (Austen, 1987: 216). For an earlier analysis see William Hance, "Africa's Minerals: What the Future Offers", *Africa Report,* June 1971.

Against common perceptions of the region—war; Marxist and nationalist governments; corruption—no investor really fears political unrest or sees it as detracting from the viability of oil production. Washington, DC has its Gulf of Guinea lobbyists spinning the rich treasures the region can bestow on brave explorers, and the benefits of supporting this government or that rebel group. But it is the following decade that will reveal the true vastness and worth of the Gulf of Guinea's oil endowment.

5

OPPORTUNITY AND UNCERTAINTY
IN THE OIL FRONTIER

What Africa could use is a number of well-distributed giant [oil] fields [...] But the prospects for such a scene, or even a modestly better scenario, appear to be dim for the foreseeable future

Robert W. Brown, "Africa's giant oilfields", *Africa* Report

The last decade's scramble for Africa saw growing rivalries between US and French oil companies, as deep-water technology opened new fields off West Africa. Swift expansion followed, as countries called in the oil majors and hoped for a big find. The result was years of fast growth, big signature bonuses and blurred borders between politics and business

"Risky money", *Africa Confidential*

Ce golfe de Guinée qui fait actuellement baver d'envie toutes les grandes firmes pétrolières du monde[1]

"Les damnés de l'or noir", *Le Marabout*

In the past decade, the unlikely has become inevitable as the Gulf of Guinea region witnesses a major ongoing reassessment of the magnitude of its petroleum reserves. The estimate is now of over 50 billion barrels, or just under 5 per cent of world proven oil reserves,[2] roughly 80 per cent of which is in Angola and Nigeria alone. Similarly to other critical

---

1 "This Gulf of Guinea that makes the world's major oil companies drool with envy".

2 Source: *BP's 2004 Statistical Review of World Energy*. 34.3 billion related to Nigeria and 8.9 billion to Angola alone; Chad, Equatorial Guinea and STP are not included. As recent downward revisions of Royal Dutch/Shell's reserves show, the issue must be approached carefully; these are estimates, not facts. As Noel (2003) notes, "it is not nature that produces reserves, it is the oil industry by making investments where oil deposits have

junctures in its history, external technological innovations—this time, in the form of ultra-deep water machinery and expertise—have ushered in fundamental changes in the outside engagement with this region of Africa and its relations with the international system. The first ultra-deep water discovery, Angola's Girassol giant oilfield, was made in April 1996. In less than three years, there were twenty-five such finds, the fastest rate in the world.[3] The excitement over the ultra-deep waters of the Gulf of Guinea was such that a worldwide decrease in capital expenditure by oil companies during the 1998-99 fall in prices had no impact there: in fact, it rose 10 per cent relative to the previous year.[4] Natural gas, still largely flared across the region, is increasingly being harnessed for production and export.[5] Moreover, even before these fields came on stream, there had been a 30 per cent boost in production between 1990 and 1997,[6] and the region had seen a steady growth in investment from the early 1990s.[7] Interest in onshore development has proceeded in tandem, despite political instability. Yet as new opportunities beckon, oil investors are facing the hazards of association with a region where decaying institutions, widespread poverty and a deadly struggle for oil revenues are all-pervading realities.

been found"; accordingly, most reserves are but "estimates of reserves" which cannot be confirmed until investment takes place.

3 "Le point sur l'exploration/production dans le golfe de Guinée", *Marchés Tropicaux et Méditerrannéens* 502, 5 March 1999 and also Bruso *et al.* (2004). Extending the analysis to deep waters (from 300 metres) there were 43 finds by late 1999, according to Harbinson, Knight and Westwood (2000).

4 See Anne Guillaume-Gentil, "Pétrole: la nouvelle donne", *Marchés Tropicaux et Méditerrannéens,* 10 March 2000 and "Deep-Water Survey: Gulf of Guinea Development Work Set to Boom", *Petroleum Economist,* 23 October 2001.

5 Only a decade ago, the region's natural gas deposits were a nuisance to oil companies, "stranded" too far from industrial markets to be worth developing and therefore "flared" (i.e. burned off) or injected back into the oil wells. This is in the process of considerable change—natural gas is now the subject of escalating interest, with ambitious developments being pursued in Equatorial Guinea, Nigeria and Angola to liquefy natural gas and transport it to the world markets. See Yergin and Stoppard (2003) for the emergence of liquefied natural gas (LNG) as a "global business".

6 Guillaume-Gentil, "Pétrole: la nouvelle donne".

7 "Finding oil in troubled waters", *Africa Confidential,* 3 June 1994.

This chapter analyses the cost/benefit calculus behind oil company involvement in the Gulf of Guinea over the last decade. It sees the investment rush in terms that can only be described as those of a new scramble for Africa. This is so first and foremost for the oil majors, but also for international lawyers in search of the profitable maritime border shenanigans, oil services providers, export credit agencies, accountancy firms, banks, tax havens, construction companies, a variety of consultants, and crime syndicates in for a share of the oil prize. The goal is to understand how companies juggle conflicting images of commercial opportunity and political risk, and how these frame "their" Gulf of Guinea. Section one exposes the phenomenal rewards that hurl this momentous historical process forward while highlighting its pitfalls. Section two concentrates on company strategies for seizing opportunity and minimising risk vis-à-vis two of the region's most recent and least experienced oil players, Equatorial Guinea and São Tomé and Príncipe. Through brief analyses of violence in Nigeria's oil-producing region and the methods of the French company Elf Aquitaine until the late 1990s, section three contrasts the "El Dorado" vision with the actual devastation wrought on the region amidst a commercial race haunted by decades of mismanagement and oil-related decline. Finally, I come to the paradoxical conclusion (already spelt out in Chapter 1, when discussing the fortunes of the successful failed state) that companies have got the equation right. Their involvement in this violent and profitable frontier is viable for the long term even if, occasionally, local political dynamics lead to the bruising of investment expectations, and international pressures demand a measure of reconfiguration in the partnership with host states.

I. TECHNICAL BREAKTHROUGHS AND THE INVESTMENT CALCULUS

Offshore exploration in the shallow waters of the Gulf of Guinea dates back from the late 1950s and was successful throughout the following decade in Congo, Nigeria and Gabon[8] as well as Angola, but existing technology and ignorance of geological conditions precluded a more accurate

---

8 The following paragraph is mostly drawn from Harbinson, Knight and Westwood (2000).

understanding of the region's potential. Research in the deep waters off Brazil and in the Gulf of Mexico from the 1970s eventually aided in understanding the South Atlantic's geological past, the processes of sand transport and deposition and the prospects for crude oil abundance on the other side of the ocean. Advances in deepwater drilling technology and seismic data collection were also crucial in allowing present-day research in water depths of more than 3,000 metres.[9] Computer improvements have in turn afforded a much better interpretation of seismic data, which can now be used to "quickly and accurately visualize, model and interpret the 3-D subsurface with resolutions that would have been unbelievable just a decade ago".[10] While existing technology is still deficient in assessing the economic viability of hydrocarbon accumulations,[11] it has made a significant impact both on the drilling success rate and on the accessibility of previously unreachable petroleum deposits. The technological complexity of the Gulf of Guinea's deepwater as well as the project management capability needed mean that only a handful of companies worldwide can fulfil the prerequisites for involvement. Equally formidable are the capital requirements of that investment, especially if compared with much cheaper production costs in Middle Eastern states, to which oil majors have limited or no access:[12]

According to the American Petroleum Institute, a whopping $10 billion in capital investment will be needed for every one million bpd increase in production. About $52 billion will be invested in deepwater African fields by 2010, with approximately 32 percent coming from the US. According to industry sources, the Gulf of Guinea region will receive the world's largest amount of offshore hydrocarbon capital investment by 2005. Angola alone could see $20 billion in investment in the next five years, according to the Center for Global Energy Studies.[13]

Companies that have long dominated upstream operations, like Elf, Chevron, Shell, Exxon-Mobil and AGIP as well as the recently returned

---

9 Drennen (2002).

10 Ibid.

11 Ibid.

12 As explained in the previous chapter, Nigeria (though a member of OPEC) did not go for all-out expropriation and remains very dependent on oil majors for investment. The name applies to Angola which gained the control in 2007.

13 Gary and Karl (2003: 11).

British Petroleum, currently play the leadership role in offshore exploration and development. The same applies for the development of oil resources in landlocked states like Chad: the upstream technology required there is not particularly innovative but the capital expenditure needed for the costly transport structure (in this case, the Chad-Cameroon pipeline) is premised on the involvement of world-class firms. This said, there are numerous segments of the regional oil economy where many more private sector players have a role to perform. In addition to the majors, other companies are entering the Gulf of Guinea or substantially enhancing their presence onshore, in the shallow or deep waters offshore, or in partnership in the operations of the majors, including the three Chinese NOCs, Petronas, Marathon and Amerada Hess and European companies such as Norsk Hydro and Statoil. Some companies have a complementary role to that of the majors, either at an early phase of development, when a province is still "frontier", or later, at a mature stage when no giant oilfields remain but plenty of niche opportunities exist. They include a number of smaller, more adventurous companies pioneering in the less inspiring corners of the region, and the firms of local entrepreneurs, often connected with dubious international accomplices and usually without much technical expertise.

Regrettably, there are no clear statements by the oil majors on the length of their "expectation and planning period"[14] for specific countries, although there is plenty of information on a project-specific basis. Negotiations over exploration licenses, the size of signature bonuses and the competition from other oil companies for promising acreage is normally dependent on the prospects for oil wealth afforded by accurate information, and those in early possession of such commercially sensitive data tend not to share it. This said, most interviewees spoke in terms of "some decades" for the major producing countries; it is the new producers that elicit vaguer or more contradictory projections. The caution in stating timeframes is due to the fact that the bulk of proven reserves are in the older and better-charted oil states: in the absence of this for the newer producers, current estimates are unsure ground for speculation. BP interviewees in London have referred to "fifty years in

14 Byé (1957: 149).

Angola".[15] The life of the Chad-Cameroon pipeline project is nominally twenty-five years, but this is based on current proven reserves. A rare interview with a Marathon executive revealed that Equatorial Guinea's "supplies may dry up in as little as a decade"[16] if they are submitted to maximum exploitation.[17]

Except for country-specific variations, there seems to be no impediment to a long-term regional presence by companies. All the major players have the capital or the capital markets creditworthiness for continued development of oil resources. Furthermore, prolonged "time horizons" are typical of most large-scale investments in the extractive industries,[18] as shown in the previous chapter. Whatever the exact projections, and bearing in mind state-specific variations that may include both decades of production and impending dryness, there is no doubt that oil company involvement *at the regional level* is being perceived in a very ambitious, long-term manner. Companies are certainly aiming at the construction and maintenance of partnerships for the long haul.

When discussing the rush of oil investors, substantial cross-regional differences must be borne in mind. Firstly, there are the territories that have occupied the centre stage: primarily Angola, Equatorial Guinea and Chad, with sizeable investment in Nigeria's offshore also taking place. The thus far not fully substantiated ebullience for São Tomé and Príncipe can also be seen in this context. Secondly, there are the stagnant or declining oil exporters: Gabon, Cameroon and, more ambivalently, Congo-Brazzaville, none of which has seen major recent investments; production in Gabon and Cameroon has actually declined considerably in the past five years. Short of major finds that cannot be ruled out but look increasingly unlikely, both will be out of petroleum in the next decade and a half. Finally, there is a third group of states that are not the focus of

---

15 Several interviews with the author since 2000.

16 See International Consortium of Investigative Journalists (2002: 2).

17 Chevron is also rumored to be reassessing its previous enthusiasm on the basis of disappointing results and Total (then TotalFinaElf) left the country in 2000. For "a mixed view of oil prospects" see *Equatorial Guinea: Country Report*, EIU, January 2004.

18 Byé (1957: 150).

this study such as Niger,[19] the Central African Republic, Ivory Coast and the Democratic Republic of Congo, as well as West African states such as The Gambia,[20] Benin, Sierra Leone, Mauritania[21] and Guinea-Bissau. Some of these are minor producers and others have seen intermittent enthusiasm about their oil potential.[22] It is conceivable that a number of them will join the petroleum club of major producers in the next decade but this has not happened yet.

In a context of rising appetites for offshore acreage, competition between companies is fierce, and so is the support they glean from home states, especially since the former *chasses gardées* have disappeared and the whole region lies open to willing investors, old and new. It seems accurate to say that companies accept the regional business climate without qualms. Indeed, while in the long period analysed in Chapter 4 different corporate cultures may be said to have played an influential role, over the last decade these idiosyncrasies have become much less relevant and company actions are better understood as the product of the context they have to engage in. As a number of corruption scandals have lately shown, no company in the Gulf of Guinea is above trying to gain ascendancy with host states in questionable ways. But theirs is a moneymaking rivalry: it is profits, and not geopolitical advantage for their home states, that drive oil firms on. This is often misconstrued by observers, who perceive firm-on-firm cutthroat tactics as driven by hidden ambitions, especially of the US and France. In reality, collaboration between companies is a common occurrence, including between American and French ones, especially in the expensive and technologically complex ultra-deep waters. The differences in outlook between the Western majors, once significant, are currently negligible. Even at the level of the home governments, it is mostly narrow commercial advantage, rather than loftier "sphere of influence"

---

19 See "Elf Aquitaine, Exxon to continue oil research in Niger for 5 yrs", *Agence France Presse*, 28 November 1995, and "Exploration Pact with Sonatrach", *Africa Energy Intelligence,* 30 January 2002.

20 "Oil found off the coast of Gambia", *The Guardian,* 18 February 2004.

21 "Mauritania to soon join growing ranks of Africa's oil exporters", *IRIN*, 10 February 2004.

22 "Chad and its Neighbors: Explorers Flock anew to Africa's Interior", *African Energy* 73, April 2004.

aspirations, that hurl the process forward. In turn, oil companies welcome the diplomatic leverage but do not partake of the political imaginings that may exist among armchair strategists in Paris or Washington.

Companies seem to accept, as a BP executive put it, that at an early phase of development of an oil province "there is heightened competitiveness; later on there is less need for such attitudes."[23] Moreover, the costs and geological risk of oil investment are such that companies seek to limit them by establishing consortiums; in many cases, companies fighting for a certain license will be pooling their resources together for joint petroleum development elsewhere. Cooperation, as well as conflict, abounds. This does not preclude companies from casting aspersions on each other, an oil industry trait that, as Geoffrey Jones pointed out, is unusually widespread since early in the last century.[24] In the course of my research, references were occasionally made by Anglo-American interviewees to "Elf tactics" being "disloyal because we can't compete with the money bags",[25] but they sound like contrived naivety by companies equally proficient in the methods needed to secure local business opportunities.[26] A majority of those interviewed also claimed that there was no clear "political leaning" in firm-firm behaviour, i.e. companies would not make decisions on a "national" basis. Alternatively, the most commonly mentioned worry with relations between firms was the fact that "old timers" (i.e. those companies enjoying a longer presence and/or a particularly close relation with the government, or elements of it) "shamelessly" use their advantage over others to acquire individual concessions and exemptions. For new arrivals this is particularly grievous at times when a "collective stance" by the oil sector is needed.[27]

---

23 Interview, Luanda, January 2004.

24 Jones (1981).

25 Interview with Chevron consultant, London, November 2000.

26 Furthermore, this plays into the Elf myth, discussed later in this chapter, that other companies have consciously peddled in an effort to differentiate themselves from the French company; such references have been less common since Elf's merger with TotalFina in 2000.

27 This was the claim discreetly levelled against Chevron by other majors in regard to the Angolan government's new Petroleum Law (interviews, Luanda, January 2004). Similar claims about Shell in Nigeria surfaced in discussions with other companies (interviews, Port Harcourt, September 2002). In most cases, there is predictable tension between settled companies that enjoy personalised relations with local rulers and those seeking to

It is important not to take these statements at face value. Firstly, when threatened with measures that are truly negative, or when they share a long-term concern, companies react in consonance.[28] Secondly, the reader is reminded that oil investors in the Gulf of Guinea are the beneficiaries of very favourable investment terms: the imposition of some added burdens on recent investors may penalise them in regard to "old timers", but still leaves ample room for profit. Thirdly, the obvious implication of this disadvantage would be that the Gulf of Guinea is a skewed playing field where seasoned oil investors thrive and newcomers fail—but the presence of so many new faces belies this. The best way to see the situation is as an investment context where first movers have clear advantages but where other companies can carve a place for themselves (by accepting more onerous costs; by creating technical partnerships; by accessing presidential insiders; by paying overgenerous signature bonuses, etc.). Furthermore, there are country-specific trends that matter here. Angola and Nigeria, because of the presence of most oil majors and a large number of other companies, can afford to promote a much more competitive environment. In turn, states such as Gabon and Cameroon are now much less attractive, and a licensing round in either tends to elicit little enthusiasm from investors.

Oil company concerns

Africa's attractiveness is lowest worldwide in foreign direct investment (about 1 per cent of the total) but oil and mining account for 90 per cent of that which takes place:[29] what applies to other economic sectors is not necessarily relevant for businesses moved by the geological imperative. Nonetheless, a survey of the politics, economies and institutional arrangements prevalent in the Gulf of Guinea shows that

---

enter. But experience shows that the recently arrived quickly become acclimatised, and are unlikely to question the nature of business practice.

28 An example is a report compiled by oil companies in the wake of a March 2002 attempted coup in Equatorial Guinea, "assessing the security situation and detailing possible options if the country were to disintegrate into violence". See International Consortium of Investigative Journalists (2002: 2). Moreover, a Working Group on Equatorial Guinea created by the Corporate Council on Africa provides a venue for mutual consultation.

29 Goldwyn and Ebel (2004: 4).

the ample rewards need to be counterweighted by a measure of attention to risk. The following are some of the most frequently cited arenas that oil companies consider worrisome in their engagement with the Gulf of Guinea.

Political risk. Almost without exception, the states of the Gulf of Guinea rate poorly as credible FDI destinations.[30] Other than the obvious drawbacks of recurrent warfare and high criminality, there is the fragility of property rights, the tenuous nature of the court system and the pervasive corruption of the civil service and most spheres of public life. Small oil and seismic companies in frontier areas face great contractual uncertainty and unreliable authorities, as the number of ongoing legal attempts at collecting debts from the region's states testifies.[31] Furthermore, operating costs are very high in customs duties, port authorities' red tape, the provision of electricity and water supplies, healthcare, etc.: more than enough, in sum, to give the most enthusiastic investor a pause to think through any commitment at all.

Yet this is not the way the major oil companies determine their involvement. Firstly, they have a long and successful experience with operations in states where property rights are weak or non-existent.[32] Expansion of foreign investment to the oil sectors of Central Asia and Russia, for instance, has taken place despite equally unappealing circumstances, although in the case of the latter companies seem to have

---

30 France's ECA Coface rates Angola, Chad and Gabon in its C category, "a very unsteady political and economic environment [that] could deteriorate an already bad payment record". Nigeria, Equatorial Guinea and STP are rated D: cases in which "the high risk profile of a country's economic and political environment will further worsen a generally very bad payment record". Only Cameroon manages a passable B www.coface. com/insurance/risk_management_consulting.htm accessed 17 February 2004). See also Collier and Patillo (2000) for an assessment of risk in sub-Saharan African economies.

31 The fact that small companies at an early phase of the oil sector run considerable political risks in addition to the adventurous character of their undertaking was usefully pointed out by Claire Pickard-Cambridge, Africa Editor of the Argus Media Group, in "Other Acreage - How and Why?", paper presented at the conference on "Oil and Gas in Africa", Royal Institute of International Affairs, 24-25 May 2004.

32 A point defiantly made by the CEO of Unocal in regard to his company's decision to invest in Burma in Imle (1999).

underestimated the hostility they would face from a resurgent state.[33] Likewise, the Gulf of Guinea has proven to be an attractive location for multi-billion-dollar petroleum investments, perceived as secure "by the sheer volume of the financial resources involved".[34] Much of the uncertainty that exists is in fact shouldered by the small oil companies that frequently open up oil provinces to the attention of the big industry players (since the rewards are potentially enormous, they are keen to enter regardless). And once the sector is established, most oil-linked players benefit from the special conditions of the sector. Interviews with businessmen outside the oil sector and diplomats from their home states showed that the concern with such issues in the services, financial and import-export sectors is much higher than in the oil sector.[35] Oil companies would like to see an "improvement of the investment climate"[36] but this is mostly phrased as a theoretical, "long-term" goal towards which some movement is expected, but which is unlikely to make or break a good business relationship.

Secondly, and following on from the previous point, companies inhabit an economic enclave within the host country that is not only physical—either offshore or in company-administered compounds—but also legal. Contracts specify international arbitration to keep the oil business outside national courts. Oil majors suffer very little from the chronic misfortunes of the rest of the host economy. Basic services are reliable, luxury amenities are provided goods are available in a timely, hassle-free manner,[37] and the police do not molest a firm's foreign

---

33 Moises Naim, "If Geology is Destiny, then Russia is in Trouble", *New York Times,* 4 December 2003.

34 Interview with oil company executive, London, April 2002. As several interviewees put it to me in Luanda, "after about $1 billion, your investment is very safe" (interviews, January 2004).

35 Interviews with businessmen (various locations and dates); interviews with senior diplomats and foreign ministry officials of several European states, Luanda, 23 January 2004, São Tomé, 14 March 2002, Luanda 6 March 2002, Paris, 27 April 2003.

36 Interview with Shell executive, Port Harcourt, 22 September 2002.

37 Oil companies in Nigeria handle their own terminals (such as Bonny, Qua Iboe and Brass) which are much better run than the rest of Nigeria's decaying ports. In Equatorial Guinea, the Luba oil port, built in 2002, will be managed by a Dutch firm for 15 years. The oil industry in Angola mostly uses the Malongo terminal in Cabinda, the Kwanda

workers. In sum, relations with the host-state take place within a framework of understanding that, for the mutual benefit of all involved, limits uncertainty to manageable proportions.

Armed disruption. Local non-state actors are perceived as much more unpredictable. The one fear that is high on company agendas is the threat of disruption and terrorism from disenfranchised populations mobilised around a mix of ethnic and economic grievances. Especially since the mid-1990s Nigerian descent into violence, companies have invested in their security arrangements and contingency plans.[38] In this sense, there is a measure of preoccupation amongst companies as to the long-term consequences for themselves of shortsighted government policies,[39] although perceptions of non-state actors as predatory and essentially non-political often lead to an underestimation of the political strength of their calls for mobilisation. While companies see state elites as ultimately rational business partners, there is great concern that the blackmail tactics of local warlords and grassroots activists may fragment the lingering political order that enables the oil economy to function. If the Cabinda insurgency has never been more than a nuisance for Cabgoc, the recurrent production halts in Nigeria by Ijaw and Ogoni activists show that such activity can take up very threatening proportions.

At the moment of writing, armed activity has not impeded involvement by companies, except for Shell's ousting from Ogoniland, which is briefly discussed in section three. This is not to say that, in the future,

---

facilities at Soyo, and the Sonil centre in Luanda and Lobito. Interviews with two port authority officials of Luanda and Lobito showed that oil companies "do not suffer from port difficulties", at least to the extent to which non-oil businesses not belonging to presidential insiders do (interviews, Luanda, January 2004). See also "Guerra no Porto de Luanda", *O Angolense*, 10-17 January 2004. General port management much improved in Angola following a contract with Crown Agents in 2001.

38 Planning for disaster is of course not a preserve of the oil companies. Most entities with a presence in the region, including the bigger NGOs, have fairly elaborate security plans for emergencies. What is peculiar to large-scale private sector operations is the growing trend of permanent private security deployment to ensure operations can function in hostile conditions.

39 A BP interviewee noted that "the biggest threat to our operations is the absence of a peace dividend"; Luanda, 18 January 2004.

certain levels of disruption may not endanger the very presence of oil firms. Chevron, for instance, reported major losses for 2003 on account of recurrent disruption in Warri.[40] Paradoxically, these emergencies demonstrate the resilience oil firms will show when it is too late to go back. Exxon claimed to have delayed its active participation in Chad because of the country's Byzantine politics and penchant for civil war, but other companies have not been deterred from investing elsewhere because of similar circumstances. It is more likely that Exxon used this argument as a way of sharing risk with partners such as the World Bank.[41]

Thus there is an acknowledgement of risk, but the response has been security-focused (if PR-laced) much more than political, a bias likely to be furthered in the next decade as military cooperation between home and host governments increases and as local disruptive actors suffer a gradual criminalisation in policy circles.[42] This will also bring about further growth in the use of PMCs, whose expansion in sub-Saharan Africa is in no small part a product of oil company security anxieties.[43] Companies are entrenching themselves for long-term involvement in a difficult setting, but will not endanger the rapport with governments by telling them what to do. Whether companies are accurately gauging the potential for firm-targeted rebellion, and whether they overrate their capacity for co-optation, is an open matter. Finally, as everywhere since the events of 11 September 2001, Islamic terrorism is regarded as a potential threat in the Gulf of Guinea. Yet the absence so far of attacks

---

40 See "Shell Rules out early return to Ogoniland", "Chevron loses N12bn to Warri crisis", *Daily Champion,* 23 February 2004, as well as section three of this chapter.

41 The case of Chad is discussed in section one of Chapter 6. Oil sector reaction towards civil war in Angola and Nigeria were already explored in the previous chapter.

42 The best example here is the contrast between the sympathetic coverage of Ogoni activism in the mid-1990s and the very critical turn of current analyses that increasingly equate Niger Delta struggles against the Nigerian state with warlordism and piracy; comparisons with formations such as Colombia's FARC are rife.

43 Oil majors typically make use of three types of knowledge to deal with security threats: headquarters strategic thinking, local subsidiary understanding of the situation on the ground (which may include collaboration with other companies), and hired expertise. A combination of the three with emphasis on the latter means that companies are increasingly aware of the risks and, to some extent, capable of confronting them.

and of disgruntled Muslim populations in most of the region has meant that companies have not considered this a clear and present risk equal to other local disturbances.

Border disputes. At a more immediate level, oil companies are concerned with the disruptive effect of maritime border disputes between the states of the Gulf of Guinea. Most maritime boundaries were sloppily defined (or not at all) during the colonial period, and the obscurity and seeming irrelevance of the issue ensured that it would not figure prominently in early postcolonial concerns. Yet increased interest in offshore (shallow and deepwater) development from the 1970s led to tensions between countries suddenly rediscovering "the sanctity of their waters".[44] The major early disputes were those between Nigeria and Cameroon over the Bakassi peninsula[45] and between Equatorial Guinea and Gabon over Mbagné island. Despite the fact that the real struggle pertains to the ownership of offshore petroleum, these disputes still related to claims over dry land. In the late 1990s, controversies emerged everywhere as to the delineation of maritime lines. All contiguous states have by now quarrelled over the subject, although some of the disputes have been under-emphasised for expedient reasons.[46]These particularly serious disagreements for long halted oil investment in some corners of the Gulf of Guinea, led to the outbreak of an inter-state conflict (the short-lived Nigeria-Cameroon armed clashes over Bakassi)[47] and fed a substantial mini-industry of Western legal consultants brought in to solve them.[48]

---

44 Interview with Cameroonian academic, Cambridge, July 1999.

45 The International Court of Justice's October 2002 ruling in favour of Cameroon was initially not accepted by Nigeria, but after negotiations Nigeria evacuated the territory on 14 August 2006.

46 The current Congolese government, for instance, has played down its maritime border dispute with Angola owing to the Angolan military aid given to President Sassou Nguesso in his 1997 successful bid for power. See *La Lettre du Continent* 359, 7 September 2000. Chevron-Elf partnerships on both sides of the border have helped in this.

47 "Nigeria/Cameroon: Blundering into battle", *Africa Confidential,* 15 April 1994.

48 For a good taste of this, see the Proceedings of "Resolving International Border Disputes for Commercial Success in the Oil and Gas Industry: Two Day International Conference London, 10-11 July 2000", in which the Gulf of Guinea imbroglio figured prominently. See also Dzurek (1999).

Companies have sought to get directly involved in these simmering disputes, as the implications for their investment prospects are immense. Without legal certainty over sovereignty, no investment can take place, or it risks being lost.[49] At best, a legal dispute can take over a decade to solve, during which nothing happens; at worst, wrongful licensing can lead to a loss of the license. The perception is that the issue is much too important to be left to governments and—in view of occasionally violent rhetoric and the Bakassi precedent—that conflicts of some sort could ensue. One leading newsletter claimed that "the new [maritime] frontiers are being negotiated in secret with the encouragement of oil companies operating in the region".[50]

At the present time, many such conflicts have been resolved. Equatorial Guinea signed a decree unilaterally adopting equidistant median lines to define territorial boundaries as stipulated under the UN Convention on the Law of the Sea;[51] Cameroon, Nigeria and São Tomé subsequently accepted the decision, with minor adjustments.[52] São Tomé and Nigeria[53] settled their dispute through the creation of a much-contested Joint Development Zone (JDZ) that is discussed in more detail below. Companies have had a prominent role in this, especially in the case of ExxonMobil in Equatorial Guinea and Total in Nigeria. The presence of the former in the Zafiro field was questioned by Elf's seismic data, suggesting that Zafiro was actually an extension of its Ekanga field; the dispute was laid to rest in April 2002 by means of a joint exploration treaty.[54] Other disputes linger on. Equatorial Guinea and Gabon's old

---

49 "A question of boundaries", *African Business,* September 2001.
50 "Gulf of Guinea: Berlin Conference for Oil", *Africa Energy Intelligence,* 11 September 2002.
51 "Country Analysis Briefs: Equatorial Guinea", Energy Information Administration, Department of Energy, US Government, p. 3.
52 "Maritime Boundaries in the Gulf of Guinea", paper by Tim Daniel, Partner in D.J. Freeman, London.
53 "Nigeria, Equatorial Guinea resolve maritime border dispute", *Agence France Presse*, 1 September 2000 and "Nigeria and São Tomé Meet on Common Ground", *African Energy,* 27 November 2000.
54 "Country Brief: Equatorial Guinea", EIA, p. 4.

conflict has of late spiralled out of control,[55] and Angola and the DRC remain at odds as to Congolese claims over territorial waters and offshore acreage at the mouth of the Congo River.[56]

Relations with the host state. Another issue that rates highly in company priorities is the imperative of a good relationship with the host state. In general, the assumption by companies is that of continuity in the ruling elite (if not the specific presidential set-up), so there is great care about fostering and maintaining friendly and durable relations. Nowhere is aggressive anti-firm behaviour expected, even in Equatorial Guinea where the leadership is presumed to be highly erratic, but also highly dependent on oil rents.[57] The initial decision to invest already factored in the risks that can be said to exist.[58] Short of major political reconfigurations that change the local political calculus, a highly unlikely event and one that has not taken place in the last decade, this is not an issue that is daily revisited by local managers. But in between improbable expropriation and a perfect rapport lies a grey area of growing assertion whereby governments may seek to impose new costs on companies, change laws or force joint ventures with local business interests. The Angolan government, for instance, has passed a new Petroleum Law that imposes such unwanted costs on companies.[59] In Equatorial Guinea, the use of a private security firm owned by the president is non-negotiable.[60] A good

---

55 "Equatorial Guinea-Gabon: UN mediates dispute over Corisco Bay islands", *IRIN*, 23 January 2004. A deal was struck at the African Union Annual Summit that provides for the "common management of oil resources" but the unpredictability of both sides, and especially of the Equatoguinean leadership, means this dispute may resurface at any time. See "Un deal pétro-diamantaire", *La Lettre du Continent* 451, 15 July 2004.

56 "Snarl-up in the Gulf of Guinea", *Energy Day*, 28 May 2001.

57 Frynas (2004) is more skeptical, stating that oil companies can "expect that the Equatoguinean government will become more assertive [...] indeed, it is not inconceivable that the government could even demand majority shareholding in the future".

58 Interview with official of the Norwegian Petroleum Directorate on Statoil's involvement in Angola, Brussels, 15 May 2001; interview with TotalFinaElf official, Luanda, 8 March 2002. The point has been made by a number of other interviewees.

59 Interviews with BP, Chevron and Exxon Mobil executives, Luanda, January 2004.

60 See International Consortium of Investigative Journalists (2002: 4) and *Equatorial Guinea: Country Profile 2003*, EIU, p. 13.

rapport with host governments may also deteriorate or be tested by how a company reacts to international community stands on transparency, democracy, human rights and governance standards. While none of these is a make-or-break concern for decades-long, billions-worth investment, they can result in substantial setbacks for firm-state relations.

International respectability. The inescapable need for a good rapport with the host-state has in the last decade come to be tempered by a (somewhat less pressing) need to prevent damage to company reputations. Well-orchestrated international campaigns and occasional media coverage, ironically heightened by the oil rush, mean that the Gulf of Guinea is no longer the unreported backwater where company dealings proceed discreetly. More worryingly, investor groups are starting to agitate for ethical investment and using annual shareholder meetings to make their voices heard. While this is of no importance to small, less known or non-Western companies, it matters for the oil majors that I am discussing here. As Chapter 6 shows, these pressures have worked in two important ways. Firstly, they have added to the usual profit motive the semblance of a progressive engagement between the international system and the Gulf of Guinea specifically centred on how oil monies are spent. Secondly, it has led to a revamping of how companies address criticisms of collusion with local kleptocracies, environmental damage and violent excess in the deployment of private security, if not to tangible changes in policy.

On the bright side of things

The previous paragraphs outlined the non-technical, political concerns that the oil rush presents companies with. But even when these have been accepted into the equation, the advantages of the Gulf of Guinea remain obvious. Companies seem to believe that the existing risks are navigable: according to UNCTAD, Nigeria, Angola and Equatorial Guinea were the three main destinations for FDI in Africa, while other more respectable, stable and comparatively well-run states like Ghana, Senegal and Tanzania do not even figure in the top ten.[61] The following are prioritised in company thinking about the region as the "plus factors".

---

61 UNCTAD (2005: 41).

Security. The politics of the Gulf of Guinea, for all its violence and authoritarianism, are without major negative consequences for oil investors[62] and their home governments alike. Transnational perils such as militant Islam are absent for the time being, and below the surface of political strife there is little agitation that is threatening to FDI—a point illustrated repeatedly by the aloofness from most conflicts experienced by oil firms, despite many lethal outbreaks in the Niger Delta and elsewhere. A trait the region shares with some states of the Middle East is that, under the guise of instability, the incidence of regime change is actually very low. In the cases of Cameroon, Angola, Gabon and Equatorial Guinea, heads of state have been in power for at least two decades. Those states that have seen a lot of instability reveal on closer examination a remarkable degree of personnel continuity, as former heads of state return to the presidency and as players of a previous regime find new roles in a reconfigured political game. As discussed in Chapter 6, the worth of a "safe" region without a history of serious antagonism vis-à-vis the West, and with local rulers keen on furthering partnerships, for strategic thinking on security of energy supply is evident. For our purposes here, the point to bear in mind is that this also advances the investment bottom-line for oil companies. They are vigilant as to potential disturbances but are in no way immobilised by them, as the empirical reality of massive continuing investment attests.

Growth potential. Investment in much of Latin America, Russia and the Middle East is either barred or considerably limited by NOCs and a lingering streak of economic nationalism. Yet most Gulf of Guinea oil producers are non-OPEC and those which are members of the cartel (these are Nigeria, and Angola from 2007; Gabon was member for two decades but left OPEC in 1996) have never obstructed private investment. This means that, bar a sudden and highly unlikely bout of conservationist zeal, Gulf of Guinea oil states will pursue the development of their production potential as far as it will go. The cupidity of local elites unlikely to think of future generations and bent on maximum short- and medium-term

---

62 A case made forcefully in Frynas (1999).

profit gives a measure of certainty to company expectations in this area.[63] And now that Nigeria's production is expected to more than double to about 4.4 million bpd by 2020, many analysts share the assumption of her deliberate break with OPEC quotas, whether within the cartel or outside it.

Contractual terms. The tax structures offered by states in the Gulf of Guinea, while varying greatly from, say, Nigeria to Chad,[64] are overall much more attractive than elsewhere, so that "companies are often able to make higher profits per barrel of oil than from other parts of the world".[65] Furthermore, the burdensome joint venture agreements between companies and states, which presuppose that states pay their share of investment and are therefore subject to delays in project development, are increasingly giving way to production sharing agreements where companies make the investment but recoup a bigger slice of the profits. According to some analysts, the new contracts also "transfer volatility away from international oil companies towards host countries".[66] These fiscal incentives are related to the Gulf of Guinea's complete dependence upon foreign upstream technology to exploit oil resources. This is so even for Nigeria and Angola: while having acquired enviable expertise in

---

63 This does not preclude pro-conservation statements. Gabriel Nguema Lima, Equatorial Guinea's new Oil Minister and previously Secretary of State for Oil, stated in 2001 that he would like to see the careful phasing of developments, but this has not taken place yet (*Equatorial Guinea Country Profile 2002*, EIU). More recently, the government denied rumours that it was about to implement production caps (Country Analysis Briefs: Equatorial Guinea, EIA, p. 4). Angola has also occasionally put forward conservationist views, but there are no signs of implementation. At any rate, because of the large capital costs involved, the development of at least some of the newer finds may be sequential rather than simultaneous.

64 Typically, the older producers like Angola and Nigeria have better contracts; Equatorial Guinea had some of the world's poorest contracts when it started oil production in the 1990s but these have been revised upward since; São Tomé and Chad still have the worst in the region, although both have agitated for renegotiation.

65 Ellis (2003: 135).

66 See N. Shaxson, "Mitigating the Disastrous Effects of Oil Price Volatility", *African Energy* 61, April (2003). According to Shaxson, in order to maximise revenue in times of high prices, the oil states of the Gulf of Guinea have foregone "flat rate" approaches that could smooth the revenue stream in bad years.

the area of negotiations over the last decades, both continue to rely on their foreign partners for the technical side of the business, and this will likely remain so for the foreseeable future.

Geography. This affects the situation in two ways. The first is West Africa's proximity to the North Atlantic industrialised states. The distance from the Gulf of Guinea to European and North American East Coast refineries makes the shipping charges highly competitive.[67] While distance is not the decisive factor (Japan and China, for example, are major buyers of crude oil in the region) it adds an important element to the Gulf of Guinea's attractiveness. The second is the increasingly offshore location of oil reserves, which means that the sector is literally insulated from the politics of the region and from most conceivable onshore threats.

II. "MAKING IT BIG" IN THE NEW OIL PROVINCES[68]

As discussed in Chapter 1, small, very weak or "failed" states have warranted a lot of attention in recent literature on Africa. Amongst the many things assumed to be missing and beyond any realistic prospect is FDI in a legal sector. However, I have argued that this is not so in the context of the oil industry which, once commercially viable reserves are discovered, will pursue its goals in the most difficult surroundings imaginable. This section deals with corporate experiences in two such settings of the Gulf of Guinea where their substantial engagement is relatively new. Equatorial Guinea and São Tomé and Príncipe are ill-reputed, distant locations where few other formal economy engagements with the international system could be envisioned or would have worked.[69] These are truly bottom-of-the-pile states with a razor-thin empirical existence. Their mainly agricultural economies were already in a shambles on the eve of oil investment and their bureaucracies are

---

67 Mañe (2005: 4) also mentions the fact the Gulf of Guinea benefits from the absence of narrow shipping maritime lanes known as "chokepoints".

68 "Those [reserves] are totally untapped. We're the first, we're making it big there!" (interview with oil company executive on oil investment in São Tomé, Washington DC, 7 July 2002).

69 The Chad experiment in externally monitored governance of a new oil exporter is discussed in Chapter 6.

corrupt, unpaid and untrained. And yet, companies have encountered an international environment keen to support their entrance into these economies, local elites vouching for the security of their ventures and risk financing that believes in their capacity to wrest profit from the violent frontiers.

Boom in Equatorial Guinea

Equatorial Guinea epitomises the risks and rewards confronting oil investors in the Gulf of Guinea. On the one hand, its postcolonial history of crime, despotism and marginality make it by far the region's least appealing investment spot.[70] On the other, Equatorial Guinea's recently found oil wealth—and the unexpectedly good relationship with companies—have marked it as one of the most sought-after investment sites in the continent.[71] Despite an earlier involvement, French and Spanish companies were unsuccessful in their quest for petroleum and left the country in the early 1990s, before major oil finds by US firms took place.[72] Therefore, in the foreseeable future Equatorial Guinea will remain "an American game".[73] The main investors are ExxonMobil, Amerada Hess and Marathon together with Chevron and Vanco; other significant presences, especially since Elf's departure, are Petronas and CNOOC.

---

70 See, for instance, FIAS (2002) for a take on investment risk in the country. In pre-oil days, a Spanish newspaper wrote that "only someone suicidal would set up a business in Equatorial Guinea today, unless it was a funeral parlour". See "Teodoro Obiang, el brujo de Montgomo", *La Vanguardia,* 21 November 1993, quoted in Global Witness (2004). Spanish investors were particularly unfortunate in their attempts to resurrect the coffee sector in the 1980s, which failed.

71 Equatorial Guinea is presently the US's fourth investment location in sub-Saharan Africa, following South Africa, Nigeria and Angola, and as already mentioned, it is the third destination for overall FDI in Africa.

72 On oil sector investment before the oil boom see Liniger-Goumaz, (2000).

73 Remarks of Bennett Freeman, "Sustainable Investment Strategies", Conference on the Growing Importance of Africa Oil Co-Sponsored by the US Department of State and the National Intelligence Council, Carnegie Endowment for International Peace, Washington DC, 17 March 2003.

Even if one writes off Equatorial Guinea's first decade of independence as a political freak occurrence, the post-1979 record remains uninspiring.[74] In fact, a seminal work considered pre-oil boom Equatorial Guinea to be one of the few states in sub-Saharan Africa in possession of a fully criminalised economy.[75] The important cocoa sector inherited from Spanish colonialism in 1968 had been quickly squandered by Francisco Macias Nguema's despotic rule, with massive expropriation and killing of the labour force the norm throughout the first eleven years of independence. A 1979 coup by Macias's nephew who is the current head of state, Teodoro Obiang Nguema, halted the more demented forms of persecution (up to one third of the population had perished or fled during his predecessor's reign) but had little impact on the downward slide of the formal economy. The decade that preceded the oil boom was characterised by an unusual degree of delinquency, with the presidential clan exploring the most farfetched forms of economic extraversion to ensure personal and regime viability.[76] John Bennett, a former US ambassador to Equatorial Guinea, described the regime as not so much a government as "an ongoing family criminal conspiracy".[77]

From this perspective, the accelerated growth of the country's oil sector since the mid-1990s—with an estimated twelve-fold increase in state revenues over seven years—is in stark contrast with previous economic partnerships and has consequently brought a measure of

---

74 For historical background see Max Liniger-Goumaz's many works on the country, especially Liniger-Goumaz (1988) and (2000).

75 Bayart, Ellis and Hibou (1999).

76 These included unsustainable forms of forestry exploitation, the drugs trade, the importation of toxic waste, smuggling, etc. The evidence remains patchy and most accounts are anecdotal. In particular, the extent to which these activities have continued and prospered into the oil era is under-assessed. See Roitman and Roso (2001) and the chronology in Liniger-Goumaz (2000). On toxic waste imports see Gorozpe (1995), O'Keefe (1988), "Anglo-US waste rivals", *Africa Analysis,* 8 July 1988, and "Bank pressure over toxic waste", *Africa Analysis,* 2 September 1988. I thank a journalist with long experience of covering Equatorial Guinea for useful information on the country's pre-oil political economy (conversation with the author, London, 14 January 2005).

77 Quoted in Dan Gardner, "Ethics and Oil: A Canadian company with a sterling image navigates a brutal regime in Equatorial Guinea", *The Ottawa Citizen,* 5 November 2005.

international legitimacy to the regime.[78] The presence of substantial FDI in a legal activity means that Equatorial Guinea is going through an important reconfiguration of how it interacts with the outside world. At the domestic level, the networks of complicity between the government and a host of international state and non-state actors have allowed the former to go on suppressing most expressions of dissent. Equatorial Guinea is by far the most repressive context in the Gulf of Guinea oil-producing region, and human rights groups of all persuasions agree in placing its government amongst the gravest offenders in the world.[79] How do companies navigate such waters and think out a long-term engagement?

	1998	1999	2000	2001	2002	2003	2004
EG Oil Production '000 bpd)	5.3	03.1	17.9	197.5	65.0	282	350

Table 8. Evolution of Equatorial Guinea's Oil Production[80]

The question needs to be asked because there is plenty to worry about. In addition to the border disputes already mentioned, Equatoguinean politics comes the threat of destabilisation from at least two domestic sources. The first is the potential for secessionist violence by the MAIB,[81]

---

78 Equatorial Guinea was the regional pariah for most of the postcolonial era. See "Equatorial Guinea: Image and Legitimacy", *West Africa,* 23 August 1982, and "Equatorial Guinea's Obiang makes bid for international respectability", *Africa Analysis,* 28 April 1989, for some of the many "attempts" by Obiang to improve his country's international standing. This has changed greatly since 1995 and Obiang is now feted as the "new emir of the Gulf of Guinea"; see "La famille Obiang au Bristol", *La Lettre du Continent* 366, 14 December 2000.

79 Among those harassed or even killed by the regime are journalists, missionaries, Peace Corps volunteers, Spanish tourists and aid workers, a French aid worker, etc. Assessments by the State Department, Human Rights Watch, Freedom House and the CIA are in agreement as to Equatorial Guinea. See also International Bar Association (2003).

80 Source: Economist Intelligence Unit.

81 *Movimiento de Autodeterminación de la Isla de Bioko.*

an underground movement claiming to represent Bioko islanders against Fang majority rule that intermittently stages armed actions. This is a minor concern for both the companies, which see few parallels with Cabinda or the Niger Delta, and the government, which nonetheless makes use of the threat of disruption to justify the tight social and military control it exercises over Bioko. (With constant road checkpoints and numerous public and private security forces under the control of power-holders, Equatorial Guinea presents a measure of internal control that is rare in sub-Saharan Africa.) The second is the threat of inner-circle squabbling over the presidency. President Obiang has been rumoured to be ill for over a decade now, and his choice of successor, Teodorin, his eldest and most unreliable son who is Minister for Forestry and Infrastructures, displeases many in elite and business circles.[82] Teodorin is very unbalanced and cannot be counted on to refrain from suicidal anti-corporate behaviour that no other leader in the region would ever contemplate. Companies have taken this threat seriously, and have elaborated scenarios for extended disruption: oil operations may be offshore, but several thousands of US oil workers inhabit Bioko's compounds. Nonetheless, the prospect of violence specifically targeting the oil industry is highly unlikely.

As can be gauged from the previous paragraphs, initial expectations about the government of Equatorial Guinea were very low: the country brings together the worst the region has to offer with the specifics of

---

82 The oil companies were probably not involved in the March 2004 foiled coup attempt which led to the arrest of 80 mercenaries in Malabo and Harare, and its spectacular failure will dampen hopes on the viability of such efforts. But there is no doubt that they face the prospect of a Teodorin takeover with dread. For coverage of the coup and the internal politics surrounding it see "Equatorial Guinea: All in the Family" and "Malabo Imbroglio", *Africa Confidential,* 19 March 2004; Basildon Peta and Andrew Buncombe, "British 'mercenary chief' faces execution in Zimbabwe", *The Independent,* 11 March 2004, and Michael Wines, "Where coup plots are routine, one that is not", *New York Times,* 20 March 2004. As the trials started, more heads begun rolling, including that of Mark Thatcher, the wayward son of the former UK Prime Minister who was allegedly involved in the plot. See Michael Wines, "Thatcher's son held in failed Africa coup", *New York Times,* 26 August 2004; "South African police arrest Margaret Thatcher's son", *Financial Times,* 25 August 2004, and Vivienne Walt, "Mark Thatcher's Coup de Grâce", *Fortune,* 7 September 2004.

a uniquely marginal, unruly elite. From that viewpoint, relations have been surprisingly good since.[83] In the more or less accurate words of the former Minister of Oil paraphrased by a leading industry source, "political and legal stability, competitive fiscal terms and a lack of red tape make Equatorial Guinea an attractive place to invest in oil and gas".[84] At the working level, companies are broadly positive about their dealings with the current Minister of Oil, Gabriel Nguema Lima, who is the president's second son. Especially since the creation of a NOC closely modelled on regional counterparts such as Sonangol, the government has become more vocal in arguing for a growth in equity stakes, joint ventures with service companies and a re-negotiation of oil contracts. But the technical and human capacities of Equatorial Guinea are so limited that it is unlikely that it can substantially break away from the current dependence on oil company expertise.[85]

Equatorial Guinea is a state unfamiliar to most people. Africanists know its absurd and violent stories but will most likely not have visited the country. This relative obscurity has been a blessing for oil companies, for defending a business relationship with the regime seems more of an uphill struggle than in any of the other cases under review. Despite a nominally "democratic" constitution since 1991-92, the domination of the presidential clan over the country looks permanent, the turmoil in regard to the succession choice of an ailing Obiang mostly an elite internal affair without implications for a real liberalisation of politics. Soaring revenues from the oil sector have merely heightened the already obvious wealth gap between the people and the elite. Furthermore, after a decade of conspicuous consumption and supposed cosmopolitanism, the

---

83 Interviews with some oil executives showed them to see Equatorial Guinea "as the very model of a modern Gulf of Guinea hotspot" and "ethical... a breath of fresh air when you go into West Africa". See Jon Marks, "Equatorial Guinea Puts its Faith in God, Geology and Governance", *African Energy* 57 (2001).

84 "Equatorial Guinea: Broadening Horizons", *Petroleum Economist*, 19 November 2002. Former Oil Minister Cristobal Mañana Ela added: "If, in the next 10 years, there is no war, no disorder or turbulence, Equatorial Guinea will be a new place. It will be a developed country."

85 That is, short of the sort of irrational politics of which the country had an unhappy bout in the 1970s; hence the fear of the oil companies about the president's choice of successor, who is widely perceived as untrustworthy and prone to violence.

manners of the latter have not yet reached the level of the sophisticated machinations of regional counterparts, eager to avoid at least some public relations snares. Even allowing for the fact that the companies dominating oil production in Equatorial Guinea are not amongst the most enthusiastic CSR voices in the oil sector,[86] the reputational damage that can be incurred by doing business in the country still matters to them. In this context, public scrutiny is resented because it can expose a corporate responsibility "problem" where none is perceived to exist by the companies. But this is in the process of happening, and Equatorial Guinea can no longer lie hidden in the closet.

Organisations such as Global Witness have come to scrutinise opaque practices in the management of the country's oil revenues and shed light over company complicity with the government's malfeasance. The examples are legion. In order to ensure the President's support, oil firms have gone even further than elsewhere to deposit oil payments in his personally held foreign bank accounts,[87] which in turn are labelled "a state secret".[88] A September 2003 investigation by CBS's prestigious "Sixty Minutes" TV show trod the same ground and for the benefit of a much bigger audience. The attention of the quality press is also homing in on the country's paradoxes, if mainly driven by compelling anecdotes. Company actions outside Equatorial Guinea are thus increasingly driven by the need to improve the international standing of their sovereign

---

86 Tough talk on Equatorial Guinea from the EU, UNDP, IMF and World Bank has had little effect on companies: according to a Malabo-based senior official from an international organisation, all attempts at bringing together companies on the subject of "governance" were "characterized by their non-attendance" (phone interview with the author, 25 August 2003).

87 See Global Witness, "Does US bank harbor Equatorial Guinea's oil millions in secret accounts? US Department of Justice must investigate", press release, 20 January 2003, and Ken Silverstein, "Oil boom enriches African leader while the people of Equatorial Guinea live on a dollar a day", *Los Angeles Times,* 20 January 2003. Reports suggest that oil companies have paid prices above market rates for Equatoguinean property held by the President or his family and deposited the money at Riggs Bank.

88 "Equatorial Guinea: Exxon probed over $ 500m scandal", *The Independent* 11 May 2003.

partner, whose poor reputation reflects poorly on them in turn.[89] A detailed investigation by journalist Ken Silverstein exposed the manner in which the oil industry, and ExxonMobil in particular, have laboured to raise the Equatorial Guinea's profile in the US. This has taken many forms, from gaining access for President Obiang to the Bush Administration through oil contacts within it (including an audience with President Bush himself)[90] to the hiring of a lobbying firm[91] and a full-time consultant to revamp the country's standing.[92] This strategy has largely worked in bringing back in a number of international actors that were formerly adamant about the badness of the regime.

The US government had until recently a sufficiently detached rapport with Equatorial Guinea to allow a dispassionate look at its politics. Especially in the early 1990s, this led to a very critical stance on human rights abuses that precipitated government accusations of witchcraft against the US Ambassador.[93] This was one affront too many from an obscure backwater and led to the closing down of the embassy in 1995. The decision would certainly not have been taken some months later, when the first truly substantial oil find was made, but the Clinton administration stood by its stance and remained derogatory of the regime. The Bush Administration, however, has operated, in consultation with the principal US investors, a decisive shift in policy towards Equatorial

---

89 See "Oleaginous", *Africa Confidential,* 29 March 1996, for the frustrated attempts by oil companies to "steer the [1996] election along more respectable avenues".

90 See "Marathon invitará a Obiang a visitar Estados Unidos", Associacíon para la Solidaridad Democrática con Guinea Ecuatorial (ASODEGUE), *Hoja* 31, 19 May 2003. This is a long report on company-state relations that charts some actions by the company to prop up Equatorial Guinea's standing. These include the initial funding of the country's Washington embassy and the hiring of Strategic Concepts, a marketing firm specialising in electoral campaigns, to help Obiang with his 1996 re-election bid. He allegedly won 97.85 per cent of votes.

91 This was Black, Maneforth, Stone & Kelly, later named Black, Kelly, Scruggs & Healey. The connection was discontinued in early 2000 and the government subsequently hired Africa Global Partners for the same functions.

92 Ken Silverstein, "US Oil Politics in the 'Kuwait of Africa'", *The Nation,* 22 April 2002, pp. 11-20.

93 For the details see Douglas Farah, "A Matter of 'Honor' in a Jungle Graveyard", *Washington Post,* 14 May 2001.

Guinea. The new energy partnership with this rather unpalatable associate is not experiencing the publicity accorded to more media-friendly states such as São Tomé and Príncipe; but the US Embassy has reopened[94] and bilateral relations are now in good health.

This is of course not unique to the US: foreign powers have mostly been "conciliatory" towards a government in control of significant commercial opportunities.[95] Spain, the former colonial power that until recently held a trenchant view of the country, is softening its criticism: a 2003 trip by the Spanish Foreign Minister to Malabo was successful and the possibility of a visit by Prime Minister Aznar and a delegation of Spanish businessmen was seriously discussed.[96] Even the IMF, erstwhile implacable critic of the regime, has much reduced its fault finding in the 2003 Article IV consultation.[97] The international networks of complicity surrounding Equatorial Guinea have matured to the extent that a decision by the UN to stop regular human rights monitoring in the country came amid a particularly violent purge of internal elements allegedly engaged in what the government described as a "diabolical" coup plot.[98] At the present time, neither outside pressure nor the brutality of the regime seems to tip the equation against further involvement; aside from fears about Teodorin succeeding to the presidency, companies have mastered internal unruliness and limited the external backlash.

A 2004 US Senate investigation report on accounts personally held by Obiang in Washington's Riggs Bank constituted the short-lived

---

94 See Michael Peel, "Oil-rich Equatorial Guinea lures back US - Claims of repression have not deterred", *Financial Times*, 4 January 2003, "US Renews Ties with Equatorial Guinea", *New York Times,* 17 November 2003, and "US reopens embassy in despotic oil-rich state", *IRIN,* 16 October 2003.

95 "Equatorial Guinea Country Profile 2002", EIU, p. 47.

96 This rapprochement soon became unlikely as Equatorial Guinea accused Madrid of direct involvement in the March 2004 coup attempt.

97 See IMF (2003) and Equatorial Guinea Country Report, EIU, January 2004, pp. 12-3.

98 See "UN attacked on rights monitor", *Financial Times,* 20 April 2002, Jean-Dominque Geslin, "Guinée Equatoriale: enfer ou eldorado?", *Jeune Afrique/L'Intelligent,* 8-14 December 2002, pp. 32-6 and Serge Enderlin and Serge Michel, "La Guinée équatoriale de Monsieur Obiang, ses palmiers et ses pétrodollars", *Le Figaro,* 17 July 2003. On the trials see Amnesty International (2002).

exception to the trend of a broader and more permissive involvement with Equatorial Guinea. Stating that Riggs Bank had "disregarded its anti-money laundering (AML) obligations and [...] at times, actively facilitated suspicious financial activity"[99] by the likes of Augusto Pinochet and assorted Saudi dignitaries,[100] the report goes on to describe in great detail the bank's "largest relationship",[101] that with Equatorial Guinea. The more than 60 accounts held at Riggs by Equatoguinean officials and their family members ranged from $400 million to $700 million at a time up to 2003. More importantly, the report notes that when analysing large transactions involving Equatorial Guinea's oil accounts at Riggs, Senate investigators encountered "a number of substantial payments made by oil companies to particular E.G. government offices, E.G. officials, their family members, or entities controlled by the officials or their family members".[102] While the scandal made it to the front pages and led to the sale of Riggs Bank,[103] the oil companies have as yet incurred no "legal exposure"[104] and, almost two years later, seem to have weathered the PR backlash. The impact on state-firm relations in Malabo has been negligible. The bilateral rapport seems equally untroubled: subsequent references by

---

99 United States Senate (2004: 2).

100 See the good coverage of the report's release, including Kathleen Day, "Record Fine Levied for Riggs Bank Violation", *Washington Post,* 14 May 2004, and Timothy L. O'Brien, "Bankers Testify on Suspect Accounts", *New York Times,* 16 July 2004, as well as footnote 87 of this chapter. The Riggs scandal includes "banking dealings since the mid-1990s that included suitcases stuffed with shrink-wrapped cash being walked through Riggs Bank's front door in Washington; offshore accounts set up to avoid regulatory scrutiny; millions of dollars wired worldwide with few questions asked about the money; and questionable transactions involving three major oil companies"; see Timothy L. O'Brien, "Senate Panel Report Deplores Riggs Dealings", *New York Times,* 15 July 2004.

101 United States Senate (2004: 37-68).

102 United States Senate (2004: 97). See pages 99-112 for the types of payments made by oil companies to these persons and entities.

103 Jenny Anderson and Eric Dash, "Chief of Riggs Bank Resigns, Citing its Pending Merger With PNC", *New York Times,* 8 March 2005.

104 Ken Silverstein, "Bank, Big Oil tied to African payments", *Los Angeles Times,* 15 July 2004, Ken Silverstein, "US investigates oil firms' deals in West Africa", *Los Angeles Times,* 22 May 2004, and David Ivanovich, "SEC looking at Marathon", *Houston Chronicle,* 4 August 2004.

Secretary of State Condoleezza Rice to President Obiang during his visit to the State Department as "a good friend" seem to show that the US Senate's indignation did not carry the Bush Administration with it.[105]

This is because, once "soft" concerns are taken out of the picture, there is not much to fault Equatorial Guinea for. While few of the country's tragic political features bear relevance for oil investors, there are economic features that make Equatorial Guinea stand out as very attractive to them. First and foremost, there is the fiscal regime. Under present agreements, oil companies keep 80 per cent of oil revenues, and the government only 20 per cent. This is up from the original 13 per cent share that was renegotiated with World Bank assistance in 1998.[106] It is also one of the world's most attractive deals; regionally, it is only surpassed by the even lower country take of Chad.[107] Furthermore, companies pay no customs taxes on their imports. Secondly, and this is a subterranean advantage no company will admit to, there is the lack of local capacity in technical and regulatory terms. Technical dependence means that the country is utterly reliant on company expertise even for comparatively simple tasks that Angolan and Nigerian NOCs can perform. It also means that companies can hold information about the oil sector that is very difficult for the government to acquire on its own. Moreover, regulatory insufficiencies result in company operations not being carefully audited, which expands opportunities for creative accountancy.

There is a history of fiscal over- and under-invoicing in the region, by private operators, service companies, state companies, etc. The issue is more serious in the smaller and poorly organised oil producers, and is of lesser importance in Nigeria or Angola where company payments to the

---

105 Paul Wolfowitz, current President of the World Bank, also claimed to have been "very impressed" with President Obiang. See Douglas Farah, "African pillagers", *Washington Post*, 23 April 2006 and Michael Grunwald, "A conversation with Amos Hochstein", *Washington Post*, 23 April 2006.

106 This "technical assistance" of the World Bank, whereby a dictatorial regime is allowed a vast increase in its revenue stream without any improvements in governance or human rights, is highly questionable. For the project team's position see World Bank (1998). For internal Bank criticism of that role, see World Bank (2002).

107 Although one should not exclude the possibility of all-time lows in São Tomé's nascent oil sector.

government are audited by international accountancy firms. Companies may not see this as the basis for long-term investment decisions or even as a plus factor, but it deserves attention as an indicator of government capacity and, therefore, of the likelihood of government assertiveness.[108] As Angola and Nigeria show, greater local regulatory capacity, while it has some advantages for the private sector, also means that the government is more of an agenda-setter. Especially in cases, such as Equatorial Guinea, where local rulers are feared to be trigger-happy, lack of local capacity can calibrate the relationship at precisely the level of harmlessness companies aspire to. "Control by the [presidential] family rather than by experts", a leading publication notes, "has been a boon to foreign oil companies".[109]

Unusual contracts in São Tomé and Príncipe

Another island micro-state, another fringe actor in the international stage: only five years ago, the name and location of one of Africa's smallest states were known by few.[110] Since then, in as startling a manner as Equatorial Guinea, it has jumped to the forefront of international concerns.[111] An ex-Portuguese colony with a socialist past and a moribund cocoa plantation economy, São Tomé has lived through most of its independence years by means of aid handouts and myriad creative strategies. These have included the sale of diplomatic passports and flags of convenience, the rental of the

---

108 The value of such discrepancies can be very important. A World Bank technical assistance team, for instance, allegedly found a gap of $53 million in company payments to the Equatoguinean government between 1995 and 1998. Interviews, Washington, July and August 2002. For the 1996-2001 period, audits of oil company payments showed that they were under the obligation to pay an additional $88 million to the government. See IMF (IMF: 11).

109 "Oiling the palm trees", *Africa Confidential,* 7 February 2003.

110 This section puts forth a very brief account of the changes faced by São Tomé and Príncipe since the onset of international interest in its oil reserves. For a fuller account based on a large number of interviews and fieldwork see Frynas, Wood and Soares de Oliveira (2003); for the authoritative account and analysis of post-independence politics see Seibert (1999) and also Seibert (2005).

111 See, among many, Anderson (2002), "São Tomé and its neighbours: what oil can do to tiny states", *The Economist,* 23 January 2003, John Vidal, "The oil grab", *The Guardian,* 9 October 2004 and Michael Peel, "Oil curses Africa's new petro-state", *Financial Times,* 27 January 2005.

national phone system for international sex calls, and the recognition of Taiwan's sovereignty, as well as an incipient tourist sector. Oil will most likely only start flowing towards the end of the decade: the Economist Intelligence Unit estimates that São Tomé may be producing about 120,000 bpd in 2009, "with the possibility of doubling after that".[112] But the excitement and speculation around it have already altered the expectations of the country's 150,000 inhabitants and fuelled bitter disputes in its democratic politics. Long an isolated outpost with just one weekly flight to Europe, São Tomé is now firmly on the international circuit of businessmen, development consultants, and journalists on the hunt for improbable tales. In June 2005, the first oil sector receipts—in the form of the country's share of the signature bonus from the 2003 licensing round—brought in, overnight, approximately $49 million, an amount similar to the typical annual state budget.[113]

The manner in which this change has taken place is a cautionary tale in state dereliction, neighbourly coercion and politicians' greed. The oil story starts in 1997, when the Environmental Remediation Holding Corporation (ERHC), a small Louisiana-based company, signed a contract with the São-tomean authorities to conduct offshore exploration and raise the country's international profile as an up-and-coming investment site. Although it is common for a small company to enter a prospective oil region in order to conduct surveys and promote its potential internationally, the advantages gained by ERHC are unheard of.[114] Nonetheless, ERHC did not fulfil its contractual obligations, as they vastly exceeded available the company's means, and the accord was revoked and subjected to

---

112 *São Tomé and Príncipe: 2004 Country Profile*, EIU, p. 22. Seismic company WesternGeco is said to have identified reserves of 14.4 billion barrels, an estimate that should not be taken for granted.

113 Before this, an estimated $20 million had been made available in the form of PGS and ERHC payments and advances from Nigeria. In turn, a substantial share of the $49 million had already been spent in JDA operating costs by the time they were received. I thank Gerhard Seibert for bringing this to my attention.

114 See Ana Dias Cordeiro, "O pior acordo da era pós-colonial", *Público*, 5 July 2003. Andrew Latham, an Africa expert for the leading oil consultant group, Wood Mackenzie, claims never to have seen "a company get a stake like ERHC obtained in Sao Tome" (quoted in Ken Silverstein, "Sinking its hopes into a tiny nation", *Los Angeles Times*, 24 May 2003).

international arbitration. In 2001, Chrome Energy Corporation, a Nigerian company owned by the well-connected millionaire Emeka Offor, acquired a 50 per cent controlling stake of ERHC and arbitration was mysteriously dropped in favour of a changed, but equally poor, deal.[115] The involvement of interests close to Vice-President Abubakar and other prominent Nigerian politicians guaranteed that the ERHC/Chrome deal would have the backing of a strong and potentially bellicose neighbour. Indeed, one of the São-tomean participants in the first meeting between the new owners and the government recalled an atmosphere of intimidation, the imposing international team of experts rallying by the Nigerian side, and the personal involvement of the Nigerian Foreign Minister on behalf of Chrome.[116]

The original contract, which had taken the form of a 25-year joint venture between the state and ERHC, and its 2001 revision encompassed, amongst other provisions, ERHC rights to 5 per cent of the state's future state revenues from the oil sector. It further included a share of other companies' signature bonuses and the acquisition of two blocks in the country's Exclusive Economic Zone without the payment of signature bonuses.[117] According to the IMF, solely on account of these clauses ERHC/Chrome's income could be more than $1.4 billion over the 25-year period,[118] an astonishing figure. There is little doubt, as my co-authors and I have written elsewhere, that there is no "similar precedent in the history of Africa's oil industry since the end of colonialism".[119] A 1998 accord with ExxonMobil for the exploration of 22 deepwater concession blocks was also unfavourable to the country, and so was an accord with PGS, a Norwegian seismic services firm.[120] Although these were also heavily criticised, the track record of both companies, and of

---

115 According to the EIU, these were "even worse terms"; see *São Tomé and Príncipe 2003 Country Profile*, EIU, p. 23.

116 Interview with the author, São Tomé, 15 March 2002.

117 For a more detailed analysis of the contracts see Frynas, Wood and Soares de Oliveira (2003: 66-8).

118 IMF (2002: 10).

119 Frynas, Wood and Soares de Oliveira (2003: 67).

120 More information on these contracts can be found in Frynas, Wood and Soares de Oliveira (2003: 68-9).

ExxonMobil in particular, guaranteed that they were welcomed in the country's nascent oil sector.

These firm-related disputes took place in tandem with Nigeria's unwillingness to accept the delineation of a maritime boundary with São Tomé. Such a deadlock prevented the exploitation of the country's oil and was unsustainable. Eventually, this was solved by the creation of the Joint Development Zone (JDZ) in February 2001, which made room for the development of the contested acreage in terms that greatly favoured Nigerian interests. A Joint Development Authority based in Abuja will administer the JDZ agreement for 45 years, during which time oil revenues will be shared 60-40 in favour of Nigeria. In addition to the JDZ, there is an Exclusive Economic Zone (EEZ), belonging to São Tomé alone, which has elicited less enthusiasm from investors but where Chrome also holds significant rights. It is noteworthy that the Chrome 2001 deal was made conditional on the acceptance of the JDZ, in a show of permeability between Nigerian private and public interests.[121]

The election of Fradique de Menezes to the presidency in 2001, however, came to upset some of these expectations. Elected with the support of the Trovoada family as a continuity figure (for President Miguel Trovoada was constitutionally barred from seeking a third term), Menezes swiftly turned against his erstwhile patrons. On the basis of an IFI-commissioned April 2002 study that exposed serious flaws in the country's oil deals,[122] Menezes damned the contracts with Chrome, PGS and ExxonMobil and promised to renegotiate all three. Throughout 2002, the contracts—especially that with Chrome—were denounced in the international press, their unique crookedness the source of outrage, and the country gained a name as a likeable dupe sitting on billions of

---

121 Frynas, Woods, Soares de Oliveira (2003: 68). Later on, when São Tomé reneged on the 2001 deal as well, Nigeria said it would withhold the JDZ licensing round until the Chrome deal was cleared. As the blocks in the JDZ were those most eagerly sought by investors, this "essentially gave Nigeria a veto over the development of São Tomé oil resources" (*São Tomé and Príncipe: 2004 Country Profile*, EIU, p. 23).

122 Memorandum, "Oil Business Between DRSTP and Nigeria, ERHC, Exxon-Mobil and PGS", Washington, DC, 12 April 2002.

dollars.[123] In the course of this, Menezes' multiplication of contacts in the US earned him the sympathy and support of journalists, lawyers, international civil servants, politicians and academic superstars willing to help with the renegotiation of the rotten deals and make a good governance example of the country.[124] Plans were made for revenue management laws, oil funds and transparency provisions. And yet, the more one peered into the subject, the more convoluted it got.

The reasons for such a bad set of deals struck by the country's politicians have mystified observers. In a rather apologetic manner, some have accounted for them by reference to the lack of experience of the country's elite and Nigeria's muscular approach to business. The same view adds that, at any rate, it would have been difficult to get a much better deal at a time when oil industry interest in the country was scant.[125] This seems a true but insufficient explanation. Another reading is that which, while factoring in these insights, gives equal centrality to the greed of São-tomean politicians and their willingness to sell access to the national resources in exchange for kickbacks. In turn, the modest size of the kickbacks when contrasted with the estimated worth of the resources involved betrays small-time dereliction unaware of how high the stakes were.[126] This is how São Tomé's public opinion and the highly vocal

---

123 Some industry voices were vexed at the vigour of public criticism, blaming the delay in getting this oil frontier going on "outside liberal, anti-oil, anti-business lobbies that are trying to derail development in this region". See "Untying the Gulf of Guinea knots", *Upstream,* February 2003.

124 These include the economist Jeffrey Sachs (now at the Earth Institute, Columbia University) under the sponsorship of George Soros, Joseph P. Kennedy (CEO of Citizen's Energy), Yale University, the law firm Williams & Connolly, etc. The Washington-based Center for Strategic and International Studies, among others, held a workshop on "São Tomé and Oil" in 2004.

125 "Mobil cherche du pétrole", *Marchés Tropicaux et Méditerrannéens*, 26 March 1999, claims that ERHC "at least had the merit" of attracting Mobil to São Tomé.

126 In this regard, a foreign official referred to tiny São Tomé's easily bought politicians as "micro-crooks" (author interview, Washington DC, 15 August 2002). The rewards of the 1997 ERHC deal to pliable politicians included university scholarships for their children, small cash handouts and jobs in the São-tomean company created by the joint-venture, STPetro. Kickbacks from the later deals are rumoured to have been far more substantial.

diaspora have understood it.[127] As it turns out, even the President, who is a vociferous critic of some aspects of the ERHC/Chrome contracts and, indeed, initiated much of the backlash against them, has benefited from opaque financing by the company. A February 2002 transfer of $100,000 to the President's company, CGI, has been unearthed, ostensibly a "campaign donation".[128] The result of enmeshed external pressures, with Nigeria in particular threatening to jam the whole process were its ambitions not fulfilled, and unwillingness by local power-holders to look carefully into deals mostly signed by themselves[129] meant that it was just a matter of time before the disputes were resolved. Some observers were quick to point out that, while there were grounds to disparage the oil agreements, potential investors might instead think that in São Tomé no contract was above arbitrary revocation. The country's credibility was undermined further by the heated internal altercations for control of future oil monies and the endlessly recomposed parliamentary coalitions—whilst São Tomé's democracy was to be commended in theory, in practice many companies privately complained of lack of direction.[130]

Amidst political instability and elite infighting, popular confusion and international criticism of the oil deals, São Tomé experienced a *coup d'état* by disgruntled army officers taking advantage of Menezes' absence

---

127 Disturbances in April 2003 that resulted in one fatality and much damage in the capital were partly caused by perceptions of lack of transparency and corruption in the oil sector. With regard to the vocal Lisbon-based São-tomean community, see the open letters that have been published since 2001 under the title *A Coisa Pública* (i.e. "Public Goods"), denouncing corruption in the country's oil policy.

128 Fax sent by First Bank of Nigeria (London) to HSBC Bank USA (NYC), 21 February 2002, 15:44.

129 Not only was there great continuity in the oil team since 1997 (ex-Energy Minister Rafael Branco, for instance, was involved in it throughout the Trovoada and Menezes presidencies and several cabinet reshuffles until the aftermath of the coup, when he finally quit), but several of those involved either have worked for ERHC-Chrome (like Carlos Gomes, the São-tomean director of the JDA) or are shareholders of the company (this was the case with the former Foreign Minister of São Tomé, Mateus Rita).

130 In less than five years, São Tomé has had five prime ministers. See Dino Mahtani, "São Tomé - where the champagne swills in before the oil gushes out", *Financial Times*, 25 March 2006.

from the country in July 2003.[131] Most analysts gave exclusive coverage to the perceived linkage with the oil contracts (a view aided by the coup spokesman's constant references to oil and social justice[132]) and all but forgot the country's coup-prone past and the older grievances of sections of São-tomean society.[133] That said, there is little doubt that the timing and the ambitions of many involved coincided with the impending arrival of unprecedented wealth. The bloodless coup was finally halted by the end of the week following unanimous international condemnation and the willingness of the country's political class to assuage some of the grievances of the plotters. This included an amnesty and provisions for greater transparency in oil-related matters. Shortly afterwards, President Menezes returned to São Tomé accompanied by President Obasanjo and his presidential guard. This was a Nigerian show of support for Menezes, but also underlined his growing dependence on his Nigerian patron, who is rumoured to have brought about an early resolution to the conflict by threatening to invade São Tomé. Nigeria's grip over the country and its oil sector was certainly tightened by the events.

By the end of the year, the conditions for going through with a licensing round were finally in place. Agreements had been renegotiated with PGS, ExxonMobil and Chrome: they remained bad deals, but some of the more shocking elements had been subtracted.[134] And finally, the argument that even a bad deal would bring forth enough resources to

---

131 See "Military coup ousts government of São Tomé in West Africa", *New York Times*, 17 July 2003 and "A coup in São Tomé: troubled waters over oil", *The Economist*, 19 July 2003.

132 In an interview given later in the year, Sabino Santos stated that "at the bottom was the oil [...] those who signed [the contracts] should be punished. They are mortgaging our future" (quoted in N. Shaxson, *Reuters News* report from São Tomé, 23 December 2003).

133 For a reassessment of the 2003 coup that stresses non-oil causes see Seibert (2003). For the 1995 coup, see Seibert (1996).

134 For instance, whereas before there were fixed limits for signature bonuses, the renegotiation stated that first movers would have to match the amounts offered by other companies in a competitive licensing round. See "Revised Deals may Pave Way for JDZ Licensing", *African Energy*, April 2003 and "Hopes rise for troubled Nigeria-Sao Tome zone", *Argus*, 30 January 2003. See also the "Memorandum of Understanding with Chrome Oil, 15 March 2003".

radically transform the country won the day. Against predictions that political unrest would make the licensing round "an even harder sell than before",[135] a number of investors flocked with offers for the country's promising acreage: they had missed neither the farcical aspects nor the ultimate harmlessness of instability in São Tomé. In fact, the biggest hassle for oil majors was not the untidiness of domestic politics but Nigeria's imposition of essentially speculative players without technical experience in deep waters, such as Chrome, Sahara Energy, Atlas Petroleum and Foby Engineering Company,[136] all of which were connected to Abuja's elite circles.

In October 2003, twenty companies, including Chevron, bid for nine blocks in the JDZ presumed to hold between six and eleven billion barrels, with Chevron offering $123 million for 100 per cent of oil block 1.[137] Chrome and ExxonMobil had preferential rights, under the condition that they matched the signature bonus offers of the other companies.[138] Exxon did so for one of the blocks[139] but the process has soured yet again since.[140] This resulted in the personal involvement of President

---

135 "Oil Fuels a Minor Coup in STP as Investors Look on", *African Energy* 65, August 2003.

136 See Ana Dias Cordeiro, "As mais bem colocadas nos melhores blocos a concurso" and "Empresas nigerianas posicionam-se para controlar petróleo em São Tomé e Príncipe", *Público*, 20 January 2004.

137 N. Shaxson, "Nigeria/São Tomé give oil rights to Exxon, Chrome", *Reuters News*, 31 December 2003.

138 "Mobil, local firm get preferential oil blocs in Gulf of Guinea", *ThisDay* (Lagos), 18 March 2003.

139 "Exxon to take 40pct of Nigeria/Sao Tome oil block 1", *Forbes,* 13 February 2004. According to Gerhard Seibert, Exxon thought the signature bonuses for the other blocks were too high, and disliked being forcefully partnered with Chrome and other small Nigerian companies. See Seibert (2004: 11-12).

140 At the same time, the government's signed a MoU on oil trading (though no oil exists yet) with a subsidiary of the mining company DiamondWorks, a company with little oil sector expertise. This agreement was subsequently revoked. See "DiamondWorks Ltd - Sao Tome oil deal formalised, expanded", Press Release, DiamondWorks Ltd., 18 February 2004, and also "Rough diamonds", *Africa Confidential,* 5 March 2004. DiamondWorks is also active in the field of privatised military services, as documented in Vines (2000) and Singer (2003: 105).

Obasanjo[141] amidst accusations of foul playing on the part of some of the key participants, including Tajudeen Umar, the JDA's former Chairman. The first round eventually closed with only one block being awarded.[142]

The second round for Blocks 2-6 that started in November 2004 was even worse.[143] Amongst the 22 companies submitting bids, the absence of most oil majors was as conspicuous as the proliferation of hitherto unknown Nigerian firms. While ExxonMobil decided not to make any bid or exercise preferential rights, ERHC-Chrome not only exercised its options on *all* the blocks, but also formed consortia with three other companies for exploration in three of them. Five difficult horse-trading months ensued, with Nigerian vested interests pushing for the rewards they had been refused in the first round. Eventually, the awards were announced on 1 June 2005 but the polemics they raised led to yet more delays in JDA negotiations with the winning firms. In December 2005, São Tomé and Príncipe's Attorney General made public a careful investigation of the Second Round.[144] Not mincing its words, it concluded that:

Following the investigation, the Office of the Attorney General found that the procedures used to select the companies to receive awards were seriously flawed, and did not meet the minimum international standards for a licensing round. The process worked to the financial detriment of São Tomé and Príncipe [...] The second round had other serious deficiencies, including the award of interest to many unqualified firms, or firms with inferior qualifications, technically and financially.

Despite these mishaps, not all expectations are gloomy. For some, São Tomé's open society and relatively non-violent politics give it a head start over the authoritarian blind alley of Equatorial Guinea. While the bidding rounds show that a verbal commitment to transparency has not transpired

---

141 "New Eldorado: Obasanjo Steps in to Take over JDZ Licensing", *African Energy* 74, May 2004.

142 This was for Block 1: Chevron (51 per cent), ExxonMobil (40 per cent) and a Norwegian company, Energy Equity Resources (9 per cent).

143 See Gerhard Seibert, "Adjudicação de blocos petrolíferos rodeada de irregularidades", *Expresso*, 5 April 2006.

144 Procuradoria Geral da República de São Tomé e Príncipe, "Press Release, 9 December 2005". See also República Democrática de São Tomé e Príncipe (2005). This impressive document sheds light over a much wider set of issues, including the conflicts of interest of Saotomean politicians and their links to Nigerian companies.

into oil sector practice, the National Assembly did unanimously pass a Revenue Management Law in November 2004.[145] The US willingness to build a non-permanent military base mooted more than three years ago, or at any rate the deepening of its involvement in the country, can only ensure its autonomy from overbearing Nigeria.[146] More importantly, though, the results of Chevron's oil drilling efforts in Block 1 are expected in mid-2006. If, as is hoped, they point to large recoverable reserves, this is likely to create a new, geologically more informed race for the acreage. It is to be expected that when faced with near certainty rather than speculation, at least some major players will be more forthcoming about getting their hands dirty in São Tomé's corrupt business dealings. If so, they would merely be following companies such as China's Sinopec, which discreetly entered São Tomé as a Chrome business partner and bloc operator in 2005. It seems safe to predict that, for the foreseeable future, São Tomé will remain a port of call for corporate actors of all kinds, with established firms securing unparalleled terms, and obscure and fictive ones capturing important business opportunities through legal loopholes, opaque contracts and local corruption.

III. PAST PRACTICES AND PRESENT WOES

Despite the optimistic outlook for the Gulf of Guinea, some aspects of regional politics present the private sector with complex, unpredictable dynamics and scenarios of decline that jeopardise its upbeat pretensions. As companies move to fresh grounds, older ones show the scars of the oil trade and the impact of state-firm partnerships that have had detrimental effects for the lives of the majority. This is particularly clear in the states where oil firms have been a jutting presence for decades, and where oil money has worked its way through the veins of politics and society. For

---

145 The law was written with the assistance of external experts. Amongst other things, the law creates an oil sector watchdog, a publicly available register of all oil sector documents, and a national oil account held outside the country with strict drawn-down rules as well as a future generations' account. For details see the information available at www.earthinstitute.columbia.ed/cgsd/STP (accessed 3 March 2006). It remains to be seen whether, as in the case of Chad, the donor pressure that has been key to the passing of the law will be as influential once oil production starts.

146 Seibert (2004: 7-8).

the private sector, whose strategies "adapted to the local setting"[147] for long guaranteed rich rewards, these are the locations where it is most difficult to re-brand a firm's image. Unsurprisingly, these are also the countries where protests have gone furthest in questioning company procedures through words and, increasingly, through violence. Herein one finds the ingredients both for unwavering company entanglement (for potential dividends are enormous) and for mounting criticism and obstruction. This section explores ways in which companies have dealt with such bleak outcomes and how they are attempting to frame them in advantageous terms through, firstly, a look at the Nigerian experience with oil sector disruption and, secondly, perusing of Elf-Aquitaine's manifold African activities.

The Niger Delta troubles

Nigeria is currently a nightmare for oil companies. On the one hand, it is the Gulf of Guinea region's, and one of the world's, biggest exporters, with expectations of a long producing life and great prospects for expansion and diversification, onshore and offshore, in crude oil and natural gas. On the other hand, the last decade has brought about a remarkable degree of civil strife, international condemnation and, increasingly, the halting of production by local protesters and criminal gangs. This compounds problems that have restrained development for decades, such as the incapacity of the Nigerian government to keep up with its financial commitments in joint ventures. While primary responsibility for Nigeria's all-encompassing decline lies with its leaders, oil companies have been key participants in the extraction of resources and environmental damage wrought upon the Niger Delta region, which produces an estimated 90 per cent of the country's oil. Their continued presence and further involvement will remain burdened by this record.

The oil-rich and densely populated Delta region, which is mostly inhabited by minorities belonging to none of the country's three principal ethnic groups, has been traditionally circumvented by the federal government. This happens both through informal neglect and a revenue allocation formula that withholds from it the lion's share of Nigerian oil

---

147 Author interview with oil company executive, Rome, 23 June 2000.

money.[148] Delta communities are thus saddled with the consequences of callous and inexpensive extraction of oil but do not partake of the wealth that comes with it. Economic opportunities are scarce, gas flaring and oil spills have damaged the land, and the provision of public goods by the state is virtually absent. Until the late 1980s, the security situation in the Delta was not of great concern to the oil sector. Criminality and occasional protests were a problem, as was the Nigerian penchant for "oil bunkering" (i.e. the stealing and smuggling of oil). But the companies seemed content with cultivating careful relationships with Nigeria's power circles and ensuring that the country's complicated politics would have no impact on foreign investors.

The equation changed as perceptions by Niger Delta populations, especially those of the disenfranchised youth, shifted from tacit toleration of companies to an ever-deepening sense of dispossession. Delta protests quickly escalated from incipient violence in the late 1980s to take on the character of a veritable war over the last decade. The Federal Government responded to local activism with the coercive might it was able to muster, ruthlessly defending its revenue lifeline. Although the repression started under the presidency of General Babangida, it was made worse by the onset of Sani Abacha's dictatorial rule in 1993, when state violence and political autism towards civil society claims were heightened.[149] The deterioration in security conditions unsurprisingly led to the strengthening of links between companies and state security forces, which were used to safeguard production sites and areas adjacent

---

148 According to the derivation formula in section 162 of the Nigerian Constitution, oil-producing states in the Niger Delta receive 13 per cent of oil revenues. See Constitution of the Federal Republic of Nigeria, 1999, p. 69 (www.nigeria-law.org/ ConstitutionOfTheFederalRepublicOfNigeria.htm) accessed 3 January 2006). While Delta protesters and politicians alike complain about this formula and would like to see it increased (up to 50 per cent of the total), oil derivation payments make some of the Delta states Nigeria's highest-spending states after Lagos.

149 The reference work on the several conflicts that beset the Delta for the better part of the 1990s is Human Rights Watch (1999). See also the follow-up report Human Rights Watch (2002).

to company premises, and to restore "normality" when local protesters threatened or committed acts of violence.[150]

This period also witnessed a strengthening of the use of private security forces within company premises. By the time military rule ended in 1999, none of the companies had "shared information on the Memorandum of Understanding on Joint Operations Agreement with the Nigerian government governing security, nor the internal guidelines relating to protection of their facilities", but the record points to tight collaboration.[151] This included the purchase of weaponry for local police forces,[152] the use of company premises and material such as helicopters for the waging of repressive campaigns,[153] and the calling in of police or military forces for disbanding protesters. The first such major event occurred in 1990, when police units were brought into the town of Umuechem at the request of Shell to stop community protests against it; more than eighty people were killed, and the town was wrecked. Despite company policies that are now much clearer and more careful in this regard, stipulating the rules of engagement for companies and local police forces, similar events have taken place throughout the last seven years. The Delta Region's experience illuminates the extent to which it is increasingly difficult to discern between violence against the state and violence against the foreign private sector. In practice, both have become interchangeable in the eyes of local actors, whether politicised activists, highly organized criminals or amateurs on the lookout for quick profit.

Far from abating, the conflict has taken on a new dimension since democratisation in 1999 with political patronage of militias and the mobilisation of the Ijaw ethnic group against the state, the Itsekiris, and other perceived enemies.[154] This challenge has been brutally but

---

150 "Nigerian troops move into Delta to put down ethnic riots", *New York Times*, 20 March 2003.

151 Human Rights Watch (1999: 4).

152 Human Rights Watch (1999: 13).

153 Booker and Minter (2003: 45).

154 The mobilisation of Ijaw ethnic nationalism went into full gear with the Kaiama Declaration of December 1998—a ten-point statement demanding oil control for the people of the region and that oil companies leave what it referred to as "Ijaw land"—and has not abated since. See Nwajiaku (2005).

ineffectually repressed by the authorities[155]—with many politicians covertly abetting armed groups for their own ends—and this has ensured continued support. As explained in Chapter 2, Nigeria's federal allocation of revenue means that the fight for the national cake, which in most Gulf of Guinea states focuses on the national government, is here replicated at the level of state and local governments, where revenue can be appropriated by power-holders without accountability. This is the case with several Delta region states such as Bayelsa, Rivers and Delta, where electoral politics, rising gun ownership paid for by oil theft, radicalised unemployed "youth" and international criminal syndicates have come together in a pattern that is now described by observers as "a high intensity conflict".[156] Its specificity and relevance lie in the fact that its ebbs and flows impact directly on company operations like nowhere else in the Gulf of Guinea and, by doing so, significantly affect the world market prices for crude oil.

The targeting of companies has two main motivations. The first is their perceived complicity with the government in denying locals an "oil dividend"; the second is the retreat of the state from tasks it was, until recently, expected to perform, and the concurrent shifting of these expectations towards companies. Until recently, protests remained predominantly of a local nature, even when articulated by ethnically based outfits such as the Movement for the Survival of Ogoni People (MOSOP), the Ijaw Youth Council and the Ijaw National Congress. Likewise their demands, which frequently pertain to an increase in job opportunities, company-related sub-contracts, and local improvements in health and educational facilities, as well as compensation for perceived or actual damage done. The methods are those of occupation of oil facilities,

---

155 In the most notorious incident, federal troops massacred around 2,000 Ijaw civilians in Odi (Bayelsa State) in 1999.

156 Alex Vines, "Oil and Human Rights in the South", paper presented at conference on "Addressing the Resource Curse: Solutions from Oil Exporting Countries in Africa and the Caucasus", Heinrich Boell Foundation, Berlin, 27-28 May 2004. On company views of the more recent disruption see Michael Peel, "Oil giants face uncertain future as tribes clash", *Financial Times,* 5 May 2004. On the Warri conflict see Human Rights Watch (*l2003*).

kidnapping of oil company personnel,[157] sabotage of oil infrastructure, and riverine and coastal piracy.[158]

This sort of serious incident is now chronic in the Niger Delta. In July 2002, for instance, 150 women shut down a ChevronTexaco oil plant.[159] Early in 2003, similar protests resulted in unprecedented disruption of oil production on the eve of national elections, whereby 40 per cent of the country's output was interrupted and the activities of all leading operators ground to a halt, costing the Nigerian government and the companies untold millions. In the event, Shell, TotalFinaElf and ChevronTexaco took the very rare decision of "shutting down their facilities in the region and evacuating staff members".[160] In January and February 2006, pre-dawn raids on oil installations led Shell to halt 455,000 bpd of production, an estimated 19 per cent of Nigeria's output.[161] Needless to say, these incidents are profoundly disruptive; but by now, they are more than that.[162] If left unchecked, these trends have the potential to undermine the profitability of operations across the Niger Delta, with the exception of recent and forthcoming deep and ultra deepwater developments. This is particularly so for Shell's operations, which are the most affected

---

157 Company premises have been assaulted on a regular basis and company employees are routinely abducted by youths. This has been a nuisance companies have learnt to deal with, as most are now heavily insured against kidnapping, but Cezarz, Morrison and Cooke (2003) cautions that this too may be changing in view of the March 2002 killing of six Nigerian oil workers. This is perhaps substantiated by the subsequent slaying of two US employees of Chevron. See "2 US Oil Workers Are Killed in Nigeria", *New York Times,* 25 April 2004, and also "Security Risk in Nigeria" in *Nigeria Country Profile 2003*, EIU, pp. 13-4. Foreign company employees were again kidnapped in January 2006 but were eventually released unharmed.

158 According to the International Maritime Association, Nigeria is one of the world's top three countries suffering from piracy.

159 "Nigerian Women, in Peaceful Protest, Shut Down Oil Plant", *New York Times,* 14 July 2002.

160 See Somini Sengupta, "Oil companies to resume work in the Niger Delta", *New York Times,* 5 April 2003 and also "Big Trouble in Nigeria Poses Geostrategic Questions for Oil Companies, Government Planners", *African Energy,* April 2003.

161 See Dino Mahtani, "Nigerian oil industry helpless as militants declare war on Obasanjo", *Financial Times* 21 February 2006 and "Nigerian militants assault oil industry", Reuters, 19 February 2006.

162 See, for instance, Pegg (1999).

because they are predominantly onshore, though the takeover of shallow water platforms by protesters shows that most oil production is at some risk or other.

The pattern of disruption seems to be recurrent rather than permanent, with "substantial disruption" in 1993, 1998, 1999 and 2003 offset by a degree of stability and even considerable growth in other years.[163] 2005, for instance, has been described as "quietish and immensely profitable":[164] within limits, companies "have adapted".[165] The worst-case scenario is that some areas may fall out of the reach of the state and companies. In Ogoniland, for instance, where oil production started in the late 1950s, the 1993-94 campaign waged by Ken Saro-Wiwa's MOSOP led to the unprecedented departure of Shell.[166] Ten years later, it has yet to resume: the landscape is littered with abandoned oil infrastructure and Ogoniland remains a no-go area for the company. Its pauper status is hardly an inspiring example for would-be saboteurs who are more likely to want to extract money from companies than to oust them,[167] but the possibility remains.

Despite being at the receiving end of local frustrations, companies have not voiced their disagreement with state policies—including those that impact negatively on companies, like the crushing poverty and unemployment of the Delta—beyond the faintly reformist space for discussion opened up by the Obasanjo Administration. The relationship between the companies and the government may be akin to a "fragile marriage", but a marriage it is, and strong public criticism does not

---

163 *Nigeria Country Profile 2006*, EIU, p. 36.

164 "A New Year Offensive", *Africa Confidential*, 20 January 2006.

165 This information was made available by John Bearman, Managing Director of Clearwater Research Services, Ltd, paper presented at the conference on "Oil and Gas in Africa", Royal Institute of International Affairs, 24-25 May 2004. Clearwater conducted two extensive surveys of oil bunkering and security for NNPC and the Federal Government of Nigeria.

166 On the Ogoni crisis see Human Rights Watch (1995) and also Osaghae (1995).

167 In a meeting I had with the Paramount Chief and notables of Bodo on 22 September 2002, the desire for a negotiated return of Shell to Ogoniland was explicitly articulated. I noted that the only substantial sign of Shell's presence in a community where it stayed for over three decades was a 2002 "corporate responsibility" calendar in the Paramount Chief's living room.

play a role in it.[168] Yet the seriousness of the situation—together with pressure from Western public opinion, particularly in the wake of Shell's perceived complicity in the 1995 execution of Saro-Wiwa —called for substantial reconfiguration of company discourse and policies. In response to these pressures, oil companies have not revised the unquestioning rapport with the state but have upped their financial commitment to community development and made some concessions to protesters.[169] As Frynas notes, company generosity is part of a broader set of "generic strategies" that also included additional repression and fairly shallow public relations stunts,[170] yet there is no denying that considerable efforts have been made. Shell has gone furthest in establishing a development company and in increasing its social spending from a mere $100,000 in 1991 to $25 million in 1996 and a staggering $69 million in 2002 with the aim of improving company-community relations.[171] The *New York Times* reports the example of Chevron and its involvement in medical care, schooling and welfare across the area where it operates.[172] A Niger Delta Development Commission (NDDC) that is partly paid for by oil companies has been created.[173] Additionally, companies have tried

---

168 Jeff Gerth and Stephen Labaton, "Shell withheld reserves data to aid Nigeria", *New York Times,* 19 March 2004.

169 For a good study of responses to oil-related dissatisfaction, see Frynas (2001) and also Frynas (2000).

170 Frynas (2001: 27-8).

171 Frynas (1999) and Frynas, "Engagement of the Local Community: the Niger Delta", paper presented at the conference on "Oil and Gas in Africa", Royal Institute of International Affairs, 24-25 May 2004.

172 Norimitsu Onishi, "As oil riches flow, poor village cries out", *New York Times,* 22 December 2002.

173 Expectations about Nigerian institutions are typically low, but the NDDC has managed to disappoint. A brief look at the Niger Delta Development Commission (Establishment etc) Act of 2000 (Act No 6, Laws of the Federation of Nigeria) shows the contours of this "bureaucratic monstrosity" (conversation with foreign official, Abuja, April 2006). The Act establishes a commission with no less than twelve directorates. These include Administration and Human Resources; Community and Rural Development; Utilities Infrastructural Development and Waterways; Environmental Protection and Control; Finance and Supply; Agriculture and Fisheries; Planning, Research, Statistics; Management Information System; Legal Services; Education, Health and Social Services; Commercial and Industrial Development; and Projects Monitoring and Supervision. It

to stem threats to their operations and facilities by contracting "ghost workers" (normally restive youths) on "surveillance contracts", making payments to communities for "allowing work to be carried out", and other direct handouts.[174] More recently, companies have lent support to Obasanjo-led transparency initiatives such as EITI.

While they show that company thinking has evolved, these changes are no longer enough to halt the worst. Many have had no effect, while some measures have actually exacerbated Delta problems. This is the case with the cash payments that have created incentives for hostage taking and other criminal activities[175] as well as fuelling conflict amongst and between local communities for access to monies.[176] Even the most sincere efforts at corporate social responsibility would be tough to implement in a terrain such as the Delta which, in addition to the oil-related predicament, is ailing from the many intractable problems that plague Nigerian politics.

The political and social evolution of conflict in the Delta over the past decade, however, has resulted in "genuine" grievances leading to local criminal phenomena that are enmeshed with the national and international activities of crime syndicates. While there was always a criminal dimension

---

adds that "the Board may, with the approval of the President [...] increase the number of directorates as it may deem necessary and expedient to facilitate the realization of the objectives of the Commission." Over six years, NDDC has been plagued by allegations of corruption over its vast resources and has in no way performed its mission to "formulate policies and guideline for the development of the Niger Delta" (Act 6, clause 7 (a).

174 Human Rights Watch (2002: 3).

175 The recent report commissioned by Shell on Conflict in the Delta comes to the same conclusion. See WAC Global Services (2003).

176 In 2002, I visited Ogbodo, a small community in Rivers State scarred by internal discord due to compensation paid by Shell for an oil spillage. Accusations that some residents had kept the money for themselves had already resulted in confrontations, death threats and one fatality. Although no violence was yet attached to it, I encountered similar concerns in the population of Toura, a small village near the Willbros (the US oil services company) compound in southern Chad. The village had reportedly received compensation from Willbros for the early dismissal of its men. According to interviewees, this was paid to the *chef de canton*, who is reported to have kept the money. See Soares de Oliveira (2002a: 10-14) and (2002: 25-6). There is extensive coverage of the impact of corporate money on communities in Nigeria (and far less elsewhere); see Academic Associates PeaceWorks (2006) for a comprehensive set of studies.

to the Delta troubles, this has arguably turned into its defining feature: in effect, oil-fuelled poor governance created a many-headed monster that has taken on a life of its own. It is difficult to say what the tipping point was. Some have argued that the "violence of March and April [2003] departed fundamentally from prior patterns",[177] in terms of strategy, sophisticated weaponry and level of organisation,[178] but there are previous instances of such activities.[179] Be it as it may, highly sophisticated criminal gangs[180] and even secret cults engaged in gruesome acts of violence[181] are now as much part of the picture as "legitimate" protesters. More importantly, militias have grown increasingly independent of their former political patrons and can no longer be masterminded with any predictability. This trend has been accentuated in the past two years by the rhetoric and the actions of groups such as Alhaji Mujahid Dokubo-Asari's Niger Delta People's Volunteer Force (NDPVF),[182] Ateke Tom's Niger Delta Vigilantes (NDV) and the more recent and still enigmatic Movement for the Emancipation of the Niger Delta (MEND).[183]

---

177 See Cezarz *et al.,* (2003) for an analysis that sees the March 2003 events in terms of a qualitative change of the security equation in Nigeria. See Douglas, Kemedi, Okonta and see Watts (2003), for a critical response.

178 Some observers claim that companies and the Nigerian state exaggerate the criminal element in order to justify a heavy-handed response that would otherwise be internationally unacceptable: "[…] These sorts of rumors and insinuations […] set the stage for yet another cycle of ethnic cleansing".

179 Davis and D. Kemedi (2006: 11) argue that the criminal dimension was there from 1999.

180 Oronto Douglas, a prominent author and activist of the Ijaw Council of Human Rights, expressed concern that the phenomenon of banditry that is now prevalent in Ijawland might lead to a "long term incidence of warlordism across the delta" (author interview, Port Harcourt, 24 September 2002). Douglas' reputation was subsequently dented by his decision to work for Bayelsa State Governor Diepreye Alamieyeseigha, arrested in 2005 on corruption charges.

181 See Kemedi and Oko (2006: 57-69).

182 Asari was one of the 1998 founders of the Ijaw Youth Council and its former president. See Ebimo Amungo, "Profile: Nigeria's oil militant", *BBC News*, http://www.bbc.co.uk/news (accessed 22 February 2006).

183 MEND is a previously unknown group that has claimed responsibility for a string of attacks and abductions of oil workers in early 2006. See Dino Mahtani, "Nigerian oil industry helpless as militants declare war on Obasanjo", *Financial Times,* 21 February

The question is, does any of this make the presence of oil companies untenable? Ultimately, it does not. On the one hand, the conflict makes the Niger Delta one of the world's most dangerous and difficult oil-producing areas. It is obvious that, by now, companies cannot manage exclusively with national-level partnerships, which cannot deliver the basic security available, say, in Angola. In Nigeria, oil companies also need local partners, even if those local partnerships are no more broad-based or "developmentalist" than their national counterparts. In fact, they are remarkably similar, having to do with empowered actors capturing rents through a privileged relationship with corporations, whether directly through handouts or indirectly via the control of local institutions. On the other hand, the chaos these local malcontents create, while extremely serious, has tended to be manageable, and the rewards of the Niger Delta remain too big to say no to. Thus recent times have seen both an increase in armed activities and sabotage and mounting investment interest by oil firms from around the globe.[184]

Throughout the first half of 2004, there were persistent and carefully cultivated rumours that Shell was finally planning to abandon its onshore holdings.[185] This coincided with the amply covered leak of a very critical internally commissioned report on conflict in the Delta in which the company is severely criticised and real options for conflict resolution are described as scant.[186] But exit is not an option, for a number of reasons. Firstly, the situation has not yet reached the level of permanent efforts at targeting oil personnel and closing down production for good. Shell and

2006 and "A New Year offensive", *Africa Confidential,* 20 January 2006. While little is known about it, some press coverage of MEND has labelled it a "new threat" fuelled by "frustration and an ideology of armed resistance". See e.g. Simon Robinson, "Nigeria's Deadly Days", *Time,* 18 May 2005.

184 Of which the most high profile is CNOOC's $2 billion-plus purchase of a 45 per cent stake in a Niger Delta oil block, in a demonstration that its "relentless demand for energy has it aggressively on the prowl in Africa, no matter the obstacles". See Vivienne Walt, "China's Africa Safari", *Fortune,* 20 February 2006 as well as "It's Definitive: CNOOC Concludes its Niger Delta Mega-deal—or so it Thinks", *African Energy* 95, February 2006.

185 Author interviews, April-July 2004.

186 WAC Global Services (2003). For coverage see "Shell 'feeds' Nigeria conflict, may end onshore work", *Bloomberg News,* 10 June 2004.

other oil companies are still seen as a soft target from which to extract money and their departure would confront would-be insurrectionists with Ogoni-style poverty and a federal army unlikely to pay for ransoms, phony contracts, and roads to nowhere. Secondly, the investment on the ground, and the prospects for return on that investment, are too substantial for retreat. Thirdly, Nigerian oil supplies are too important for industrialised economies for the possibility of indefinite disruption to be contemplated.[187]

And finally, as mentioned earlier in the chapter, a growing characterisation of violent protest in "criminal" or even "terrorist" terms, while empirically accurate, will aid in its proper and legal repression. Shell's 2004 threat of departure can thus be read as a warning to the government to take stock of the seriousness of Delta disruption and act accordingly, rather than a real acknowledgment of defeat.[188] The usefulness of a worst-case scenario lies precisely in the way it illustrates that oil companies will not be driven away from the Delta. The intensification of repression in partnership with the Nigerian federal state is the much more likely outcome of enhanced disruption that really and indefinitely jeopardises operations, which is not yet the case. Whether the Nigerian armed forces can competently repress an armed challenge in the Delta is a different question.[189]

Elf "africaine"

I will now look at the record of private sector involvement in the region, not from the host-state perspective, but from that of one of the major, and until recently least understood, oil companies. While others have engaged in similar activities, most of the available information about corporate malfeasance other than Elf's is anecdotal. In the Gulf

---

187 It is noteworthy that the need to secure Nigerian oil flows in the face of prolonged disruption is now a priority for the US military's Gulf of Guinea strategy.

188 Asian companies have not balked at building explicit military partnerships with the Sudanese government (see section one of Chapter 6). In the prospect of a retreat by Shell, the technically simple Niger Delta onshore deposits would certainly be attractive to a number of Chinese, Malaysian and Indian firms.

189 The Nigerian armed forces are in many respects more poorly equipped than the militias and have declared that no military solution to the conflict is possible.

of Guinea, there is no individual case of carefully documented corrupt practices involving the major Anglo-American companies, let alone the structural corruption that characterised Elf's *modus operandi* in Africa and elsewhere. Nonetheless, the recent surfacing of similar accusations against Chevron in Kazakhstan and Halliburton (the leading oil services company) in Nigeria[190] should caution one against believing in morally opposed methods of securing contracts and good relations with governments in the Gulf of Guinea. It should be noted that the Foreign Corrupt Practices Act of 1977, which forbade American companies to pay bribes to secure commercial opportunities abroad, resulted in them having to proceed far more discreetly, whereas European companies could do so with impunity until very recently. This did not eliminate corrupt practices by US firms but made them exceedingly careful. In fact, those implicated in the Elf scandals have been adamant in stating that these practices are used by "numerous other companies".[191] It is therefore not the specificity of Elf's mission or methods, to which reference was made in Chapter 4, but rather the wealth of available information and the suggestion that all corporate actors are pursuing similar strategies that explain a closer look.

Until the forceful retraction of some of its more colourful activities in the last few years, Elf was both an arm of the French state for deeds good and ill and an "informally privatised" company that benefited "top management, its own related networks, and French political factions and parties".[192] According to the French political scientist Jean-François Médard, "we have at our disposal enough information to construct a plausible picture of Elf as a delinquent organization".[193] Corruption and intelligence were as much part of the company's operations as the more orthodox tasks of an oil giant.[194] As our knowledge of these dealings grew,

---

190 Simon Romero, "Inquiry Widens to Bid-Ridding at Halliburton", *New York Times* 2 March 2005.

191 André Tarallo, quoted in "J'ai eu à gérer l'"indivision africaine" du groupe Elf", *Le Monde*, 24-25 October 1999. The editorial of the same edition of *Le Monde* notes that "M. Tarallo is unfortunately right" about the widespread nature of corrupt practices by oil multinationals.

192 Médard (2001: 1).

193 Médard (2001: 1).

194 See "L'étrange interpénétration entre les services d'Elf et de la France", *Le Monde*, 28 September 1997, for the close links between Elf and the French intelligence

a number of analysts expressed concern with the Elf "*fantasmagorie*";[195] they claim that myth-making overrates Elf's importance and misses the complexity of the company's rapport with local rulers as well as the extent to which the latter have often had the upper hand in (and have actively constructed) the relationship. These remarks are correct but beside the point here. It is the backlog of company activities and the way it circumscribes and defines its current policy horizons, rather than its impact as weighed against host-state impact, that is of concern.

A non-exhaustive list of these activities is breathtaking. Elf was involved in oil for arms deals; it tampered with elections; it financed different sides in civil wars; and it cooked the accountancy books. In order to buy off political support, it was *de facto* intermediary for oil-backed loans for some of the countries where it operated, from which it covertly benefited handsomely.[196] It under-invoiced oil sales through its subsidiary Elf Trading and pocketed the difference.[197] And mostly, it did these things with the complicity of heads of state and important politicians in oil-rich states. Indeed, Elf maintained a slush fund from which many were remunerated on a permanent basis through the payment of fictitious salaries: it is rumoured that this included more than 160 prominent individuals across the region.[198]

---

community.

195 See for instance Assemblée Nationale (1999: 129-37).

196 The facilitation of oil-backed loans, which the former Elf CEO Le Floch-Prigent claimed was meant to "balance the accounts of the producer states, to allow them generally to pay civil servants' salaries and avoid revolts", figures prominently in the 600-pages-long *Elf Indictment*. This is summarised in the Congo-Brazzaville section of Global Witness (2004).

197 Global Witness (2004: 20).

198 As Elf's key producer in the region until recently, Gabon was the centrepiece of Elf's regional strategy: a leading analyst has described it as "an offshore financial turntable for generating hidden accounts". See Shaxson (2004: 69). The Banque Française Intercontinental (FIBA), where both Elf and President Bongo held important stakes, was noteworthy for its role in paying heads of state for favourable business treatment. An exposé of FIBA can be found in Smith and Glaser (1997). See also "Les secrets africains de l'affaire Elf", *Le Monde*, 24-25 October 1999. For colourful, more anecdotal accounts to be taken with a pinch of salt, see "Elf et ses amis africains", *La Lettre du Continent*, 325, 18 March 1999 and "Les salaires d'Edith Bongo et Georges Rawiri chez Elf à Genève",

Although Elf's policies in Africa had always been subject to a degree of public criticism, only in the early 1990s did sustained questioning of its methods surface. Despite privatisation in 1994 and merger with TotalFina in 2000, and the unwillingness of France's elite to give free rein to investigations, the Elf scandal grew throughout the 1990s. It eventually encompassed corrupt practices by the company in Africa as well as China, France, Germany, Russia, Spain, Taiwan, the United States and Venezuela[199]—a system articulated through a vast network of accounts in offshore banks—and led to a trial without precedent in France or elsewhere.[200] Radical NGOs sought to undermine the legitimacy of the French Republic on the basis of the findings[201] and the ample coverage by the satirical weekly *Le Canard Enchaîné* lampooned Elf's dubious connections with the French political class.[202] The final verdict in November 2003 convicted thirty of the thirty-seven executives on trial. These included the CEO of the early 1990s, Loïk Le Floch-Prigent, and two of Elf's top old Africa hands, Alfred Sirven and André Tarallo.[203]

The pitfalls of company attempts at domination are best illustrated by one of Elf's principal engagements in Africa, in Congo-Brazzaville, where the company was for decades the towering economic entity. It paid the government most of its tax earnings, conducted infrastructure works, operated banks and ministries and arranged for loans when the country's mismanaged accounts gave way, amongst many other jobs normally deemed the preserve of the state. Congo, while not the proverbial banana republic, did provide Elf with much more leeway than

---

La Lettre du Continent, 366, 14 December 2000. The Elf "salaries" of these 160 individuals (who jointly amount to the "who's who" of the region) were deposited in Crédit Suisse and Paribas accounts, two banks that figure prominently in all manners of business with the Gulf of Guinea, including the oil-backed loans discussed in Chapter 2.

199 Shaxson (2004: 68).

200 A good memoir of the case is by the senior magistrate in the case, Eva Joly (2003).

201 Pride of place should be given here to François-Xavier Verschave and his association, Survie. See Verschave (1999) and (2000). Both works have sold very well but are marred by substantial factual and analytical weaknesses.

202 See, for instance, "Les dossiers du Canard Enchaîné 67: Elf, l'empire d'essence", April 1998, but also coverage until 2004.

203 There is no space here to delve deeply into the decade-long scandals. See "The Elf Affair: Who's Who", *Financial Times,* 15 April 2003 for a good guide.

it had in other regional partnerships; it also heightened its visibility and the potential for resentment. As explained in the previous chapter, the relationship developed between Elf and President Sassou's regime,[204] while occasionally rocked by the boom-and-bust pattern of the Congolese economy and criticism of Elf's overbearing domination, was manageable.

This was severely upset by the 1991 holding of a "national conference", the unseating of Sassou, and the installation of a caretaker government before general elections could be held. The government led by André Milongo asked consultants Arthur Andersen for an audit of the Congo subsidiaries of Elf and AGIP. Neither was forthcoming with the accounts, and the effort flopped.[205] Hostility towards Milongo on the part of Elf was such that there was discussion of supporting a coup d'état against him.[206] Sensing that the tide was changing against itself, Elf supported Sassou all the way to the elections, but to no avail. Pascal Lissouba, a former Marxist minister with a UN background, was elected to the presidency in 1992, and duly pressed Elf for improved contractual terms and cash handouts to finance his political machine and the debt-ridden state. This was done through blatant threats of alternative alliances with US firms, and did not work. Elf's time-honoured readiness to advance money to Congolese authorities was halted, and the deals Lissouba pursued with Occidental

---

204 Sassou was extensively involved at the regional level on behalf of Elf and reportedly facilitated the company's acquisition of Angola's Block 17. This allowed him great assess to wealth (which he partly invested in his personal militia) during his time outside the presidency, via deposits in Gabon's FIBA Bank.

205 A Congolese senior official of the period claimed that the lead auditor was threatened with death. He also claimed that, in November 1991, Prime Minister Milongo was received by President Mitterrand of France and asked him to get Elf to play by the rules and pay Congo what it should. Mitterrand reportedly answered that he was going to do all he could provided Milongo accepted that the past would not be delved into, i.e. that the audit should not uncover anything unbecoming. Retelling this story, the Congolese senior official added, "I was not aware of either the difficulties [Congo] was going to face or the perception of all-out challenge that the audit created". Interview with the author, Brazzaville, 9 November 2005.

206 Numerous interviews, Brazzaville, November 2005, and conversation with Congolese academic, May 2003. See also Shaxson (2007), Chapter 6.

Petroleum ($150 million in exchange for future oil production[207]) were highly unfavourable to the country. After repeated bankruptcy, Lissouba approached Elf with a number of short-term strategies for cash. Elf profited from the country's dire position and Lissouba's willingness to plunder it further by buying out the 25 per cent share of Elf Congo that belonged to the government for a staggeringly low $50 million, a move that would in time create complications with the post-1997 regime.

Despite equivocal attempts at a thaw on both sides, as the political situation in the country deteriorated, so did the rapport with Elf. The company went on to make donations to Madame Lissouba's foundations and continued the development of the Nkossa offshore field, then one of the most talked-about in sub-Saharan Africa.[208] It also backtracked on its maximum control and maximum exposure precedent by selling a 34 per cent stake in Nkossa to Chevron and by allowing Marathon Oil and ExxonMobil in. This reflected "a desire to hedge against the political and other risks of investing in the country".[209] Finally, a miscalculated attempt by Lissouba's militias to capture Sassou in June 1997 led to all-out war.[210] By mid-October, and at the cost of thousands of lives but no disruption to the oil sector, Sassou was back in the presidency, and Lissouba was off to exile. The portrait of "Elf sabotage" painted by Lissouba after his forced ousting is inaccurate, for Elf was actually doing what it had done elsewhere: it was simultaneously sponsoring all major political forces.[211] Sassou was especially displeased when the summer of 1997's oil rents were paid to Lissouba instead of himself,[212] supposedly Elf's favourite. In Congolese politics, Elf holds the status of a mythical demon. Most local accounts seem to credit it with far more power than it has ever had

---

207 Elf and AGIP later bought the loan from Oxy.

208 According to the EIU, high expectations on Nkossa were not fulfilled: production, which was expected to reach 110,000 bpd, attained only 62,000 bpd in 2000. See *Congo Brazzaville- Country Profile* 2003, EIU.

209 *Congo Brazzaville Country Profile 2003*, EIU, p. 32.

210 "And across the river", *Africa Confidential*, 20 June 1997.

211 The winning side certainly did not treat Elf as an ally: in an interview three years later, Sassou confirmed that relations with Elf had remained sensitive since the end of the civil war. See François Soudan, "Sassou a-t-il changé?", *Jeune Afrique/L'Intelligent*, 26 December 2000/8 January 2001.

212 See "A dictator returns", *Africa Confidential*, 24 October 1997.

while putting Congolese politicians in a reactive, subservient role.[213] As John Clark explains, it is ultimately to the endogenous logic of Congolese politics in general, and the greed and shortsightedness of its leading men in particular, rather than to an external bogeyman, that the thrust of our analysis must point.[214] But that should not obscure Elf's central role in the trajectory of Congolese politics.[215]

When the Elf scandals finally broke out in France, the arguments marshalled in defence of the company's activities were similar to those used by Western companies caught red-handed around the globe:

In oil circles, one speaks of "bonuses". There are official bonuses, which are foreseen in contractual arrangements [...] other bonuses, which one could call "parallel", can be paid in order to ensure a stronger chance of getting a block. Such payments are inscribed within a continuum, in the framework of long-term relations established between the company and the country, and within a climate of trust towards the leaders of that state. One of the missions of [Elf] was to entertain such relationships and ensure the fulfilment of commitments.[216]

---

213 In Congo, I found a pervasive tendency to repeat baseless and often fantastic rumours and a general belief in foreign plots and conspiracy theories, especially in regard to the role of France in Congolese politics. While external complicity in Congo's governance record exists, excessive emphasis on it tends to portray *both the Congolese and their government* as dupes of foreign companies. This often results in the attenuation of the foremost relationship of accountability—that between the people and their government—to the extent that many critics of the government seem nonetheless to think that "there isn't enough money" because "companies are fooling Congo" (conversations with the author, Brazzaville and Pointe-Noire, 5-12 November 2006).

214 Clark (2002: 186).

215 The opaque nature of Elf operations in Congo remains. In 2003, Elf finally settled with Sassou on claims of compensation for President Lissouba's hasty sale of the state share in the Elf Congo joint venture in the early 1990s. Elf's package —the relinquishing of its stake in the Linkouala oil field—was very unfavourable for the state as it was worth less than one third of the $500 million initially demanded by Sassou. It was made acceptable by the fact that the equity was taken up by Linkouala SA, a previously unheard-of company with Sassou links that in turn paid the Congolese state a mere $70 million. See "Cleaned out", *Africa Confidential,* 5 December 2003.

216 André Tarallo quoted in "J'ai eu à gérer l'"indivision africaine" du groupe Elf", *Le Monde,* 24-25 October 1999. However apologetic, this seems to me a very candid description of the way things are done.

The very politicised Africa strategy of Elf, as the Congolese example shows, was reconfigured but not fundamentally displaced by privatisation in 1994. The government kept a "golden share" of stocks that gave it de facto control over the company, and despite CEO Philippe Jaffré's attempts at corporate cleanup—which he dubbed a "Cultural Revolution"[217]—old practices were slow to die. A two-pronged change in the late 1990s would finally undermine it. This was, firstly, the political disengagement of the Jospin administration, which was reluctant about paying too much attention to Africa and continuing the notorious Foccart-style networking; and secondly, a successful merger bid by TotalFina, which set about diluting the more conspicuous elements of the old Elf.[218] This included a clear re-centring of the new conglomerate's strategy around the wealthier and more promising states of the Gulf to the detriment of the old and declining Francophone producers, and the very contentious abandonment of Elf's participation in the Chad-Cameroon oil and pipeline project. The company had always been lukewarm about involvement in Chad and had pursued this for political reasons.

Since then, Total has engaged in laborious attempts at a facelift, but as with other companies, success is limited, even while it has discontinued many of the Elf practices. Firstly, it lacks sincerity.[219] It seems to see the 1990s scandals fundamentally in terms of a media and public relations setback and not to have accepted the fundamentally criminal nature of its track record in governance and transparency, among other dimensions. As I experienced in contacts with the company in Nigeria, it seems to be much less concerned with corporate responsibility when it feels the spotlight is elsewhere.[220] Secondly, it knows that the business

---

217 Dominique Gallois, "Philippe Jaffré défend son bilan à la tête d'Elf", *Le Monde*, 29 October 1999.

218 On the many articles written on the merger, see for instance Dominique Gallois, "Les destins croisés d'Elf et de Total", *Le Monde*, 9 October 1999, "Total/Elf: Adieu les affaires?", *La Lettre du Continent*, 333, 15 July 1999, and also "Elf: end of "petropolitique" in sight?", *Africa Analysis*, 12 November 2002.

219 See Peter Gumbel, "Operation Total Makeover—Can a tainted oil giant change its spots?", *Time*, 8 December 2003.

220 I accompanied Father Kevin O'Hara from the Port Harcourt based-Centre for Social and Corporate Responsibility (CSCR) to a meeting with a Public Relations Officer at Elf's Port Harcourt Headquarters on 24 September 2002. The stated purpose was to lobby in

dynamics of the Gulf of Guinea's oil sector—created, to some extent, by itself—while not cast in stone, are structured in a way that precludes major changes. This is accentuated by the fact that the Gulf of Guinea, which accounted for 60 per cent of Elf's production and 70 per cent of its reserves in the late 1990s, remains of overriding importance to the new conglomerate's global ambitions. With Petrofina's and Total's assets factored in, the new company (the world's fourth largest in the oil sector) is in relative terms no longer as dependent on the region, but in view of new finds, its engagement in absolute terms should increase exponentially. And thirdly, the "old" Total segment that apparently has the post-merger upper hand, while claiming to want an end to Elf tactics, has brought into the merged firm its own governance problems that it has also failed to own up to.[221]

Conclusion: "Long-termism"

This chapter has argued that multinational oil companies perceive themselves to be on the threshold of a long-term regional involvement that is a different scale from that which has existed since the late 1950s, when production in the Gulf of Guinea started. The region has a terrible and well-deserved reputation for violence and for being commercially unreliable. Yet the track record of the oil industry, in the region and elsewhere, leads it to think that it can cope with the challenges that may arise. It does not take these challenges lightly. Beyond the technical and geological difficulties of the new developments, a topic I have not broached, there are political risks such as the Niger Delta uprisings,

---

favor of 144 employees who had been summarily fired in 1999. (The company alleged that they were merely sub-contracted workers, despite evidence, such as Elf badges, salary deposits in local banks and company Christmas postcards congratulating them for the "team work", that substantiated their claims; the legal process is still pending.) In the event, the two-hour meeting pertained to Elf Nigeria's labour, security, environmental, and social responsibility agendas. What struck me most was the lack of sophistication of the policies, both in substance and in PR terms, especially in comparison with the much more advanced Shell programmes I had visited days before. See Soares de Oliveira (2002: 7-10). A recent audit of the company's discourse and practice in this arena concluded that, for now, its efforts are insubstantial. See ECON (2004).

221 The prime example is Total's involvement in Burma.

unpredictable leadership twists in Equatorial Guinea, and a host of less probable scenarios which companies have certainly not neglected to consider. Companies will weather some of these out if they have already sunk their investment in a country but may be discouraged from moving there in the first place. Western companies, in particular, have been slow to move into the outright hazardous frontier states of unproven oil value, have mostly kept away from war zones such as Sudan, and have on rare occasions left contentious locations, although there are many exceptions on all three counts.[222] But in light of the willingness of empowered local actors to exploit the existing oil resources and benefit from their sale, as well as their own resilience and the ever tighter nature of the world oil market, companies have decided that the region-wide investment equation favours them. As an interviewee candidly put it, "once we are in, only the state can kick us out and they are unlikely to be stupid enough to do that; they depend on the oil more than we do".[223]

Oil companies, contrary to some press coverage, are not developing offshore resources simply because they are escaping inland conflict. The construction of a 1,060-km pipeline from turbulent Chad to the shores of Cameroon disproves this reading, as does ongoing production or exploration in Uganda, the northeastern DRC and Sudan.[224] In locations such as the Niger Delta, there is hesitation on the part of companies to commit much more of themselves onshore, and great relief at being able to discreetly move over into the sea. But the reason why oil companies are pursuing their huge investment in the Gulf of Guinea waters is *because petroleum is known to be there*, in very large quantities, and they have the capacity to extract it. Were it in the hinterland, in equally mammoth

---

222 Exits have had to do with internal company decisions rather than objective political limits that affect all operators in the same way. For instance, Texaco left Burma, but Total and Unocal did not; Shell and Elf left Chad at a time of great PR fragility for both companies, but Exxon did not. Canada's Talisman left Sudan but Total is biding its time to take possession of its Block 5 in the south of the country. A consistent feature of these three cases is the subsequent entrance of Malaysia's Petronas into the picture (see section one of Chapter 6 for an analysis of Asian companies' agendas in the Gulf of Guinea).

223 Comment made by oil executive in conversation with the author, Houston, 18 November 2001.

224 On the oil sector's involvement in Sudan, see the landmark 754-page volume researched by Jemera Rone, Human Rights Watch (2003).

quantities, the industry would move there, too; and it probably will. This said, it is clear that the shift towards the Atlantic Ocean brings in a level of attractiveness that messy onshore considerations would preclude. It is safer and easier to secure, and the invisibility of companies prevents them from being too conspicuous a factor in the internal politics of states.

The fact that working conditions (corruption, crime, poor infrastructure, warfare, etc.) in the region are not exemplary need not be laboured further; for a company's image, too, the Gulf of Guinea is very trying. One can accept that companies such as BP genuinely want to go beyond certain cruder forms of partnership with thieving rulers.[225] Companies, reform-inclined or not, would also ideally like states to perform some tasks better. State neglect in the Niger Delta means that locals direct expectations to firms; the recurrent Nigerian incapacity to pay the NNPC share of joint venture investment means that projects are uneconomically delayed, sometimes for years. Such nuisances do not make for either good business practice or friendly investment settings. But oil companies alone cannot enact structural changes. They are the prey of dynamics they have co-created but which they cannot easily change: were companies to push much further, managers think, the result would be to alienate host state allies. As this work has explained, many of these "local impediments" are not pathological deviations from the norm. Rather, this is the manner in which business has been conducted in the region and in many other areas of the developing world since the early postcolonial era; it has its own "historicity", as Chapter 4 showed.

Almost simultaneously with the surge in corporate and strategic interest in the Gulf of Guinea's oil, a different set of constituencies mostly based in the industrialised countries has grown interested in the region, yet it is not an El Dorado that it sees.[226] There are many facets to this critical turn, but all share the willingness, fuelled by scandals such as Elf's, to reconfigure the private sector's policies in the developing world. Companies are embattled by this new discourse on their purported

---

225 That does not mean that companies are keen on the transparency agenda, as most continue questionable or at least obscure accounting practices. Chevron's Angola subsidiary the Cabinda Gulf Oil Company, for instance, is registered in Bermuda and has published no annual report in four decades.

226 See section one of Chapter 6 for a discussion of the progressive agendas.

responsibilities in human rights, poverty reduction, sustainable development, etc., which is globally aimed but seems to have focused on the Gulf of Guinea as a particularly sordid nexus of politico-commercial impropriety. It is not always antagonistic and it may have some gains for companies.[227] Indeed, a degree of sophistication and double-talk can take companies a long way in improving their image, even when the behaviour on the ground has not fundamentally changed. But ultimately, there is something structurally wrong with the oil sector in the Gulf of Guinea, its rapport with governments, populations and the environment, that spin alone will not explain away. Moreover, the depth of the problems is such that even (partly) genuine reform cannot do the trick. As Shell's repentance and subsequent expansion of social investment in the Niger Delta show, even a well-intentioned company will fail at reform and not be able to eliminate external faultfinding.

The efficacy of this limited reformist agenda, alas, cannot be taken for granted. Current discourse seems to point to its mainstreaming but this can easily change, with national security or bilateral priorities overriding business ethics (a view developed further in the next chapter).[228] Furthermore, there is little evidence that the public in industrial nations can keep up a sustained interest in corporate actions in Africa, or that it would proceed to boycott oil companies on that basis.[229] It should be noted that oil companies are used to being unpopular and that this is not a great concern as long as it does not

---

227 This was the case of the World Bank involvement that made possible the Chad-Cameroon pipeline discussed in section one of Chapter 6. In exchange for a few painless concessions, ExxonMobil managed to share risk and political responsibility with a leading development institution.

228 A sobering example from the US: the US Supreme Court recently considered the repeal of, and proceeded to curtail, the Alien Tort Claims Act, an eighteenth-century piece of legislation that allows non-citizens to, among other things, sue American companies for misdeeds abroad. In this regard, the Bush Administration is intervening in favour of companies such as Unocal and Chevron, which stand accused of complicity in gross human rights violations in Burma and Aceh (Indonesia), on the dubious grounds that national security is threatened by these efforts at redress. See Dolly Filártiga, "American courts, global justice", *New York Times,* 30 March 2004 and "Legal actions over foreign misdeeds", *New York Times,* 30 March 2004.

229 Some campaigns, especially against Shell over its support for apartheid in South Africa and the Saro-Wiwa execution, were successful but short-lived. These campaigns may achieve change at specific policy levels (e.g. the 1995 backtracking by Shell in regard

affect their bottomline. A real risk for companies is the growth of ethical investment funds and the agitation by progressive shareholders favouring reform. This is more feared than any smear campaign, but it is too early to speculate on its impact.[230] The same applies to transnational litigation against companies.[231] In sum, reputational risks exist, but they are manageable, and graver risks have not yet materialised in a sufficiently serious way.

By and large, oil companies accept the Gulf of Guinea region as it is, do not want to change it, and resist attempts at doing so if these imperil relations with ruling cliques. This conservative take on local realities gives rise to a wholly different set of preoccupations from those of their critics. The goal is not to reduce poverty and increase democratic participation, accountability and broad-based prosperity; it is the successful extraction of crude oil for sale in the world market with the smoothest and least bothersome relationship possible with the state that owns it. These are the overriding concerns of commercial entities that are not meant to "enact social change", as a senior Western diplomat apologetically put it.[232] Company policies may be changing in some regards, with investment in philanthropy,[233] partnerships with NGOs[234] and the signing of voluntary codes of good conduct, but this is damage control, not reform.

With that in mind, one can more accurately perceive the strategies that the oil industry is and will be pursuing. In the field of research and development, companies will not enter bitter confrontation with each

---

to its earlier plan to dump the Brent Spar oil platform into the North Sea) but are of limited use when it comes to enacting structural changes in methods or partnerships.

230 I thank Father Kevin O'Hara for sharing his experience of work with Royal Dutch/ Shell shareholders with me.

231 See Frynas (2004).

232 Author interview, Luanda, 14 January 2004.

233 This new corporate political correctness can be pursued to a ludicrous extent: witness, for instance, ExxonMobil's hiring of "a team of anthropologists to communicate (sic) with Chadian residents" affected by the Chad-Cameroon pipeline project. See the report of the 25 January 2002 symposium, Institute of Advanced Strategic and Political Studies (2002: 12). See also "Grands groupes étrangers cherchent ethnologues", La Lettre du Continent 355, 22 June 2000. Exxon-Mobil's Corporate Citizenship Report (2003) includes a photograph with the caption "Anthropologist Ellen Brown consults with villagers in Chad about the Chad-Cameroon project construction".

234 See, e.g. Gereffi, Garcia-Johnson and Sasser (2001) and footnote 130 of Chapter 6.

other; they will wrestle opportunities from rivals in the way it is done, but when the costs or risk are too high, they will come together. All over the region, there are many instances of partnership between BP, Shell, AGIP, ExxonMobil, Chevron, Total and other companies, including the Asian newcomers. In the field of security, companies collectively think out solutions to disruptive scenarios and have elaborate mechanisms for protecting their investment and people.[235] Collaboration with state forces and local and international private security firms, as old a process as private sector investment in the region, will be deepened.[236] Finally, in the field of links with local elites, careful friendships will continue to be cultivated and so will the cravings of particular politicians. In some cases, off-budget deposits will continue to be made, as will generous and "no-questions-asked" donations to regional "development commissions" and "charitable foundations" close to the presidency. Everywhere, companies will act as important international allies for host governments, lobbying for access to powerbrokers in the West, fixing oil-backed financing and polishing the image of the more dubious ones.

The new partnership is thus being thought of in terms of unswerving support for the status quo. Corporate attitudes towards the many state-firm problems that exist are enacted within the framework of not jeopardising mutual relations. If villagers are unruly, hand out some cash, build a higher fence, and work more closely with security forces; if joint venture agreements are fatally flawed and cannot respond to the investment needs of the new offshore frontier, change to production sharing agreements. An end to the opacity of the oil sector, which has hardly ever held companies back, will not be a priority for investors that are long-established and well networked. Oil companies are a key constituency whose support is pivotal for the deployment of a reform strategy in the oil sector. This has been rightly noted by a number of advocacy groups that have concentrated their attention on company

---

235 An analyst noted that companies have "robust contingency planning" for even the most extreme scenarios (author interview with Kevin A. O'Brien, RAND Corporation, Cambridge, 10 May 2004).

236 On this point, I call the reader's attention to the huge profits made by international lawyers from the above-discussed maritime border disputes. There are few dimensions of "uncertainty" as sketched in this chapter that are not amenable to profit.

performance as a site for policy change. But on the face of the evidence, it seems that most oil firms either are indifferent to the prospect of reform or stand to lose from its accomplishment. Companies will not budge, especially at a time of rising international demand, tight supplies and minimal spare capacity, and all-time highest prices. The frontier of oil is a frontier of fierce exploitation by all means necessary, because three sets of actors (oil companies, oil-producing states, and oil-importing states) have invested everything in it.[237] The Gulf of Guinea El Dorado will be a violent affair.

---

237 A number of recent contributions on corporate actors and conflict assume that the private sector does not understand the dynamics of conflict in which it inserts itself. See, for example, Litvin (2003) and Banfield, Haufler and Lilly (2003). My research has pointed to a different picture, namely, that companies are frequently aware of the conflict potential of their actions and partnerships and act whenever possible to stem it. When conflict erupts, it is because of structural contradictions in their local involvement that they cannot address because they form the very tissue of their presence on the ground.

6

LOCKED IN: THE GULF OF GUINEA, OIL AND THE INTERNATIONAL SYSTEM

Access to energy has thus emerged as the overriding imperative of the twenty-first century

Paul Roberts, *The End of Oil*

La vie du monde du pétrole est ainsi faite que la morale n'y aura jamais sa place

Jailed former CEO of Elf Aquitaine Loïk Le Floch Prigent[1]

When it all goes, the oil I mean, sometimes we think: will anything be left?

Chadian resident of Bebedjia, Logone Province, Chad[2]

Oil production in the Gulf of Guinea is of increasing relevance for the global economy and dwarfs all other trade flows between the region and the rest of the world. It is one of the few exceptions in an Africa that matters less to the world economy now than at any time since the fifteenth century. Far from waiting on the sidelines, the Gulf of Guinea is playing a role in the vast process of enhanced global interconnectedness. This role is only obliquely related to the "new industrial revolution"[3] of IT and the knowledge economy, despite the important technological innovations that have allowed the more recent leap in production.[4] The

---

1 "Life in the world of oil is such that morals will never have a place in it"; quoted in Friends of the Earth (2001).

2 Author interview, Bebedjia, Chad, 2 October 2002.

3 Gilpin (2000: 29).

4 See section one of Chapter 5.

oil nexus I have described—a mutually rewarding relationship between firms with the technical know-how, states with legal ownership of mineral resources and consumer states—is a long-established one that has recently deepened but has not mutated significantly for decades. The new machinery fortifies, rather than challenges, the Gulf of Guinea's role in the hydrocarbon economy. In the age of "postmodern" intangible flows, the oil economy of the Gulf of Guinea is familiar and modernist in its provision of the fuel of industrial civilisation.[5]

This chapter analyses the way in which the political economy of oil of the Gulf of Guinea is embedded in the global political economy and the structural reasons and contingent motivations that sustain it. Section one focuses on the Gulf of Guinea policies of the key players of the international system. It seeks to understand the decades-long preponderance of oil in shaping outside interactions with the region in the context of much more recent concerns with good governance and transparency, made more pressing by the oil boom described in the previous chapter. The section argues, firstly, that mild strands of a progressive agenda have made inroads—in cases such as Chad and in the foreign policies of some Northern states—but that the oil rapport remains dominated by a *Realpolitik* agenda that sidelines softer concerns. Secondly, that the actual policies pursued by external actors show incentive structures for qualitative improvement not to be in place.

Section two investigates the oil nexus from the perspective of the oil states, for which this is the external link accounting for regime prosperity, the guarantor of international respectability and, more often than not, the lifeline. Starting from an analysis of this dependent relationship with the international system, I explain the international strategies of oil states and their achievement in convincingly presenting themselves as bona fide sovereign entities despite evident empirical shortcomings. After acknowledging this success—which, as noted in Chapters 1 and 3, is of almost exclusive benefit to a tiny elite—I inquire as to the political possibilities open to these societies within the constraints and opportunities

---

5 As *Le Monde* recently noted, "the world's conjuncture finally depends more on the Najaf combats and Iraqi oil production than on the floating of Google [...] There is something [of the 1970s] in this return to the 'physical' economy". See "Les leçons du pétrole", *Le Monde,* 23 August 2004.

created by oil. My assessment is very pessimistic. Oil extraction has taken place under the sign of state brutality, mismanagement and private sector connivance and the outcomes for the vast majority are devastating; economic diversification seems implausible; and the nearing end of oil in some of the region's older producers questions their very viability. I argue that the domestic political arrangements and ways of being on the world stage that were established on the basis of oil will be radically unmade when it draws to a close.

I. THE WORLD AND THE GULF OF GUINEA:
PROGRESSIVE VERSUS *REALPOLITIK* AGENDAS

The significance of the Gulf of Guinea for the world economy rests on its oil and this is, fundamentally, the prism through which it is seen by outside actors, especially since the end of the Cold War. This does not mean that other economic and political aspects do not count or that some actors do not deal with it on entirely different grounds. Bilateral relations with former colonies have a history and a (dwindling) importance that partly go beyond the economic calculus; there are non-oil legal and illegal trade flows; and international civil society organisations, for instance, certainly look beyond oil wealth into the bottomless poverty of the region's masses. But ultimately, it is the oil that counts; it is oil that makes it the exception to the neglect of Africa by the rest of the world.

This importance can only increase. Oil, once taken for granted in the pre-1973 world, is no longer perceived as inexpensive and abundant. Through the creation of the International Energy Agency, strategic stockpiling, an emphasis on energy conservation, and the search for oil outside the preserve of OPEC, importer countries have striven to break the latter's grip over oil pricing and supply. A measure of rebalance was achieved from the early 1980s too the late 1990s, but prices have mounted steadily in the past few years and the prognosis is an extremely expensive $50 a barrel (2005 dollars) by 2014.[6] The long-term prospects

---

6 This is the *Annual Energy Outlook 2007* forecast of the generally upbeat Energy Information Administration of the US government, which now foresees a rise to $59 per barrel by 2030. If anything, this is an overly optimistic assessment and most other estimates point to even higher prices; OPEC, for its part is aiming at a $60 a barrel for

are therefore dire, with a growing market share again falling in the direction of the OPEC producers that control the bulk of the world's reserves and spare capacity, severely curtail private sector oil activities, and are apt to cause price hikes in their own fight for market share.[7] Overall supplies will probably peak within two decades and the "easy oil" (i.e. the non-OPEC oil that companies can exploit and the readily accessible oil that presents few technical quandaries) may be in the process of peaking already. The debate is on between pessimists who see an impending zenith and those who believe the industry will go on providing the technological innovations and fabulous oil finds needed to cope with ever-rising demand in the medium and even long term.[8] On their part, oil companies—despite record profits—are certainly having difficulties in replacing production,[9] let alone adding to their reserves, as the 2003 Shell scandal surrounding inflated reserves showed. Indeed, as the *Financial Times* notes, "oil majors are quietly panicking because they cannot find sufficient new projects or fields big enough to make material difference to companies of their size. So [instead] they are spending whatever they can on getting access to assets in such countries as Russia and Libya that have largely been explored already."[10]

To make matters worse, the emphasis in many importer states, especially outside the EU, is on supplying rising demand through available

---

the near future. In contrast, "oil prices in real terms [...] averaged below $20 in the 15 years to 2000". See "World to become more dependent on Mideast oil", *Financial Times,* 27 October 2004.

7 OPEC currently produces about 40 per cent of world output. According to the International Energy Agency's *World Energy Outlook 2000*, OPEC will have a 51 per cent market share by 2020. According to BP's 2004 *Statistical Review of World Energy*, OPEC holds an estimated 881.6 billion barrels of the 1146.3 billion barrels of proven worldwide reserves.

8 For a review of the debate see Chapters 7 and 10 of Roberts (2004). For optimistic assessments that stress natural gas, non-conventional oils (like tar sands) and the continued technological prowess of the industry in coping with demand, see Daniel Yergin, "Towards a $7-a-gallon future?", *New York Times,* 24 November 2003 and Claude Madil, "Oil: is the sense of crisis overdone?", *Financial Times,* 19 October 2004.

9 See Alex Berenson, "An oil enigma: production falls even as reserves rise", *New York Times,* 12 June 2004.

10 "Use less or save more", *Financial Times,* 29 July 2004.

means rather than urging demand-side restraint. As Paul Roberts writes, the 9/11 attacks have caused a reaction that is the opposite of that elicited by the 1973-74 hike in oil prices.[11] Rather than seeing an opportunity to ease reliance on oil—through conservation, research on renewable energies, etc., like in the 1970s and early 1980s—the Bush Administration has put forth a security of supply strategy premised on close relations with producers and the pumping of more oil into world markets. Other states are following policies of variable aggressiveness to secure their own supplies.[12] It is therefore likely that hydrocarbons will remain at the centre of the world economy. Their continuing centrality will be boosted, firstly, by the high cost of renewable alternatives and their capacity to address only a minority of total energy needs; secondly, by the marginalisation of the nuclear sector as an alternative, which will either diminish in importance or stay at the same level, but will not grow substantially;[13] and thirdly, by the unwillingness of the major importers to conserve energy or proactively work towards a new energy economy through investment in renewable energies. This is not to assume that there will not be a non-oil world economy, but that its advent will be the result of resource depletion and climate change rather than enlightened

---

11 Roberts (2004).

12 The politics of oil imports and the nature of competition between importer states are too broad a subject to be engaged here. Generally speaking, there are two attitudes: firstly, those of states that are protective and competitive and seek to secure their own supply first. Examples include not only the present-day aggressive behaviour of Asian states but also that of France until about a decade ago (in 1973-74, France refused to join multilateral efforts to deal with shortages and did not enter the IEA). The second is that currently prevalent in the OECD states, which is premised on maximum production for the world market and diversification of energy sources. The confrontation of these mercantilist and market-based understandings of oil provision will shape the future of energy supply.

13 This is certainly the tendency in most EU member-states with the exception of Finland and France. However, one cannot exclude a reassessment of the nuclear sector in the light of unsustainable hydrocarbons dependency in the near future. See Andrew Taylor and Sheila McNulty, "The nuclear option: creeping back into political favour", *Financial Times,* 10 August 2004 for this view. In the Asian economies, strong investment in the nuclear sector is already taking place but it is unlikely to have a taming effect on hydrocarbon needs for a long time.

policy change, and that it will happen later rather than sooner. It is likely to be traumatic.

In this context, the Gulf of Guinea matters immensely, despite the fact that it will never be a swing supplier or price-setter or replacement of the Persian Gulf. In the context of a strategy of propping up non-OPEC output in under-exploited oil provinces such as Central Asia, frontier areas such as Siberia and parts of Alaska, and regions such as Mexico, where foreign operators are still not allowed in, the Gulf of Guinea plays a considerable role. OPEC membership of its two largest producers notwithstanding, it is company-friendly, and its instability is of a kind that operators feel they can cope with. To that effect, the states in the region are courted and the symbolic (legitimacy, international support) and material (financial, technical) bases of political order are provided to them by the international system. The export credit agencies (ECAs) of the industrial states provide ample political and commercial risk insurance and loans—well in excess of that provided by all other multilateral lending institutions combined—that make energy investments in the Gulf of Guinea possible, and indeed attractive, to companies.[14] Western commercial banks continue to provide finance to these states in exchange for mortgaged future oil production.[15] The World Bank Group and its private sector arm the International Finance Corporation lend generously towards extractive industries projects across the region and, more importantly, put their political clout at the service of investors.

The progressive agendas

This essentially realist engagement with the Gulf of Guinea clashes with the proclaimed underpinnings of present-day Western foreign policies in the fields of good governance, human rights and poverty alleviation,

---

14 ECA activities are the subject of an important paper by Rich, Horta and Goldzimer (2001). According to the authors, ECAs are "governmental or quasi-governmental entities that subsidize and promote a country's exports and investment abroad" (p. 1) and their support is essential for large-scale projects in the developing world. ECAs are virtually unknown to the public and their loans are often not subjected to the "environmental and social policies and standards" (p. 3) that organisations like the World Bank have to fulfils.

15 See discussion on oil-backed loans in section two of Chapter 2.

goals pursued not only by dissident voices in the activist community but often by OECD governments themselves.[16] In the past ten years, the growth of distinct but quickly juxtaposed "progressive" agendas on transparency, the responsibilities of corporations towards "stakeholders", and the consequences for the poor of investment in extractive industries have jointly eroded the space for self-interested, security of supply-driven policies by (Western) companies and states. Policy suggestions vary from the voluntary and faintly reformist to the radical call for tough regulation, but everywhere business practices across the developing world, and in the energy sector in particular, are being scrutinised for needed amendments.

The first call for change pertains to the lack of transparency of business transactions and the attendant corruption. Though colossal graft in developing nations has been a concern for decades, the broad-based recognition of the need to tackle it is a recent one. Partly on the tracks of the "blood diamonds" campaigns of the late 1990s,[17] activists closed in on the theft of public resources by the elites of oil-exporting states and the contributory role of oil companies in the process. Despite attention being paid to oil despots in other regions, the Gulf of Guinea, and Angola in particular, was in the spotlight from the start. The transparency effort was aided by the mounting anti-corruption trend, visible in the 1997 approval by the OECD of an anti-corruption convention, the high profile of organisations like Transparency International, the IMF's growing interest in huge discrepancies in member state accounts, and pronouncements by the G8 and other international organisations. In particular, the Publish What You Pay (PWYP) campaign, which now includes more than 200 associations around the world, has called for a legally binding requirement

---

16 Jagdish Bhagwati (2004: 169) points to the paradox of progressive actors now rallying not against multinational corporations' involvement in domestic politics, as in the old days, but in favour of (positive) corporate meddling in domestic politics. "Some NGOs today", he writes, "seem to accuse multinationals of neglect, rather than intervention, when it comes to advancing their own agendas."

17 This focused on the role played by diamond smuggling in the financing of rebel forces across Africa, particularly in Angola, Sierra Leone and the Democratic Republic of Congo, and eventually led to the UN-backed, peer-reviewed "Kimberley Process", whereby the diamond industry and external observers supervise the provenance of gems.

that companies divulge the amounts they pay governments of oil-rich countries as a precondition to being listed in Western stock exchanges.[18] The argument is that the civil society of these countries can only call their rulers to account if they know the location and amounts of money earned by the sale of their natural resources. In response, the British government (with Prime Minister Blair's personal involvement[19]) has put forth an Extractive Industries Transparency Initiative (EITI, launched in eptember 2002) that calls on both governments and companies to render public their transactions on a voluntary basis.[20] The transparency issue is today at the forefront of all actors' dealings with the Gulf of Guinea.

The second arena for change is the responsibilities of firms themselves. In tandem with its attempts at responding to the transparency agenda, the private sector has conjured a discourse on Corporate Social Responsibility (CSR). It appeared as an endeavour by companies to both deflect criticisms on the environmental and social impact of their activities and prevent the creation of putative transnational regulatory frameworks to tackle them. About ten years later, CSR is an industry that employs thousands within and outside large corporations who produce social audits and community relations advice, but is often (and mostly correctly) dismissed as glorified PR of varying degrees of honesty and effectiveness.[21] The noteworthy innovation of CSR lies in the fact that companies now admit as "their responsibility" tasks and consequences that were until recently seen as

---

18 I thank Henry Parham, the coordinator of the PWYP campaign, for discussing his work with me on two occasions in February 2003 and May 2004.

19 The UK is at the forefront of governmental and corporate attempts at thinking out a "progressive" compromise for the oil sector. This includes EITI and the initiatives of companies like Shell and BP, which are noticeably more involved in such efforts than their American counterparts.

20 A measure of how easy it has been to fit into a voluntary scheme with vague performance criteria is the fact that Gulf of Guinea oil states have some sort of rapport with EITI. São Tomé and Príncipe, Nigeria and Congo Brazzaville are said to be in the process of "implementing" EITI principles. Cameroon, Gabon and (incredibly) Chad and Equatorial Guinea are listed as states that have "endorsed" EITI, which is defined in the EITI website (www.eitransparency.org access 12 March 2006) as governments "that have endorsed EITI and are presently considering how they will implement it". Angola was an "observer" to the process.

21 See Christian Aid (2004) for a recent critical evaluation.

situated beyond their profit bottomline. Most high profile companies have since adopted non-binding good behaviour codes and can cope with criticism of their operations with less of the old awkwardness. Home states have taken to the CSR concept with equal enthusiasm,[22] as have international governmental organisations.[23] In particular, the UN has put forth with great fanfare its "Global Compact", a voluntary forum where member companies publicise their CSR good performance in the hope of creating benchmarks for industry best practices.[24]

Finally, the third progressive agenda is interested in the extent to which investment in the extractive industries can lead to poverty alleviation.[25] As discussed above,[26] this dubious link was the subject of increasingly disparaging research from the 1980s. It has finally been called into question by the work of NGOs[27] and, grudgingly, by the IFIs themselves. This went to such an extent that the World Bank itself was compelled to commission an independent Extractive Industries Review (EIR). The resulting EIR report made a number of important recommendations and acknowledged that the extractive industries have not only brought about disruption to the environment and local communities the world over, but failed to alleviate poverty as well. Most importantly, it called on the Bank to stop supporting extractive industries altogether.[28] Unsurprisingly, this set of proposals was not accepted, but it succeeded in shedding further

---

22 See e.g. www.societyandbusiness.gov.uk for the British government's CSR activities.

23 See in particular European Commission (2001).

24 For the activities of the Compact see www.unglobalcompact.org; for an evaluation after four years see McKinsey & Company (2004). It is telling that McKinsey establishes from the outset that "an attempt to assess the Compact's impact on the ultimate goal of promoting inclusive globalization would have been spurious, given the many variables that affect it" (pp. 1-2). See Ruggie (2002) for a spirited defence of the Global Compact by its foremost intellectual advocate.

25 For a typical defensive statement on the positive role of the extractive industries, see McPhail (2000).

26 See section two of Chapter 1.

27 Amongst many, see the work of Catholic Relief Services, Oxfam USA, Environmental Defense, Friends of the Earth, Christian Aid and Human Rights Watch.

28 The six-volume report is available at www.eireview.org (accessed 2 September 2004); see also the op-ed by the head of the EIR, Emil Salim, "The World Bank must reform on extractive industries", *Financial Times*, 17 June 2004.

light on the disruptive effects of extractive industries and in putting the latter on the defensive.

This, then, is the challenge faced by companies and *Realpolitik* government strategists: to make security of supply self-interest and the now mainstreamed progressive agendas, if not compatible, then mutually co-opting through inclusive partnerships in which the goals of multiple actors are partly and incrementally fulfilled. The former pertains to the very survival of industrial society; the latter, to expectations of business behaviour by companies and states that is geared towards broad based prosperity. Chad, the newest oil exporter in the Gulf, is where the most ambitious attempt at bringing the two together is being enacted. This is an innovative and frankly experimental framework in which NGOs, IFIs, oil companies and foreign governments, in addition to the Chadian state, all played along. I will briefly review the experience to try and weigh its success in bridging these, seemingly opposed, sets of expectations.

Doing the right thing in Chad[29]

As late as 1994, a survey of oil politics in sub-Saharan Africa, commenting on increased excitement about the continent, noted with some bewilderment that "even Chad has attracted interest".[30] Exploration for oil in the country started in the early 1970s but the escalation of the civil war from 1979 led to the suspension of petroleum activities as well as a proposed plan to build a pipeline from Lake Chad fields to a new refinery in N'Djamena, the capital. Although Exxon found the commercially viable Bolobo field in 1989,[31] the security situation of the country was still deemed unmanageable. To the frustration of cash-strapped President Deby, the investors' consortium (then comprising Shell and Elf, as well

---

29 I thank Catholic Relief Services, and Ian Gary, Nicole Poirier and her family in particular, for their help during my visit to Chad in late September and October 2002.
30 "Finding oil in troubled waters", *Africa Confidential*, 3 June 1994. For the background on Chad's pre-oil politics see Buijtenhuijs (1998) as well as other works by the same author and also Foltz (1995).
31 The original consortium included Exxon, Royal Dutch/Shell, Conoco and Chevron. Conoco (the consortium operator) left in 1981 and Exxon assumed operations.

as Exxon) was cautiously assessing any further involvement,[32] and it was only in November 1996 that it signed a MoU with the governments of Chad and Cameroon. This provided the terms for the development of the Doba fields in southern Chad and the construction of a 1,078 km-long pipeline to transport the oil to a terminal on the Atlantic coast. The $4.2 billion project—the most important onshore investment in Africa—came to involve the consortium, the governments of the two countries and, crucially, the World Bank. The consortium was resolute that it would not pursue the project at all in the absence of World Bank engagement, seen as a protective device against the many attendant risks.[33] The long period of gestation was given a further twist when Shell and Elf suddenly revised their commitment to the project and decided to pull out. The void was filled by Petronas (35 per cent) and Chevron (25 per cent) with ExxonMobil remaining as the operator (40 per cent), but still sent shock waves through the oil sector.[34]

---

32 "Chad: more pipe dreams", *Africa Confidential,* 2 December 1994. In an 1990 interview, Deby's predecessor Hissein Habré had already gone out of his way to reassure "all foreign investors" that their investment was safe in Chad and that the "doors were open". See "Le président Hissein Habré fait le point de la situation politique et économique", *Marchés Tropicaux et Méditerranéens,* 2 November 1990. See also the article "Energie: les espoirs pétroliers" in the same special issue on Chad.

33 Chad has few of the technological complexities of the Gulf of Guinea's ultra-deep offshore but is nonetheless dependent on the involvement of big players for the exploitation of its oil wealth. This is because it is a landlocked country: the transport structures needed to export the oil are so expensive that only the majors can afford to build them. In turn, with its unenviable reputation as one of Africa's earliest failed states, Chad is a context for which only global players with impeccable credentials can muster financing in international capital markets.

34 It also further dented Chad's credibility: the fact that two oil majors with a reputation for not shirking tough assignments thought the country to be beyond the pale did not augur well for the project. Although the reasons for doing so were never fleshed out, it is thought that the environment, governance and human rights-related criticism that the project was bringing forth was too much for the two companies, then going through bruising public relations flops over their African operations. In the case of Elf, the impending merger with TotalFina added some urgency to the distancing from Elf's traditional grounds. See "Chad: Project in Trouble", *Petroleum Economist,* 21 December 1999, "Tchad: Retrait de Shell et Elf du projet petrolier", *Marchés Tropicaux et Méditerranéens,* 19 November 1999, and "Un mort dans une manifestation contre Shell et Elf", *Les Echos,* 17 November 1999.

The World Bank's board of directors finally approved its involvement in the petroleum and pipeline project in June 2000.[35] In response to the many concerns voiced in regard to the project, the Bank made its presence conditional on a willingness and ability to direct revenues to social and economic infrastructure.[36] The Law of Petroleum Revenue Management (the so-called Law 001) adopted in 1999 as a precondition for Bank involvement included scrutinised offshore holding of direct revenues in escrow accounts, a "Future Generations" fund, and the earmarking of most direct revenues for "priority sectors". The law left only 5 per cent of revenues at the discretion of the government.[37] A nine-member Revenue Oversight Committee (the *Collège de Contrôle*) with four civil society representatives was created to oversee the use of the oil set aside for priority sectors. The International Advisory Group (IAG), a five-expert team appointed by the World Bank in February 2001, monitored the project through periodic trips to the region.[38] In addition, the External Compliance Monitoring Group (ECMG) was under a contract with the International Finance Corporation to specifically monitor the consortium's compliance with the Environment Management Plan.

The Chad-Cameroon project became the most closely observed World Bank project ever, decried by NGOs and hailed as a "model" by industry and government,[39] and was never out of the headlines during construction. Important elements of the oil pipeline project were readjusted over four years as the result of the (often adversarial) rapport

---

35 This consisted of a modest contribution: $93 million from the World Bank and $200 million from the International Finance Corporation and the IDA. The financial importance of the World Bank participation lay in its giving the "green light" to other sources of funding. Within a month, IFIs such as the European Investment Bank and ECAs such as the US Export-Import Bank and France's Coface, as well as 18 commercial banks, had approved additional loans.

36 See World Bank (2000). For a briefing on the project see Guyer (2002).

37 80 per cent is earmarked for social spending in priority areas (education, rural development, health, infrastructure and environmental resources); 10 per cent for the future generations fund; and 5 per cent for the oil-producing region.

38 I thank Prof. Jane Guyer for discussing her work in the IAG with me (conversation with the author, Washington, DC, 23 June 2003).

39 See the hearings held by the US House of Representatives (2002) for widely divergent views on the project. See also Norland (2003).

between a myriad constituencies permanently jostling over many issues. The involvement of the international community thus brought some tangible benefits in addition to the revenue management framework, many due to the critical on-the-spot presence of civil society networks.[40] At the environmental level, the pipeline technology is substantially better than that common in the Niger Delta,[41] and an exhaustive (if not flawless) 19-volume environmental impact study meant that the impact, while severe, is not what was feared. At the level of compensation over land and felled trees, the initial trivial amounts were upgraded to more substantial sums. More importantly, the project kept the spotlight on the country, allowing a flourishing of civil society activity beyond the narrow confines that had been previously tolerated.

In spite of World Bank rhetoric on "partnerships", however, this was never one that was accepted by sectors critical of the very existence of the pipeline project. Rather, the rapport was mostly confrontational. There was never a consensus about ideal outcomes, and the pipeline project was mired from the start in criticism of the consortium, the government, and the Bank's actions. The government grudgingly played along because it yearned for the revenue but was unenthusiastic about the intrusive conditions[42] and poorly implemented most of the 250-odd agreements that are the basis of the project. Presidential elections in 2001 were widely seen as fraudulent.[43] The *Collège de Contrôle* was under-funded, harassed

---

40 See Environmental Defense (2002) and Friends of the Earth (2001) for critical civil society views midway through the construction period.

41 "The Chad-Cameroon Pipeline is buried and insulated and it uses lined pits and ponds and oil-spill detectors. Gas produced with the oil will be used for local power generation rather than flared". See "Chad: digging a hole", *Africa Confidential,* 30 August 2002.

42 An early sign of lack of commitment was its use of a $4.5 million share of the $25 million signature bonus to acquire arms. Although, strictly speaking, not a violation of the rules set forth, the action clearly questioned the government's commitment to poverty alleviation and institutional reform. See "Le bonus pétrolier en question", *N'djaména Hebdo* 613, 19-22 September 2002 and "This Clean Oil Deal is Already Tainted", *Business Week,* 5 July 2001.

43 See Stephen Smith, "L'élection présidentielle au Tchad est entachée de multiples fraudes", *Le Monde,* 24 May 2001 and "Old president and new oil for Chad", *Financial Times,* 28 May 2001.

and denied important information from the start.[44] The consortium was accused of using the high-profile World Bank involvement to deflect criticism[45] while basking in the easy finance and diminished risks that came with it. And far from assuaging detractors by its presence, the Bank bore the brunt of the criticism that Exxon was able to avoid, including a high-profile 2002 dressing down by its own Inspection Panel.[46]

The Bank is accused of rushing ahead with support for the project without prioritising Chadian capacity building, both civil society and state, and of not guaranteeing a good enough revenue share for Chad.[47] Furthermore, the commendable project framework was before long known to be riddled with loopholes. Revenue allocation under Law 001 could be legally changed after five years of adoption, i.e. in 2004, soon after oil started to flow. Moreover, indirect oil revenues (customs duties, taxes, etc, amounting to almost half the total) did not fall under the ambit of Law 001, and neither did further oil developments outside the Doba fields. The fact that the consortium soon found five additional fields and other companies acquired the right to prospect over much of Chad meant this was a serious gap.

Political developments since the oil started to flow in October 2003 did nothing to mitigate these earlier criticisms.[48] A critical radio station,

---

44 I thank *Collège de Contrôle* member Thérèse Mékombé for interesting talks on the many political and logistical obstacles the *Collège* has faced (Kribi, September 2003, and London, May 2004).

45 "In Chad and Cameroon, Exxon saw early on what was coming, and it deftly interposed the World Bank between itself and the NGOs. [...] Despite the multiple layers of supervision, it is far from certain that the agreement will work as intended. [...] But it is already working for ExxonMobil, because the World Bank now finds itself in the hot seat bearing the brunt of criticism if money is diverted or the system fails". Ottaway (2001).

46 See World Bank (2002) and World Bank (2002). See also Alan Beattie, "World Bank Team Attacks Own Project", *Financial Times,* 19 August 2002.

47 As already noted, this is currently a modest 12.5 per cent of revenues, i.e. the Gulf of Guinea region's lowest share.

48 The best work on Chad's experience with oil (taking the story up to early 2005) is Gary and Reisch (2005). For the almost immediate slippage in Chad's fulfilment of its obligations, see also Ken Silverstein, "Chad sees first trickle of cash from pipeline", *The Nation,* 26 December 2003, Somini Sengupta, "The making of an African petro-State"

FM Liberté, was closed down;[49] public executions returned, and an attempted coup d'état was staged.[50] The president, though ill, changed the constitution so that he could stay on for a third term.[51] Forty per cent of the "social spending" was set aside for road building (compared with only 5.1 per cent for education and 3.3 per cent for healthcare), that being an area that is "particularly vulnerable to corruption and mismanagement".[52] Of course, these setbacks are only relevant for the goals of economic growth and poverty alleviation through investment in the extractive industries that the World Bank set out to achieve, for the oil project itself stands on much firmer ground. The oil producing region has been mostly pacified since 2000 despite rebellion raging elsewhere in the country. Banditry that still exists does not threaten operations and the pipeline is permanently policed by private security contractors. Exxon is enthusiastic about the country and committed to stay and invest much more,[53] even if relations with the government are occasionally prickly, and other companies are following suit. Conversely, the project's "developmental" success seems to be limited by both contingent reasons (i.e. the lack of commitment of those involved) and more serious structural limitations. For despite attempts at de-politicising the oil and pipeline project, it is being carried out in a context of state decay and authoritarian renewal,

---

New York Times, 18 February 2004, and Emily Wax, "Oil wealth trickles into Chad but little trickles down", *Washington Post,* 13 March 2004.

49 "Independent radio station closed, director beaten up", *IRIN,* 18 February 2004.

50 Korinna Horta and Delphine Djiraibe, "Africa's dangerous treasure", *Washington Post,* 10 March 2004.

51 *Chad Country Report,* EUI, May 2004, pp. 6-7.

52 Ibid., p. 8. Indeed, the greatest beneficiary has been President Déby's brother Daoussa: of the estimated $51 million spent in public works until late 2005, $48 million were assigned to a partnership between his company and a foreign construction firm. See Lydia Polgreen, "Chad's oil riches, meant for the poor, are diverted", *New York Times,* 18 February 2006.

53 Remarks by André Madec, Manager, Corporate Public Affairs, Exxon Mobil Corporation, at the conference on "Oil and Gas in Africa", Royal Institute of International Affairs, 24-25 May 2004.

lack of institutional capacity and absence of a strong civil society that could somehow capitalise on the oil boom and make it a good thing.[54]

From the start, critics pointed to four major obstacles that made it unlikely that Chad would provide a satisfactory, or even less unsatisfactory, compromise between, on the one hand, the thirst for oil and, on the other hand, the commitments to development and high ethical goals that I have outlined. The first is the very relevance of the "Chad model". Chad's recent history would caution against pioneering any scheme in it: the circumstances that made it acceptable to the leadership in N'djamena were unique and unlikely to be replicated. It is impossible to wrestle analogous commitments and degrees of foreign involvement from oil giants such as Angola or Nigeria—São Tomé is perhaps the only example of a state that is powerless enough to accept such a degree of external impingement. Companies, too, are unwilling to lobby for or implement similar policies elsewhere.

The second obstacle lies in the degree of influence that the IFIs and the international community in general can go on exerting when a project comes on stream and the revenue starts flowing. This is primarily so in regard to the host state, which may ditch the deal as soon as it can afford to do so, but it applies to the private sector as well. The Chad consortium, critics predicted, could not be assumed to side with the Bank in pressuring Chad to comply to previously defined regulations, as this might jeopardise its own position. The third obstacle is that the private sector gets more out of this scheme than it gives: World Bank involvement gets companies cheaper credit and a high-profile institution with which to share blame if things turn out badly—in short, it makes projects happen. But in return for this boon, companies can take refuge in a business-only perspective that puts the Bank at the centre of public responsibility. And finally, there is the limited extent to which the goal of poverty alleviation can be fulfilled by oil revenues. The revenues still

---

54 The extent to which this affects the consortium calculus is moot. Under the leadership of Exxon, it has been able to attract the support of Western governments, import-export agencies and development institutions to a project that did not seem feasible only a few years ago. Moreover, within the confinements of its corporate culture, Exxon has been able to adapt slightly to a discursive realm heavily dominated by development themes that were not familiar to the company.

accrue to the state, fuelling unproductive rent seeking, and the "road-building" frenzy shows that it is possible to neuter many good governance provisions.

This is to say that the writing was on the wall from early on and that the Bank and its defenders cannot pretend that the more recent developments come as a huge surprise. Faced with a mounting insurgent challenge from the east (partly a Darfur conflict spill-over but decidedly fuelled by Déby's own oil-hungry clansmen), the President concocted a solution that to his mind would bring in the additional resources needed to shore up the regime and fight the rebels. Law 001 was to be modified and the pesky progressive clauses dealt away with. The rubberstamp parliament obliged him on 29 December 2005, despite the vociferous opposition of the Western donors and the World Bank. The modified law was a shadow of its predecessor.[55] The Future Generations' Fund was scrapped and the $36 million in it gobbled up; defence and administration were added to the priority sectors list, whose share of revenues fell from 80 per cent to 65 per cent; and the national budget was to receive 30 per cent, as opposed to 5 per cent, of revenues.[56] The World Bank under its new president, Paul Wolfowitz, reacted with rare outrage by freezing an estimated $124 million in loans to Chad as well as all oil payments to the government.

The stalemate continued into April 2006, with President Déby telling the Consortium to hand over an estimated $100 million in oil revenues or face the shutdown of the pipeline[57] and, Chadian officials claimed,

---

55 See Chip Cummins, "Exxon oil-fund model unravels in Chad; government breaches deal requiring it to spend royalties on development", *Wall Street Journal,* 28 February 2006 and Lydia Polgreen, "Chad backs out of pledge to use oil wealth to reduce poverty", *New York Times*, 13 December 2005.

56 Probably with the goal of not stirring up too much trouble in the oil-producing region of Southern Chad, its relatively modest share of 5 per cent of direct revenues was left undisturbed. See Alain Faujas, "Tchad, la malédiction de l'or noir", *Le Monde* 17 March 2006.

57 Chad's Oil Minister Mahmat Hassan warned that "either the World Bank unblocks the offshore account and the frozen revenues enter into that account, controlled by Chad, or the oil companies make direct payments". See Daniel Flynn, "Chad confident of World Bank deal by end of April", Reuters, 18 April 2006. See also Stephanie Hancock, "Chad threatens to halt oil output", *BBC News*, 20 April 2006.

"alternative oil deals" with the likes of China.[58] By then, there were indications that the Western critics were ready to flinch.[59] Its tough tactics not having worked thus far, the World Bank was mortified about losing face completely, which would happen whether the consortium accepted Déby's terms or the project got cancelled altogether by a vindictive government. The Bush Administration, for its part, sent in Deputy Assistant Secretary of State for Africa Donald Yamamoto as a troubleshooter. Paul Wolfowitz himself had earlier seemed bent on a principled stand but was now amenable to a less than morally tidy outcome. "I think to stand back and say the whole thing is a dirty business and we in the World Bank don't want to have anything to do with it is very shortsighted," he said.[60]

In short, the World Bank and the international community "blinked first".[61] The ensuing compromise accepted most Chadian demands. The Fund for Future Generations was eliminated and the $ 36 million kept by the government. Chad revised its 2006 budget and the Bank will recommence its loans and release the oil revenues it holds. An estimated 30 per cent of direct revenues will go to the discretionary budget. A minor victory for the Bank is the inclusion of indirect revenues of the oil project into the Revenue Management Law. But few are presenting these stopgap measures as much more than that. The April 2006 agreement is only valid till 2007, when the whole deal will again be up for grabs. More recent developments point to continuing unpredictability: claiming the non-payment of taxes, President Deby went after Petronas and Chevron with threats of eviction.[62]

The Realpolitik *agenda*

Chad illustrates how far the progressive agendas have travelled, and the structural limitations they face. On the one hand, an unprecedented experiment was enacted. A sovereign state seemed to accept major,

---

58 David White, "The 'resource curse' anew: why a grand World Bank oil project has fast run into the sand", *Financial Times,* 23 January 2006.

59 Some never supported Wolfowitz's confrontational tactics. This was the case of France, which took up the defence of Chad from the start (telephone interview with Ian Gary, 12 May 2006).

60 Lydia Polgreen, "Chad's oil riches".

61 Phone interview with Ian Gary, 12 May 2006.

62 "Chad tells 2 oil firms to pack up", *International Herald Tribune,* 28 August 2006.

sustained external involvement in social policies, budgetary processes and macro-economic management. Recalcitrant companies had to compromise on stringent transparency measures and environmental standards. A huge degree of constant scrutiny by the Western press forced all involved to be moderately responsive to some demands and criticisms of local and international civil society groups. On the other hand, this "model" was put at the service of turning the possibility of oil production under a dangerous and remote dictatorship into reality. Pressures to slow down the project and build local capacity failed. Pressures to democratise Chad, or make political liberalisation a precondition for investment, failed too. And there is now near-certainty that the money will be wastefully spent. In short, while the contraptions put together by the international community could not save Chad from poverty and repression, the benefits for companies, the Chadian elite and importer states are immediate. From their joint perspective, the project is a winner already, even if the World Bank's reputation has been badly bruised in the process.

It is this *Realpolitik* agenda—the same that has cemented the political economy of oil for many decades—that ultimately shapes oil policy, though it is now tempered by a lot of progressive rhetoric (and some progressive practice) unthinkable a decade ago. By way of illustration, I will now look at the Gulf of Guinea policies of the United States, the newly industrialising economies of Asia with emphasis on China, and the two key international financial institutions, the World Bank and the IMF.[63] These are, respectively, the largest importer of oil in the world, the second largest oil importers and all-out challenger of the progressive agenda, and the pivotal international governmental organisations in the areas of finance and economic management. In different ways, all point to the fact that, despite the importance of the progressive agendas, it is oil and the intertwined interests of oil exporters, oil firms and importing states that still take precedence.

The United States. "This is not a policy driven by commercial or economic interests, it is driven by political and humanitarian factors [...] I wanna

---

63 Lack of space precludes an in-depth analysis of the policies of EU member states, and of France and the United Kingdom in particular.

make that clear".[64] Thus US Ambassador to the UN Richard Holbrooke, outside Luanda's Futungo de Belas presidential complex in December 1999, in an attempt to assuage criticisms of oil *Realpolitik* by the world's largest consumer. And yet it is difficult to claim that the US's recent interest in the Gulf of Guinea—which, together with South Africa, is the exception in a continent that does not register much in US geopolitical concerns[65]—is not directly connected with the size of its oil reserves. The Bush Administration is now pursuing the firmest multilevel, long-term engagement ever with the region.[66] Particularly since 9/11, the Washington think-tank scene is awash with conferences and workshops on the subject. Visiting Gulf of Guinea heads of state are given access to the White House and feted by lobbyists and oil companies, while policy documents stating the region's importance for the "national interest" are amply discussed and given wide coverage in the press.[67] Proposals typically concentrate on existing and potential US bilateral policy tools (e.g. AGOA, the Millennium Challenge Account) and multilateral collaboration (e.g. through the G8 and the IFIs) to arrive at long-lasting relationships with oil exporters. They typically stress both realist US concerns and commitment to African development.

Despite this enthusiasm, US policy co-ordination on the region is disappointing. The authors of an important report on the energy stakes in Africa note that the State Department "is not currently organised, empowered or resourced to lead a robust and informed US energy policy in Africa".[68] Furthermore, there does not seem to be a great deal of articulation taking place between companies and government, and there are reasons to question whether this ever happened in the past, at least in the familiar format of, say, Elf/France and ENI/Italy.[69] That said, it is

---

64 "Diplomatic Briefings: Luanda", *Africa Analysis* 342, 10 December 1999.

65 For US-Africa relations see Schraeder (1995), Morrison and Cooke (2001) and also Zartman (2001). There is no real literature on the Gulf of Guinea but see Valle (2004).

66 The State Department is increasing its staff and intelligence capacity on the continent. See "US/Africa: Spinning the Continent", *Africa Confidential,* 30 May 2003.

67 See IASPS (2002) as well as the very balanced Goldwyn and Ebel (2004).

68 Goldwyn and Ebel (2004).

69 While it is simplistic to say that US companies are independent from government pressures to the extent that they can pursue policies that are contrary to the national

undeniable that the present-day appetite by oil majors towards investment across the Gulf of Guinea is much to the liking of the US government, and fits its undeniably ambitious strategy of escalation in involvement at the regional level. According to an analyst, the widespread government belief is that "the market will provide",[70] i.e. that the oil firms' interest in the region can be taken for granted and needs no micro-management or interference beyond the creation of hospitable conditions like investment access, legal frameworks, etc. Note that in the last two decades, the US energy policy has been that of securing and extending market access to oil supplies, not the state-centric security of national oil supplies in the manner of France until recently or China at the present time (see below). Its greatest concern is therefore to open markets for the private sector in order to enhance the oil flow in free market conditions.[71]

From this perspective, then, US private sector and governmental policies (the "Houston vision" and the "Washington vision"[72]) are both autonomous and mutually reinforcing, not partaking of the same lines of reasoning but jointly pushing for a decades-long presence in the region. The "Washington vision" for the region cannot be characterised as simply warped by the thirst for oil, as assumed by most critics. Important sections of the government are advancing a "carrots and sticks" approach that stresses issues of good governance, transparency and human rights.[73] US bilateral relations with oil producers frequently touch upon these matters,[74] and US officials in the region have consistently and publicly voiced their concerns and worked towards addressing them (certainly

---

interest—as claimed of Gulf/Chevron from 1976-93—there is certainly great autonomy that precludes a forceful juxtaposition of strategies.

70 David L. Goldwyn, Former US Assistant Secretary of State for Energy, paper presented at the conference on "Oil and Gas in Africa", Royal Institute of International Affairs, London, 24-25 May 2004.

71 See Copinschi, Noel and Soares de Oliveira (2004: 38-57) and Noel (2003).

72 Copinschi, Noel and Soares de Oliveira (2004 : 114-9).

73 Author interview with senior State Department official, Washington DC, 25 June 2003. See also James Lamont and Michael Peel, "Human rights issue in focus as US interest in African oil surges", *Financial Times,* 29 October 2002.

74 Vines *et al.* (2005: 35) for instance, note that "The United States has been involved in robust programmes of governance and forthright public statements" in Angola.

far more than most of their European colleagues).[75] Its vibrant NGO community has been central in highlighting and mainstreaming the Gulf of Guinea's governance problems and in constructing strong partnerships with its regional counterparts.

The worthy aims have, nonetheless, ultimately taken the back seat relatively to more material concerns, and this is likely to continue for the foreseeable future. There are four main reasons for this. The first and simplest is the economic imperative of securing the region's oil, particularly in the face of disruption in the Middle East and perceived rivalry from importer states that "do not think in market terms".[76] The second is the Washington bipartisan drive for enhanced partnership with the oil-producing states, which includes an unholy alliance ranging from the Black Caucus in Congress to Republican oil sector-friendly politicians.[77] The third is the considerable leeway that some of these states have in regard to accessing financing (if at a high cost for them),

---

75 A good example is that of Tawfik Ramtoolah, the US Treasury Department adviser to Chad's Revenue Oversight Committee. His role was invaluable.

76 Comment made by a former Clinton Administration official in explicit reference to China (conversation with the author, London, 25 May 2004).

77 This Washington DC support for engagement with African oil producers is reinforced by the subsequent career paths of politicians and diplomats with regional expertise, which often includes the setting up of consulting firms custom-tailored to pro-Gulf of Guinea regime lobbying. At least two former Assistant Secretaries for Africa have been active on behalf of oil producers. Witney Schneidman, a Deputy Assistant Secretary for Southern African Affairs during the second Clinton Administration who had earlier been a prominent Angolan government lobbyist through Samuels International Associates, advised Assistant Secretary Susan Rice on Angolan affairs. Three senior diplomats with ample Angola experience in the 1990s continue their engagement with the country in their private sector incarnations. For a slightly outdated take on the Angola connection see "The business of US diplomacy in Angola", *Africa Analysis,* 17 November 2000; for an example of US lobbying for Nigeria during the infamous Abacha years see Lucy Komisar, "The people who peddle Nigeria", *The Nation,* 25 December 1995. See also "Nigéria: les lobbystes attaquent", *La Lettre du Continent* 341, 25 November 1999, for the reconfiguration of DC lobbying brought about by the Obasanjo election. Paul Beran, "Nigeria's dictator, his Ally, and their access to the US", *Christian Science Monitor,* 18 December 1997, accuses Abacha and his business associate Gilbert Chagouri with spreading "millions of dollars" across Washington, including some for the Democratic National Committee, to curry favour with the Clinton Administration.

which diminishes pressure on the sovereign debt; they are simply not as easily pressured as other developing states. Finally, there is the basic contradiction between building oil partnerships and forcing political change. Numerous policy proposals on these matters claim that a pursuit of the "right policies" (enhancement of democracy, equality, human rights, transparency, etc.) is "not only [...] the right thing for Africa, but will also protect [US] interests in the region".[78] But this claim, so frequent in US foreign policy statements of intent, does not necessarily hold true: long-term investment by US companies has never been prevented by the venality and despotism of the region's rulers, or the poverty of its people, or war. In short, bad governance and destructive politics have never been defining issues, whereas the disruption to oil relationships that can result from pushing reformist agendas too strenuously is much more obvious.

This means that, while it phrases its engagement in reformist terms, and indeed, purses some reformist policies very firmly, the mainstay of US policy will be the preservation of a good rapport with oil-exporting states. The result will be a muddled and poorly implemented compromise on a number of governance areas that may change the way some things are done but not the structure of the relationship. US companies will keep a very low profile in contentious arenas, concentrate on making friends and ensuring business success, and leave most of the "stick" approach to the US government, which will in turn use it sparingly. Multilaterally, transparency will be a key theme, but outside regulatory frameworks. At the bilateral level, critical engagement will be pursued, with the oil relationship tempered by occasional references to good governance.[79] Nigeria and Angola, and to a lesser extent Equatorial Guinea, will remain the US priorities in the region. To the concern of the two would-be regional powers,[80] there has been talk of a potential build-up of US

---

78 Goldwyn and Ebel (2004).

79 Equatorial Guinea, to which a US diplomatic presence has recently returned, is a case in point. The US chargé d'affaires is outspoken about the country's problems but his mere presence highlights the fact that the policy of ostracising the regime has been abandoned.

80 Antony Goldman and James Lamont, "Nigeria and Angola to discuss US plan for regional military base", *Financial Times,* 4 October 2002.

military presence in the region, including a unified command for Africa.[81] This was partly the result of visits to São Tomé by US high-ranking military and civilian officials at the height of the country's tensions with Nigeria, and a general increase in military cooperation due to fears of Al-Qaeda activity in the region.[82] There will certainly be a growing military component to the US presence in the region.[83]

China and other Asian States.[84] Interest in the Gulf of Guinea is also arising from Asia's booming economies, with their unquenchable thirst for oil and fragile energy security.[85] China's mounting demand has taken her from self-reliance before 1993 to the status of the world's second importer of oil a decade later and Sub-Saharan Africa—already the stage for the prolonged China-Taiwan diplomatic tussle for recognition— is one of the key growth areas.[86] Only in 2004, China's daily imports swelled from 1 million barrels to 1.7 million bpd and Africa now provides an estimated 25 per cent of that. China has become one of the key economic partners

---

81 For a good, if slightly alarmist account of US military involvement in the Gulf of Guinea and Africa more generally see Barnes (2004). See also Volman (2003), Jean-Philippe Rémy, "Les Etats-Unis renforcent leur dispositif antiterroriste dans plusieurs pays africains", *Le Monde*, 5 March 2004, "US military commander to visit African oil producers", *IRIN*, 24 August 2004, and Richard Wilcox, "An American proconsul for Africa", *New York Times*, 15 October 2004.

82 In Equatorial Guinea, for instance, MPRI, the leading security company, is on a Pentagon contract to help the country build its currently non-existent coast guard.

83 On the enhanced US presence in West Africa, see Morrison and Lyman (2004: 110-14).

84 This sub-section concentrates on the Chinese experience but is also of relevance for understanding the actions of companies from other Asian states such as India and Malaysia.

85 The reference work on this subject is Ebel (2005). See also Xu (2002) and Manning (2000).

86 See "China's Demand for African Oil Helps Boost Gasoline Prices", *Bloomberg News*, 6 May 2004, Jasper Becker, "Why the world feels China's growing pains", *The Independent*, 8 May 2004, and "China's oil companies: drilling for the Party", *The Economist*, 24 May 2003. The most important growth area remains inland Asia: China was involved in a bitter struggle with Japan to gain the right to build an oil pipeline from Siberia. A leading analyst claimed that "relations between Tokyo and Beijing may sink to its lowest, potentially most dangerous, levels since World War II". See Paul Roberts, "The undeclared oil war", *Washington Post*, 28 June 2004.

of West Africa, behind France but ahead of the US.[87] China's approach to oil is based on a preference for ownership of resources rather than their purchase; this mirrors "a distrust of international energy markets [and a focus] on direct political and economic action by the government and the state companies".[88] In additional to acquisition of crude oil, China's three major national oil companies, Sinopec, CNOOC and CNPC, are rapidly expanding across the Gulf of Guinea's upstream. At an early stage, they took up acreage that is too politically explosive for Western multinationals, as in Sudan, or too "frontier", as in Niger, but their interest in areas such as the Nigerian onshore, Gabon, Angola and Equatorial Guinea is growing.[89] In addition, Malaysia's Petronas is now involved in exploration in 16 African countries and is an equity partner in Chad and Sudan. India's strong economic growth has recently brought it to the Gulf of Guinea[90] and South Korean and Taiwanese companies are also present.

A number of analysts have overstated the "Asian threat" in Africa. At least at the present time, the general standard for oil company activity is so low that one cannot claim there are major differences between the newcomers and the Western companies. Yet it is undeniable that they are not frontrunners of the reform agenda: while Western operators and their home governments can be accused of hypocrisy, the policies of Asian companies are positively illiberal. Asian companies pay generously, have no governance preconditions for engaging with the Gulf of Guinea, and no vocal civil society and shareholders criticising their African partnerships;[91] they are unconcerned with their "corporate image" and rarely bother to talk to the press.[92] In the absence of international regulation, they will use all weaknesses of their richer and more technically able Western

---

87 "Quand la Chine réveille l'Afrique!", *La Lettre du Continent* 457, 28 October 2004.

88 "China: Surging oil demand changes energy scene", *Oxford Analytica*, 26 February 2004.

89 This interest now includes the crisis-ridden refining sector. See "China aims to expand in Nigeria", *Bloomberg News*, May 2004.

90 "The East needs oil", *Africa Confidential,* 28 May 2004.

91 Howard W. French, "China in Africa: all trade, with no political baggage", *New York Times,* 8 August 2004.

92 Daojiong (2006).

counterparts to access commercial opportunities in the Gulf of Guinea and elsewhere. Theirs is a full embrace of the *Realpolitik* agenda. The successful running of Sudan's upstream by Indian, Malaysian and Chinese companies[93] amidst Sudan's civil war has shown their willingness to transgress the limits binding Western investors.[94] It has also shown—for instance, through China's unabashedly pro-Sudan stance in the UN Security Council discussions on Darfur—the lengths to which China will go to protect its investments in a client state.[95] Furthermore, though ultimately wanting to access international capital markets—which, as of now, demand no progressive preconditions anyway—Chinese companies benefit from their government's unswerving political support and are unlikely to be broken by international progressive demands.

They are thus perfectly situated to capture an important share of oil-related economic activity in the Gulf of Guinea—especially the technically unsophisticated onshore production—should the rapport with Western investors and donors become overly dependent on reforms that are not sustainable from the viewpoint of local elites.[96] Already Sonangol, the Angolan NOC, has opened up a Singapore-based Asia oil trading office, and although commercial reasons were pointed out,[97] at least one Angolan

---

93 These are India's Oil and Natural Gas Corporation Limited (ONGC), the China National Petroleum Corporation (CNPC) and Petronas.

94 Canada's Talisman, the only Western firm involved, left last year. Some discordant voices have been sceptical about assuming an inherent clash between the interests of Asian and Western companies. They point to the fact that big players like Shell, BP and Exxon have in the last few years taken to buying stakes in the internationally floated subsidiaries of these companies. See, e.g. "Oil's new world order faces its first test in postwar Iraq", *Financial Times*, 12 May 2003, "China's oil companies: drilling for the Party", *Economist*, 24 May 2003 and "China paves way for 14bn pounds BP oil stake", *Observer*, 26 February 2006. Perhaps Asian companies will in the future (so the argument goes) come to advance oil majors' agendas without the same risk of a Western public relations backlash.

95 Stéphanie Giry, "China's Africa Strategy", *New Republic*, 15 November 2004 documents the manner in which "quietly but steadfastly, China's Ambassador to the UN [...] has helped defang US-sponsored drafts against Sudan".

96 David L. Goldwyn, Former US Assistant Secretary of State for Energy, paper presented at the conference on "Oil and Gas in Africa", Royal Institute of International Affairs, London, 24-25 May 2004.

97 Angola is now China's largest supplier of oil, ahead of Saudi Arabia. See "Oil empires", *Africa Confidential*, 21 March 2003.

senior official referred to the "strictly business" attitude of the Asians as an incentive.[98] Certainly the tying in of the Asian importers to the region can only be to the benefit of local elites in staving off of conditionality or sanctions.[99] An example of this is the significant credit lines opened up by China for Nigeria and Angola which, in the Angolan case, will allow it to remain autonomous from IFI and Western donor conditionality. At the time of writing, the Asian takeover scenario remains hypothetical: Asian companies simply will not be able to afford the estimated $43 billion of oil investment projected for the region until 2010, all but a small part of which will come from the US and Europe. There is even the view that China's global push for resources is not a sign of strength at all, but "an acute awareness of their vulnerability".[100] But it certainly weighs in the planning horizons of Western companies that expect to be in place decades from now, as well as in those of their home states.

International financial institutions. The World Bank and the IMF are the cornerstones of international community action towards the oil producing states of the Gulf of Guinea.[101] This is only secondarily on account of their lending of last resort, technical assistance, and project finance, for oil states do not consistently require IFI money, at least to the extent that other states may. (They need it before the money starts flowing, as Chad did, and later, once mismanagement or resource depletion has led to heavy indebtedness, as in the cases of Gabon and Congo, but not at

---

98 Author interview, Luanda, 23 January 2004. In a February 2004 visit to Gabon (part of the ongoing courting of the region), Chinese President Hu Jintao guaranteed that the strengthening of bilateral ties with the region came "without political strings attached"; see *Angola Country Report*, EIU, June 2004, p. 17, and also "Gabon: Oil pact with China to boost falling reserves", *IRIN*, 5 February 2004. In the aftermath of this visit, President Bongo approvingly said that China's cooperation comes only "with mutual respect and regard for diversity" (quoted in Giry, "China's Africa Strategy").

99 "Friend or forager? How China is winning the resources and the loyalties of Africa", *Financial Times,* 23 February 2006.

100 "China's splurge on resources may not be a sign of strength", *New York Times,* 12 December 2004.

101 For a more detailed discussion of IFI relations with the Gulf of Guinea, see Soares de Oliveira (2003b), a longer version of Soares de Oliveira, "IMF and World Bank Relations with the Gulf of Guinea's Oil-Producing States", paper delivered at the Royal Institute of International Affairs, 20 June 2003. See also Gary and Karl (2003: 44-9).

the height of the oil economy and especially not with current oil prices.) Alternatively, the IFIs matter considerably on two accounts. The first is their role as gatekeepers providing "good housekeeping seals of approval" for bilateral donors and creditors, other international financial institutions and potential investors. Oil states are frequently unwilling to implement their proposals but nonetheless keen to be seen engaging with the IFIs; some respectability sticks to them even if no reforms are taken on board. The second is the way in which unintended developments within the IFIs have fed into the progressive agendas. These range from the Bank's stated goal that its activities must "reduce poverty" and fight corruption, the dramatic findings of its own Research Department on the nefarious management of resource-based economies,[102] to the IMF's mounting concern with fiscal stabilisation and the impact of oil price volatility on economies. In theory, this has led to realignment between the IFIs and some of their fiercest critics. Many civil society actors, while strenuously attacking IFI policies and conditionalities in other areas, are particularly keen on having IFIs flex their muscles and exercise pressure on oil states and oil companies to reform.[103]

There are numerous examples of attempted IFI positive engagement with the region as a whole, in addition to its role in Chad described above. The World Bank has provided funds for NOC audits, improvement in local technical and managerial expertise and introduction of sustained efforts at tackling the environmental and social costs of oil (though nowhere to the same degree as in the case of Chad).[104] The IMF's Article IV consultations and Staff-Monitored Programmes, whether published or leaked, have become the chief source of information on theft and mismanagement by

---

102 This includes Gelb and Associates (1988) as well as a number of publications by Paul Collier and his team on the links between resource extraction, poverty and civil conflict.

103 Catholic Relief Services, for instance, urged IFIs to "use all their leverage, with both companies and states, to strategically promote transparency, fair and accountable revenue management and allocation, and respect for human rights" (Gary and Karl, 2003: 3). Human Rights Watch notes that "whatever one thinks of the IMF's economic prescriptions, its efforts to promote transparency [...] have been an important source of leverage" (Human Rights Watch, 2004: 2).

104 See in particular the work of the Energy Sector Management Assistance Programme (ESMAP).

Gulf of Guinea governments, and the IMF's call for economic reform is often the shrillest around. Both institutions now see the management of oil revenues by producer states as a key governance issue and no longer present the extractive industries as unproblematic. Moreover, there have been policy-specific successes in a number of individual cases.[105] The IFIs are nonetheless caught in many of the same contradictions as other external actors engaging with the Gulf of Guinea's oil economy, as well as some of their own.[106] The recognition of the need for systemic reform therefore "does not appear to have affected operational policy",[107] at least in a consistent manner. Let me point to four fundamental problems.

Firstly, the IFIs cannot afford to threaten oil states with disengagement, for they need the region. As already noted, the Gulf of Guinea is the only area of Sub-Saharan Africa bar South Africa where meaningful private sector involvement and strong economic growth are taking place.[108] The region is thus exceptionally good business for the Bank, especially through its private sector lending arm the IFC. Calling off the dialogue with the worst offenders would have a minor impact on most oil states and investors—Equatorial Guinea and Angola, for instance, have managed fine despite acrimonious relations with the IFIs—but would risk sidelining the IFIs as the key development institutions in the Gulf of Guinea.[109] This means that the rapport between the IFIs and the

---

105 These include the slow acceptance of basic transparency measures by the oil states that are discussed below.

106 For the case of Chad see Uriz (2001).

107 Gary and Karl (2003).

108 The exclusive IFI focus on growth that predominated until the 1990s has been tempered by a focus on poverty alleviation policies. Yet the World Bank still labours under the assumption that "growth must be reckoned to be the principal force in alleviating poverty". See Bhagwati (2004: 64-6) for a typical statement of this argument. This is of course right in most instances, but is wrongheaded when accepted uncritically, particularly in contexts where exponential growth has had no ameliorative impact on poverty. Legal, foreign-led investments that lead to double-digit growth figures are almost inherently worthy of the Bank's attention and support.

109 "The carrot that the IMF and the World Bank hold out isn't big enough for the government", an observer of Angolan politics is quoted as saying. "The IMF is holding out the carrots, and the industry is holding out T-bone steaks". See "Angola forced to come clean", *Financial Times*, 2 October 2003.

oil states is premised on the imperative of continued engagement. This is enacted through a "ritual dance of reforms"[110] that encompasses the opening of dialogue and the upbeat prospects of all involved, followed by disappointments and breakups, followed by renewed dialogue, in which oil states reform little but keep the other side "talking for as long as possible".[111] When states renege on commitments in a confrontational manner, IFIs may react in a steadfast way, as Paul Wolfowitz initially did in the case of Chad discussed above. But pressures soon mount for lenient backtracking: following disreputable revelations about widespread corruption in Congo's oil sector,[112] the Bank threatened to postpone discussions of debt relief, only to grant it a month later without significant further concessions on Congo's side.[113] The Bank did include references to "triggers and conditions", but if these were taken seriously, the agreement would not have been signed to start with.

The second contradiction is that the IFIs resist the adoption of any objective or binding criteria for assessing potential partnerships, be it with states or companies; there are only voluntary codes for partners and broad guidelines for the IFIs themselves. A "black list" of bad corporate actors the Bank will not do business with, created only after a major corruption scandal in Lesotho,[114] has been de-fanged soon after being

---

110 The phrase by Thomas Callaghy (1995) describes the never-ending structural adjustment negotiations of the 1980s and early 1990s.

111 *Angola Country Report*, EIU, May 2002, p. 19.

112 A UK high court judgment on a case brought by holders of Congolese debt revealed that Mr Denis Okana, CEO of SNPC, was the secret owner of two oil trading companies benefiting from the under-priced sale of Congolese crude oil. See Kensington Ltd vs Republic of Congo (2005) EWHC 2684. See also "In oil-rich nation, charges of skimming; Congolese officials said to reap benefits", *Boston Globe,* 25 November 2005, for an investigation of substantial malpractice in the sale of Congo's oil. Another news report linked the "loot of the Congolese national economy" with "Oil for Food-like corruption"; see Claudia Rosett, "Dollars for dictators", *Wall Street Journal*, 19 October 2005.

113 World Bank, "Republic of Congo Reaches Decision Point Under the Enhanced HIPC Debt Relief Initiative", News Release No: 2006/301/AFR, World Bank, Washington, DC, 9 March 2006.

114 The landmark Lesotho court ruling, whereby some of the world's major engineering companies were convicted of engaging in serial corruption in the context of the World Bank-supported Lesotho Highlands Water Project, was instrumental here. See "Plucky

established.[115] There is no commonly agreed threshold beyond which a state no longer deserves finance or assistance. This means that regardless of often stringent-sounding prescriptions, IFI practice vis-à-vis the Gulf of Guinea is muddled and compromised, ridden with inconsistency in implementation, loopholes and exceptions.[116] Contradictions exist not only between discourse and practice but between different practices as well. Let me give some examples. Despite paying lip service to CSR, the Bank went on to bestow an extraordinarily favourable partnership in Chad on three companies (ExxonMobil, Chevron and Petronas) that were amongst the least forward-looking in the sector and have cultivated a deprecatory non-engagement with efforts such as the Global Compact.[117] There are also major contradictions in policy towards different states. The intrusive measures put forward in Chad have not been enacted elsewhere: in fact, states such as Equatorial Guinea have benefited from very empowering technical assistance without any real reforms demanded as a precondition.[118]

The case of IFI relations with the Angolan government shows serious contradictions between the Bank and the Fund that culminated in a 180-degree policy change. The notoriously difficult relations with the country, which never went through a SAP and only first engaged with the IFIs in

---

Lesotho Floors Corporate Goliaths, More Heads still to Roll", *African Energy*, 74, May 2004.

115 The Bank is to offer a conditional amnesty to companies that "voluntarily admit" to corrupt dealings on bank-funded projects. "Under the terms of the new Voluntary Disclosure Programme [...] the bank will not take action against companies that come forward with information about past wrongdoing". Information made available in this way will be granted anonymity. See, e.g. Krishna Guha, "World Bank to offer amnesty over corruption in its projects", *Financial Times*, 7 August 2006.

116 The lack of consistent criteria to measure slippage is much underlined by the literature on structural adjustment and the Bretton Woods reform visions. The result is that, whereas theory points to stringency, practice is much more permissive. See especially Hibou (1998b).

117 The fact that this promising scheme was established with these companies rather than some of the more reformist ones (e.g. Shell, BP) sent a clear sign to the industry that the absence of reforms would in the medium-term not really damage their prospects with the IFIs.

118 See footnote 106 of Chapter 5.

the late 1990s, because of a sharp drop in oil prices, were made worse by the sharply differing IFI approaches. The Bank, which only in 2000 had left the country in fury, changed tack the following year and lobbied hard for improvement of relations. The IMF, on the other hand, kept up a fairly draconian posture towards the financial impropriety of Angola's rulers. By March 2003, the Bank was disavowing the IMF's stance and approving a $50 million loan to the government despite the fact that the latter had implemented none of the preconditions for reengagement set out by the IMF.[119] After a period of mutual hostility over the Angola brief, the IMF caved in, citing the belated acceptance by the government of some measures suggested four years ago, and relations with Angola are now more cordial, despite the almost total non-implementation of the reform agenda by the government.[120]

The third obstacle is the IFI technocratic mindset that depoliticises what can only be described as their very political engagement with the region.[121] Challenges are perceived in exclusively managerial or technical ways and success defined in an issue-specific rather than a holistic, outcome-based manner. In some cases, oil states are lauded for having introduced new petroleum and mining codes, strengthened property rights, accepted civil service reforms, and fought fiscal evasion, regardless of whether the impact of these often apocryphal measures on broader issues such as human rights or poverty alleviation is tenuous or even counterproductive. Formerly very critical of Equatorial Guinea, for instance, the IMF became much less so once the government accounted for its oil monies in the national budget: never mind that these will still

---

119 World Bank, "Angola: Transition Strategy, Technical Assistance, and Postconflict Support", Press Release 2003/274/AFR, World Bank, Washington DC, 27 March 2003.

120 The improvement of relations with the IMF also coincided with the nomination of former IMF official José Pedro de Morais as Finance Minister. As in the past, the government gets a lot of "good will" from the IFIs because of this "reformist" choice, though actual improvements are tokenistic. To say that no actual programme has been agreed upon between the IFIs and Angola is to miss the point of relations with the IFIs from an Angolan viewpoint: Luanda needs to be seen in dialogue with the financial institutions; it does not actually need their financing.

121 Ferguson (1994).

be misspent by this unusually nasty family dictatorship.[122] Issues such as political clientelism and leadership failings—frequently one of the key explanatory factors behind the region's poverty— are mostly off their radar.[123] So is the fact that the reforms they peddle do not necessarily fit the interests of either companies or states, and therefore are unlikely to be taken up by either in the long run.

A final obstacle is the fact that the IFIs are no more empowered than other oil players to single-handedly change the political economy of oil in the Gulf of Guinea. It could be argued that the IFIs are not trying hard anyway: that far from having more ambitious reform blueprints than anyone else, IFI policy prescriptions are remarkably tame.[124] But the point here is that their influence is limited. IFIs cannot be seen in isolation from the interests of leading member states: if these do not want substantial progress, it does not take place.[125] Institutionalists tell us that international organisations may be the creation of great powers but eventually gain a life of their own. This "stickiness" may be a compelling explanation in many instances but it seldom applies when the interests of

---

122 See section two of Chapter 5 and also IMF (2003) and *Equatorial Guinea Country Report*, EIU, January 2004. Compare this with the press coverage of the very critical and never divulged 2001 Article IV consultation.

123 Consultants working on the Angola oil diagnostic, for instance, as explicitly put forth in a contractual provision begging for quotation, "shall <u>not</u> (sic) be expected or required to consider or investigate or conduct any form of enquiry into the conduct, practices, honesty, integrity or standards of, or nature or quality of work performed by, any person who has or may have, any involvement in or connection with, directly or indirectly, the facts, matters or circumstances or events which shall be diagnosed [...]", quoted in Human Rights Watch (2002: 4). The series of studies on the Nigerian oil sector, in turn, hurriedly establishes that its aim is not that of identifying "particular instances of fraud and mismanagement or [assessing] the efficiency of the utilization of funds by ultimate recipients", just in case one should think that to be the crux of the issue. See World Bank (1993).

124 Certainly the pro-business agenda of the IFIs and the instinctive identification of staff with private sector woes in Africa mean that tough prescriptions are rarely targeted at the latter.

125 See Wade (2002) for a general perspective on the US grip over key elements of IFI policy. The turnaround on Angola policy in 2003 was reportedly linked to US pressures in the wake of Angola's support for the Iraq invasion (numerous author interviews since March 2003, Washington DC, London, Luanda).

leading states are at risk. Even if the IFIs were bent on radical reform, it would never take place while key member countries did not identify with it.

At any rate, this issue barely arises because IFIs are *not* bent on radical reform. While the view of the extractive industries as harbingers of prosperity has taken a good beating, the two institutions have not fully accepted that extractive industries often create poverty and exacerbate previously dysfunctional political trends.[126] Why the progressive talk, then? Firstly, because sections of the IFIs are truly committed to reform, though from a warped non-mandatory perspective that much underestimates the complexity and urgency of the issues at stake. Secondly, and more importantly, because IFIs cannot afford to do otherwise, besieged as they are to an unusual degree by militant civil society groups scrutinising every statement that comes out of them. Like the private sector, which responded to NGO criticism by creating its own CSR mantra, the World Bank and the IMF have been reforming their discourse, and to a minor extent their practice, to better face their detractors. But the IFI strategy seems primarily to be a matter of damage limitation and attempted co-optation. Their goal, and particularly that of the Bank, is more that of saving a role for oil in development (and a role in the oil sector for itself) than seriously facing up to the problems. IFIs are caught between NGO pressures, their own research, and common sense, on the one hand, and the imperative to keep channels open with the unsavoury oil economies of the Gulf of Guinea, on the other. They thus adopt much of the progressive rhetoric; progressive actions that are not backhanded, though, seldom materialise.

The limits of benign engagement

The bungled Chad experiment and the contradictions and limitations of bilateral and multilateral policies do not mean that nothing positive will happen. The joint if contradictory Western push for more transparent politics in the Gulf of Guinea has sent ripple effects across the region since 2003. Angola published the Oil Diagnostic study and disclosed the payments

---

126 See section two of Chapter 1 and Soares de Oliveira (2006).

made by Chevron for the extension of rights to the country's Block 0;[127] secretive Sonangol now has a website and the Finance Minister claims the company is no longer used for quasi-fiscal operations.[128] Congo published the KPMG audit of its national oil company, SNPC;[129] Cameroon's NOC has been doing this since 2001. Gabon deposited money into its Future Generations Fund for the first time; Nigeria signed up to the EITI, and so did the Nigerian-São-tomean JDZ. The judiciaries of western states are investigating some corruption allegations involving their companies and anti-corruption legislation has been passed by most. Change is in the air—but will it lead to qualitative change in the web of relationships I have charted in this study, and will it mean a measure of improvement in local lives and human possibilities? In short, will it work?

I am inclined to answer negatively. There will certainly be changes and some practices will be reshuffled. Many actors across the spectrum—including NGOs—are willing to go down the road of reform lite.[130]

---

127 See Carola Hoyos and John Reed, "Angola forced to come clean", *Financial Times*, 2 October 2003 and Heather Timmons, "Angola set to disclose payments from Big Oil", *New York Times,* 13 May 2004.

128 "Oil no longer used for loan payments", *Angola News* 113, April 2006.

129 The improvement of IFI-Congo relations has led to the recent approval by the IMF of a PRGF that goes a long way to normalise relations with international creditors. See "IMF Executive Board Approves US$84.4 million PRGF Arrangement for the Republic of Congo", IMF Press Release No. 04/262, 7 December 2004. The expulsion of Congo from the Kimberley Process shows that the government may be less than sincere in its overall commitment to good governance and transparency. See Nicol Degli Innocenti, "Congo is ejected from Kimberley Process", *Financial Times,* 10 July 2004.

130 There are many civil society organisations willing to work with the corporate sector. The case of the Angola Partnership Initiative (API) is illustrative. The API is a Chevron co-financed $50 million nationwide scheme "to support education, training, and small business development in Angola" and is one of the most important actions in the field of corporate-sponsored community development (author interview with Chevron official, Luanda, 16 January 2004). Launched in late 2002 by the CEO of Chevron himself, its partnering organisations include USAID, the Government of Angola and UNDP, as well as the two top US development NGOs, World Vision and Care. Both have become noticeably less critical of the relationship between Chevron and the Angolan government (personal observation and interviews, Washington DC, June 2003, Luanda, January 2004). In Nigeria, too, Shell has used sympathetic NGOs to implement some of its programmes. In Congo, Total sought to have vocal NGOs change their focus from the

In the 1980s, international pressure did lead to the liberalisation and privatisation of the continent's economies, even if the results were dismal; and in the 1990s, international emphasis on human rights, good governance and universal suffrage did lead to the mutation of the old autocracies into more sophisticated ones.[131] Companies, too, have updated their ways of dealing with the region. One should expect similar reconfigurations of some political and economic behaviour as a result of the current stress on reform. The point is that there will be changes galore, but no tangible improvement.

There is no reason to think that international pressure will be any more successful in achieving aspired outcomes with the new themes, as opposed to a farcical rendition of them, than it was with structural adjustment or good governance in the 1980s and 1990s. Empowered local actors have their own agendas that will not be foiled by (ambivalent) external tampering alone. Oil states will accept reforms when there is no alternative or when their benefits override costs. Some of these apparently accepted reforms will then go unimplemented; others will be partially implemented and will create perverse opportunities through regulatory loopholes. States will participate in voluntary partnerships because they will feel an opportunity to look good without much concrete commitment having to be delivered: this is why all states in the Gulf of Guinea are now engaging with EITI, albeit with varying degrees of depth and seriousness. Furthermore, one should question the "irreversible trend"[132] in transparency compliance: in due course, many of the states that currently play along with the reformist agenda will have seen their revenue flow grow exponentially, and may be less keen on it. In sum, the progressive agenda risks becoming a flawed record of partial adoption of some measures that will go a long way to construct a drawn-out political process, which will in turn amount to very little.

---

national sphere to the "oversight" of the company's on-the-ground activities—for which they would reportedly receive company funding. This was perceived as an attempt at co-optation and was refused (author interview with Brice Makosso, Congolese civil society activist, Kribi, Cameroon, September 2003).

131 See Brumberg (2002).

132 Author conversation with World Bank official with responsibilities in oil sector transparency, Washington DC, 28 June 2002.

This is because oil overrides governance concerns; indeed, it overrides most other sectional concerns. Oil has had a unique grip over the international system of the last century and it will probably continue to have over that of most of this century as well. This piece of received wisdom, accepted for decades, may have been lying low until recently. Consider this excerpt from a "state of the art" volume of essays from early 2001:

The book argues that the chief concern about oil over the next twenty years will be its acceptability, not its availability. [...] It argues that there is unlikely to be any sustained increased in fossil fuel prices [...] Traditional oil issues such as security are giving way to "new" issues such as environmental and social behaviour [...] energy security has moved off the title page and is at best a footnote to today's and tomorrow's global security issues.[133]

Affordable oil; available oil; no concern about security; significance of "new" concerns: needless to say, from the perspective of a post-11 September 2001 world, much of this talk is greatly exaggerated, and most is wrong. New concerns such as climate change and the negative impact of oil on development *should* be central; but politicians utterly convinced that short- and medium-term economic growth and political stability depend on access to oil continue to relegate them to a secondary role. This is particularly the case with fast-growth economies such as China's where preoccupation with the progressive agendas seems as trivial as suggestions that their growth should be underpinned by sustainable energies instead of hydrocarbons. With oil prices reaching an all-time nominal high of $78 in July 2006 and topping the agenda at the G8 Summit held then, security of energy supply is *the* central geopolitical question, not a footnote. This is heightened by crises ranging from the Russian government's arbitrariness towards the private sector and insecurity in Venezuela and Nigeria to the terrorist threat in the Middle East. Stating the centrality of such old-fashioned concerns does not presuppose a Hobbesian view of the international system that, we have reason to believe, may indeed be progressing into deeper forms of co-operation; quite the opposite. The likelihood is that access to oil will remain an arcane "realist" survivor amongst a host of globalised, multilateralised questions: in effect, a trump card that counteracts progressive efforts that

---

133 Mitchell *et al* (2001: 1, 176).

might succeed, were oil not the issue. In this, it shares great similarities with the "national security" partnerships that the US government built with Central Asia's regimes after 11 September 2001.[134] They too evoke overriding priorities that win over softer governance concerns.

Ideas of "energy independence" by importers have been abandoned long ago.[135] It is a matter of consensus that energy provision is to be achieved in the world market. Although the fortuitous presence of oil deposits in some of the world's least appealing locations has been a source of apprehension for many years, attempts at modifying that are a policy non-starter: one cannot change the geological distribution of mineral resources. In ten years' time, energy discussions by the world's decision makers will most probably not be about transparency, or corruption, or good governance: they will be about dwindling supplies, high costs, and importer state rivalry. And perhaps also about a way out of hydrocarbons, if the environmental and political costs of oil start to become unbearable, although one is presently far from such an admission. Reform of the oil economy is coterminous with reform of the world economy, i.e. beyond today's conditions of possibility. Oil is not like sweatshop manufacturing, which can be denounced and reformed without much structural change for itself or the world around it: oil is what it is, with in-built patterns of political and economic interaction; and it is those patterns that the progressive agendas want to change.

Headway is of course not inconceivable. The pursuit of energy alternatives and the lessening of importers' dependence on oil producers, together with the introduction of well-delineated regulatory frameworks, could rein in the perverse nexus of interests that now binds together the main oil players.[136] This would create the context for across-the-board, positive change as opposed to company-specific targeting and

---

134 The most notorious of which was the alliance with President Islam Karimov of Uzbekistan. Following the Andizhan massacre in May 2005, American criticism led President Karimov to oust US forces from the Karshi-Khanabad base.

135 In different ways, both main candidates in the 2004 US Presidential Election resurrected the rhetoric of energy self-sufficiency. See Cassidy (2004) for an analysis of the rhetoric of "energy independence".

136 See Nicholas Shaxson, "Voluntary codes alone will not limit corruption", *Financial Times*, 26 June 2004.

state-specific "pilot projects", which have thus far achieved little. It would allow importers (assuming they care) to pressure producers to distribute the benefits of their oil more equitably. Furthermore, through the creation of, say, transparency preconditions for company listings on Western stock exchanges, this might come to include Asian companies as well, for they too will eventually need access to Western capital markets in order to pursue their investments. Unfortunately, all of this is very unlikely: the reasons highlighted above militate against adoption of these reforms. The engagement of mainstream actors with the progressive agenda seems to be about co-optation much more than real reform: note the penchant at almost every stage for unenforceable, vague, long-term, voluntary "codes", and the explicit refusal of the few measures that could guarantee the outcomes that are supposedly sought. It is nothing short of a leap of faith to think that these timid efforts will develop, in real time, into the framework needed to tackle the festering problems of oil governance in the Gulf of Guinea or elsewhere.

II. THE GULF OF GUINEA AND THE WORLD:
DEPENDENCE, OPPORTUNITY, LIMITS

Commenting on the advent of oil, President Fradique de Menezes of São Tomé and Príncipe said, "this is not going to change the world, but it's going to change São Tomé's world".[137] This remark encapsulates the role of oil in the Gulf of Guinea. Oil and all that stems from it are the region's firmament: the basis for domestic politics as well as external relations. Because of it, and contrary to the criminalisation hypothesis,[138] oil states are marching into the twenty-first century while continuing their insertion in the international legal economy. This takes place through formal institutions, albeit vested by informal goals, and necessitates intricate webs of international complicity. Within these, oil states are no pawns in the hands of foreign powers or investors: the outside may

---

137 See "Nigeria-São Tomé and Príncipe: oil companies bid over $500m for offshore exploration rights", *IRIN*, 28 October 2003, and also Fradique de Menezes' "Mensagem do Presidente da República Fradique de Menezes por ocasião do fim do ano de 2003", 31 December 2003.

138 See section three of Chapter 1.

be "constitutive" of the politics of these states, but it does not stifle their space for manoeuvring. On the one hand, the outside matters because it has sometimes sustained or even saved the internal political order; on the other, it is secondary because "the external resources have been subordinated to domestic political control" and used to advance the goals of domestic political forces.[139] States in the Gulf of Guinea have been capable actors in the international sphere, skilfully employing its resources to boost their standing. In an era when the failed state is a widespread reality and mere juridical statehood no longer merits full international support and assistance, Africa's oil states have managed to keep their international status intact. This section explores how this is accomplished before examining the attendant limitations.

Dependency

The dependence of the oil state on the international system is difficult to overstate; without it, the oil state would simply not exist in any recognisable way. It is from the international system that the state derives the partnerships and legitimacy to keep standing. It is from the international banking sector that oil states derive the eleventh-hour oil-backed loans. It is on the basis of external technology[140] and finance that oil can be extracted. International fads in economic management and good governance are often pressed upon oil states as the only externally acceptable reform routes available. And state reliance on oil alone leads it to depend wholly on this notoriously volatile internationally traded commodity: price changes can wreck havoc or bring in immense wealth, yet both lie outside the power of the state. In these and so many other aspects, it is at the receiving, rather than the initiating, end of the international economy.

None of this is unique to oil states, for relations between developing and developed nations are still marked by asymmetry at the technological,

---

139 I am adapting a point made by Christopher Clapham in a different context. See Clapham (1988: 220).

140 Many observers have suggested that, over the last decade, the technological gap has deepened even further because of Africa's exclusion from the Internet revolution. Castells (1998: 92-5) writes of the "disinformation of Africa at the dawn of the Information Age".

financial, organisational, ideational, cultural, and many other levels.[141] This much had been argued decades ago by the proponents of dependency theory. Partly inspired by the work of Lenin and Rosa Luxemburg,[142] a number of intellectuals, many of them Latin American, postulated that there were fundamental differences between a core of developed nations and an impoverished periphery. These structural, long-term limitations related to an international division of labour and technological differentials that prevented peripheral states from graduating up the ranks.[143] Foreign capital would only develop the part of the economy it needed, such as communications, banking, mining and plantations, "leaving the rest of the economy and society virtually unaffected and unimproved".[144] This resulted in what Celso Furtado called "a dualistic economy" that was self-perpetuating.[145] In turn, it was assumed that long-term decline in the prices of the primary products exported by the periphery meant decline in the terms of trade for developing countries.[146]

Dependency theory provided important insights on the reasons for economic and political marginality in the international system.[147] Its principal weakness was the deterministic nature of the line of

---

141 Badie (2000).

142 See Lenin (1993) and Luxemburg (1972). The turn-of-the-century Marxist critique of imperialism gave prominence to the "desperate" search for raw materials to feed the capitalist machinery. This proved particularly popular with 1960s intellectuals trying to understand the insertion of primarily extractive industry-based economies in the world economy.

143 A lengthy discussion of dependency theory is beside the point here. The reader can count on a number of critical reviews and summaries such as O'Brien (1975) and Cardoso (1977).

144 For a good discussion of dependency theory and other "pessimistic" (sic) readings of the international economy, see Fieldhouse (1999: 32-67).

145 C. Furtado, *Development and Underdevelopment* (Berkeley: University of California Press, 1964) quoted in Fieldhouse (1999: 59).

146 See for instance Singer (1950).

147 As Cardoso explains in a valuable meditation on dependency theory and academia, the clichéd dependency critics decided to engage with was "a straw man easy to destroy", and its dismissal eliminated from the mainstream many pertinent issues on inequality within the international economy. See Cardoso (1977: 15).

argument,[148] which failed to recognise that actor agendas could possibly surpass the structural constraints it identified, and consequently denied the possibility of change. It also failed to give the postcolonial state the centrality it holds as mediator of the relationship between the internal and external realms and recognise the empowerment this brings. These are major faults, for though the dependent relationship is "unequal" in theory, any careful analyst of the last thirty years of oil politics in the Gulf of Guinea knows that the view of local politics as remote-controlled by foreign interests is nonsense.

There are four elements to the relationship between oil states and the international system that disprove the idea of a fixed, economically given dependence and recognise the leeway available to local actors to construct political relationships on their own terms. The first concerns other (wealthier and more powerful) states with which the oil state may entertain a dependent relationship and what Albert Hirschman calls "the disparity of attention" between the wealthy country and the poorer country. Hirschman believes the latter may seek to "loosen or cut these ties" or, at any rate, use them for its purposes, while the former may lack the will for enforcement owing to a pressing concern with more vital interests, such as relations with other great powers. Ultimately, he argues, "the [dependent] country is likely to pursue its escape from domination more actively and energetically than the dominant country will work on preventing this escape".[149] This does not eliminate dependency, but may lead into a "relation of considerably reduced asymmetry", especially if the client state is pliable in the few arenas where the more powerful state holds expectations. Witness the still fitful (if growing) attention to the Gulf of Guinea paid by the US, and compare it with the centrality of US acquiescence for the international standing of these states. Or take, for instance, the case of Gabon and its long-serving president, for long considered France's paradigm neocolonial client, as a

---

148 The exceptions within the dependency theory canon, if one can claim it to exist, are the historically grounded works of scholars such as Fernando Henrique Cardoso, who himself berated the "economic determinism and mechanistic analysis" of "vulgar Marxism". See Cardoso (1977: 8-12) and Cardoso and Faletto (1979).
149 Hirschman (1978: 47).

good example from the other side of the equation.[150] Bongo has used the French alliance to advance his personal interests, both financial and political, ward off challenges for four decades, guarantee the prosperity of his family and clique, and shore up the regional prestige of Gabon. The friendships, connections and parallel party financing provided by Bongo over many years allowed him astonishing access to French policy makers and the capacity to influence outcomes in Paris. Gabon shows how a relationship of economic dependence can give rise to a much more balanced political rapport.[151]

A second element pertains to state-firm relations. As we have seen, firms get better deals out of host states with little know-how or finance. Moreover, at least in theory, the consequences of a falling out are more serious for states than for their business partners: faced with a hostile government, companies could conceivably leave and the state itself would be incapable of replacing them;[152] at most, it could bring in a different set of companies which would play a similar role. Much of the literature on small countries and large companies takes this asymmetry for granted instead of investigating it. In practice, such a straightforward equation is misplaced: once billions of dollars' worth of foreign investment is sunk, companies rarely brandish the threat of renunciation. Oil production is not an outsourced call-centre that can be dismantled and sent off in a few days should the workforce prove unruly; before any oil flows,

---

150 Yolla (2003) provides a good overview of Gabon's foreign policy since 1960, paying particular attention to the role of the President in policy formulation and Gabon's relatively high profile in international politics.

151 For a typical reading of the relationship in terms of French hegemony, see Reed (1987). Most writings on the Franco-Gabonese relationship until the 1990s stressed Gabon's dependence on France. It took the Elf scandals described in the previous chapter to reveal the extent of Omar Bongo's empowerment in the relationship. See in particular Smith and Glaser (1992) and (1997) for a more complex perspective.

152 This is not so in contexts where there is enough local expertise to run the oil industry. Addressing the case of mineral resources in the Latin American context, Hirschman brings up the sinking of foreign investment and the training of a local technical and managerial class ("a group that could run those installations in the event of takeover") as instances of greatly enhanced host state power. But as I have shown in Chapter 4, the threat of a successful takeover is not a credible one in the Gulf of Guinea. See Hirschman (1978: 49).

untold amounts have already been spent.[153] Gulf of Guinea states, in turn, seldom create the political conditions that trigger this zero-sum reasoning on the side of companies, as explained in Chapter 4. There are occasional attempts at getting a more favourable deal or imposing new joint ventures, which oil firms ultimately accept because the starting point is so favourable to them: but nothing that would send the companies packing. In short, the asymmetry between companies and states is visible in two instances: the type of contractual arrangements and the (as yet unheard of) occasions where exit is an option or necessity. But most of the time it is politically meaningless and does not curb the choices of power holders in most areas.

The third element, which stems from the previous one, is the very different goals of disparate external players. Most are uninterested in whole dimensions of international or national policies of the oil state. Oil companies, for their part, have business success as their only goal. Except in a narrow, directly impacting way, as with security or infrastructure maintenance, they have no host state domestic preferences. Other actors such as the IFIs or Northern states do have policy goals in a number of arenas (not all) that impinge upon domestic political choices, though often in terms of form, not substance. But their attention is erratic, their policy emphases can be arbitrary, and the manner in which they seek to bring them about is compromised. Democracy and sound economic policy can be strenuously prescribed, but in most instances they have resulted in autocratic renewal and privatisation in favour of insiders, not the hoped-for outcomes. Rarely does the reformist drive by outsiders have the intended consequences and, as shown in the previous section, it is often ambivalent anyway, with progressive policies getting sidelined by realist concerns. The upshot is that the oil state can, within limits, pursue a variety of policies that are laden with consequences (positive

---

153 Malcolm Keay makes the point at length. "Resource activities typically involve heavy upfront costs in the form of significant capital expenditure and sometimes signature bonuses to gain exploitation licences. Just as normal market disciplines may be absent from the initial location decisions, so they have little impact on ongoing activities; once the initial costs are sunk the effective surplus from the ongoing operation becomes enormous. *Almost any imposition will be accepted to ensure continuation of operations.*" See Keay (2002: 3), my italics.

and negative) for those living under its writ, and in which international actors will be uninterested. It is impossible to list these as they include most key areas of state intervention, including the long-term waging of war, as in the case of Angola.[154]

The fourth and final element is the running of extremely diversified foreign policies. This remains available to oil producers though it is thought of as no longer part of the international policy menu of most African states, at least until the recent arrival of China as a major player in the continent. It includes playing off foreign partners (companies, states, banks, consultants, etc.) against each other or, at least, juggling with different allegiances. In the bygone days of the Cold War, states of the developing world used their affiliation to either or neither of the two camps to squeeze out more resources, but this source of rents dried up with the end of the conflict. Suddenly the leanings of, say, Somalia or the Central African Republic no longer mattered; Western foreign ministries reduced their staff, their aid and their attention span across Africa. This has not happened in the Gulf of Guinea, which is one of the few areas in Africa where talk of "rivalry" and a "great game" is not anachronistic,[155] though the stakes are commercial, not political. The Francophone states in particular use the ambitions of old (French) and new (US and Chinese) patrons to their advantage. Equatorial Guinea has used France's presence to tame the criticisms coming from Spain; the recent Asian interest in Africa is welcomed as a break from the governance demands of the West; and so on. A brief look at the foreign policies of any of the oil producers shows how diversified their bilateral dealings are.[156] While many African countries that do not have oil languish in international oblivion, oil states do not. At the peak of their wealth, oil states can afford assertive gestures difficult to envisage in other circumstances. Angola and Gabon, for instance, can threaten French commercial interests because of displeasure over criminal investigations into arms deals and presidential accounts

---

154 Especially after the 1998 return to war, the international community tacitly or explicitly endorsed the government's search for a military solution for the conflict. See Messiant (2004).

155 Schraeder (2000).

156 The case of Angola is discussed in Soares de Oliveira (2001).

abroad.[157] Presently amongst Africa's chief investment spots, Chad is allowed gestures unimaginable a few years ago: witness, for instance, the manner in which the government declared the French Ambassador *persona non grata* following a spat in 2000. This once-heretical measure was taken for several reasons—resentment at Elf's abandonment of the pipeline project, perceived links of the French Ambassador with the opposition, and President Deby's dislike of the said figure—but was primarily a statement of Chad's standing vis-à-vis France, the former patron and king-maker.[158]

International strategies

The privileged international status of the Gulf of Guinea state stems from its ownership of oil. This said, it is important to understand how often-skilful managing of the international sphere heightens this status, as well as the rewards that derive from it. There are of course considerable differences between the oil states. Other than the size and demographic weight of Nigeria that make it a unique case, two variables seem to account for Gulf of Guinea states' dissimilar international treatment: the size of oil reserves and the size of the country's debt. The bigger the amount of oil, the longer is the list of allies. Conversely, the heavier the debt burden, the more troubled is the hold on power, the more demanding are external actors and the less leeway there is to avoid painful economic reform.

---

157 On Bongo's snubbing of Elf in favour of Energy Africa in the wake of the company's corruption scandals (and his exposure), see "Pas si le Joli", *Africa Confidential*, 9 May 1997. On Angola's tempestuous relationship with France after Angolagate see Fabrice Lhomme, "L'affaire Falcone pèse sur le voyage de M. de Villepin en Angola", *Le Monde,* 19 July 2002, Stephen Smith and Antoine Glaser, "Les hommes de l'Angolagate", *Le Monde,* 13 January 2001 and Fabrice Lhomme, "L'enquête sur L'Angolagate dévoile l'ampleur du système Falcone", *Le Monde,* 24 January 2001. The dispute rages on, with President dos Santos bent on "punishing French companies" in order to pressure Paris to end the investigations; see "Total, Bouygues, Air France... en transe!", *La Lettre du Continent,* 451, 15 July 2004 and "Toujours le chantage sur Total", *La Lettre du Continent,* 2 September 2004.

158 See Frédéric Fritscher, "Le Tchad expulse l'ambassadeur de France à N'Djamena", *Le Monde,* 14 March 2000 and Pierre Conesa, "Le Tchad des crises à répétition" *Le Monde Diplomatique,* May 2001 for critical coverage of the incident.

According to these variables, the oil state seeks to attain the same goals as other fragile states. These are, firstly, that of surviving; and secondly, that of managing good international relations that provide political and material resources for their prosperity with the least possible upset to the governance choices of the elite, which presupposes the boycotting of the progressive agendas.

High profile visits to the West, and especially to Paris, London and Washington DC, are essential elements in the search for international credibility; over the last couple of years, Beijing has come to be included in this list. What strikes the eye is the amount of attention the average Gulf of Guinea delegation gets when in town. A typical Washington sojourn,[159] for instance, will include breakfast with the President or the Secretary of State, meetings with the Black Caucus and perhaps a few Senators as well as fairly acrimonious encounters with the IMF. It will be followed by individual meetings with senior officials (or even the CEOs) of oil multinationals already in the country or those of prospective business partners. It may include lunch with oil traders wanting to access the substantial share of oil output that is marketed by the national oil company. Of late, the afternoon of the shrewder oil presidents has come to comprise meetings with "civil society" actors as well.[160] The day will end with a dinner sponsored by the Corporate Council on Africa where businessmen will reportedly pay untold amounts of money for a few minutes with the President.[161] The whole trip will have been minutely

159 The paragraph is based on about ten years of coverage by the newsletters *Africa Confidential* and (particularly) *La Lettre du Continent*, which have minutely tracked the wanderings of oil presidents, national oil company heads and other top officials of the oil state in the power centres of the West.
160 During his 2000 US visit, for instance, President Sassou Nguesso met the head of the NGO Africare and the head of the International Foundation for Electoral Systems. See "Bel été indien pour Sassou II à New York", *La Lettre du Continent* 360, 21 September 2000.
161 A Corporate Council of Africa-rn event reportedly charged truly outrageous fees for such opportunities: 15 minutes alone with President Bongo cost $50,000; 7 minutes $25,000; and the privilege of seating at the same table "within hearing distance", $15,000. And this is Gabon's head of state, presiding over quickly diminishing commercial opportunities. One can only imagine the fees for an Eduardo dos Santos or an Obiang. See "Bongo, Bush, L'Irak et le FMI", *La Lettre du Continent* 448, 3 June 2004.

prepared by high-profile lobbyists and by the omniscient oil companies, who will use their own impressive DC operations to give a further boost to the event.[162] The effort on the African side is to be reliable throughout: in regard to the companies, contractually; in regard to states, as dependable allies. While seeking a good word with the IFIs, HIPC concessions, AGOA status or the rescheduling of Paris Club repayments, Gulf of Guinea officials will assure their counterparts of their full support for whatever policy issue is at stake.[163] They are often successful in wresting exceptionally favourable treatment out of their interlocutors.

Most of the time, it is external respectability and debt wriggling-room that are the concerns of petro-leaderships, but on occasion it is their very survival that is at stake, or at least, that of entrenched interests that must be protected. In those cases, oil states will exercise their own "oil weapon". When confronted with a string of military setbacks against UNITA, the Angolan government "marketed equity stakes to companies specializing or linked to defense and security services", in effect "tying new investments in its oil sector to arms procurements".[164] Furthermore, it made sure to mobilise broad international support for its war option and bring about sanctions against UNITA that were "a real contribution to the Angolan war effort".[165] Eventually, the sheer superiority in resources available to Luanda ensured victory over the rebels. In the aftermath of his messy takeover in 1997, President Sassou made use of all his old contacts and the distribution of available commercial opportunities to cement his contestable claim to the throne.[166] And the importance of Nigerian

---

162 This is another element of the "discharge" of sovereign functions alluded to in Chapter 2, with states buying the human resources they do not possess to engage in lobbying, diplomacy and litigation.

163 In the case of relations with the US, two recent issues have been the signing of bilateral treaties to exempt American servicemen from ICC prosecution and support for the Iraq invasion of March 2003.

164 "Oil-fired warfare", *Africa Confidential,* 14 May 1999.

165 Messiant (2004).

166 Sassou effectively consolidated his international credibility to the extent that, by the time of the 2002 presidential elections, the 89.41 per cent of votes he claimed and the conditions under which the opposition had to compete were not the object of excessive external criticism. See FIDH (2002) for a critical report on the elections. An EU electoral observation mission listed many problems with the elections but fell short of denouncing

oil was enough to convince Western states not to prod too deeply into questionable practices during the 2003 elections[167] or, for that matter, punish the Abacha dictatorship too harshly in the aftermath of the 1995 Ogoni executions.[168] In key, fragile moments, the alliances built around oil have proven instrumental in ensuring durability.

Other than securing new business opportunities or furthering existing ones, the high-profile trips and the massive and permanent PR investment by Gulf of Guinea governments are aimed at securing their second most important goal: the circumscribing or foiling of the progressive agendas that potentially endanger their governance styles. These can be at best a nuisance and, at worst, highly disruptive. In the first instance, the reaction of most leaderships in the region has been that of invoking "national sovereignty" to cope with rising external demands; the exception here is the Obasanjo presidency, which seems partly committed to transparency. As shown in the previous section, oil states have a penchant for avoiding "the full rigors of external pressure".[169] Yet in the last two years, most leaders have acknowledged that paying lip service to reform is essential for healthy relations with some of the Northern states. Petro-elites have

---

them as a sham. See also European Union (2002). This report contains an unusually detailed disclaimer whereby the European Commission "neither adopts nor approves any of the report's conclusions" (p. 2).

167 According to official data, President Obasanjo's election results in the Niger Delta region (where he is particularly unpopular) were remarkable: he won 96 per cent of the vote in Bayelsa State, 93 per cent in Rivers State, 94 per cent in Delta state and 98 per cent in Cross River State. The victory of the Governor of Bayelsa State (belonging to Obasanjo's party, the PDP), was achieved by 131,335 valid votes, thus exceeding the 129,535 registered voters. The Delta region provided an estimated third of Obasanjo's victory margin over his rival and fellow former dictator, General Buhari. Eyewitness accounts of Niger Delta inhabitants conformed to this blatant fraud (conversations with the author, various locations, 2003-05). I particularly thank Father Kevin O'Hara for a vivid account of his experience. International observers noted gross irregularities but accepted the validity of the elections. For an excellent discussion of the elections and the international reaction see Peel (2005: 3-4, 13).

168 Nigeria was suspended from the Commonwealth and faced minor sanctions by the EU and other Northern governments(hardly behaviour-changing measures, as Zimbabwe shows)but an oil embargo, while amply discussed in the press, was never seriously considered by Western policymakers.

169 Clapham (1996: 204).

understood not only that this new trend must be indulged but also that mere non-binding "progressive" gestures will be enough to satisfy the external constituencies that matter, at least for the time being. Painless gestures such as "endorsing" EITI, for instance, have allowed for a respite in international criticism as Western reformists seek to engage in "constructive dialogue".[170] Typically, this goes nowhere but years can be spent in the process. Furthermore, they can count on a number of sympathetic actors (such as oil corporations, lobbyists) who are keen to secure *Realpolitik* partnerships and will labour to neutralise the scope of reform.

As tends to be the case in Africa, economic and political relations with partners in the industrialised countries are far more important than inter-state relations at the regional level, which are often characterised by incipient rivalry, border disputes and mutual distrust.[171] But efforts have been made over the years to resolve altercations, especially in the resolution of border disputes and, more surprisingly, the transmission of managerial and financial knowledge about the oil industry. (A number of cross-border marriages, especially in the Francophone case, have also increased collaboration.[172]) The mutually detrimental bickering over borders started to be addressed by the creation of a Gulf of Guinea Commission in November 1999. Founded with the explicit purpose of providing a forum for the resolution of oil-related disputes, the Commission's long-term goal is none other than to ensure that, in the words of Omar Bongo, the exploitation of oil resources is "the object of constant cooperation at our level".[173] While it has been suggested

---

170 It is not rare for critics of bad governance in the Gulf of Guinea to be charmed by the smallest of overtures on the part of local regimes and to recoil from overt fault-finding for a period of time lest the "reforming zeal" of their interlocutors waivers.

171 See section one of Chapter 5.

172 Pride of place must be given to the "marriage strategy" of President Sassou Nguesso, whose daughters Edith and Claudia have respectively wedded President Bongo of Gabon and the son of the prominent Gabonese politician and ex-Finance Minister Jean-Pierre Lemboumba. Relations with Gabon in particular have flourished since. In 2004, there were even (false) rumours that President Joseph Kabila of DRC had married Sandrine, another of Sassou Nguesso's daughters.

173 At the same event, President Obasanjo of Nigeria called for the need to avert "the potential sources of conflicts generated by unhealthy competition that could arise

that the Commission is a spurious construct (most breakthroughs have occurred bilaterally or with the support of oil companies, and political actors across the region make scarce references to it in everyday politics), it encapsulates the will to transcend outstanding problems concerning oil. Of course, the real action takes place informally and bilaterally rather than multilaterally, in particular at the level of national oil companies and their sharing of all manner of oil-related information.[174] The upshot of regional relations is that, outside border disputes, the degree of inter-state friction is not particularly high—sovereign neighbors have least to fear each other.

Commentators have noted the penchant of oil states for megalomaniac ambitions in the international sphere, with Nigeria craving great power status, Angola aiming to topple South Africa as regional hegemon, and Equatorial Guinea (rather improbably) aspiring to lead Central Africa. Yet few have noted that oil states do not just peddle their empowered status. Paradoxically, their international (oil-induced) status is so assured that, rather than hiding the weaknesses of their internal sovereignty, power holders seek to access new sources of funding and new chances of discharge on the basis of them. Their partners in the international sphere ask them to be believable external entities and do not question their credibility on the basis of internal outcomes. Therefore, Gulf of Guinea states explore the *à la carte* possibilities this allows. These include the joint, non-contradictory embrace of the language of sovereign empowerment and external recognition, on the one hand, and the bonuses of failed statehood such as the sub-contracting of welfare tasks to NGOs and UN agencies, on the other. For long a medium-income country on account of oil rents and a small population, Gabon tries to present itself as much poorer than it really is in an effort to earn the advantageous HIPC status and its possibilities of debt relief. The Angolan government pursues *Machtpolitik*, regional intervention and great power pretensions while

---

between us". See "Gulf of Guinea Leaders Meet on Peace, Stability, Oil", Agence France Presse, 19 November 1999.

174 This is discussed in section two of Chapter 2.

expecting the WFP to feed almost two million of its citizens on an everyday basis almost three years after the end of the civil war.[175]

Political possibilities and the end of oil

It is undeniable that the political economy of oil has been full of consequences for the survival of the state, its international standing, and the types of accumulation, political staying power and elite reproduction at the disposal of leading political actors. It has also contributed to limiting the progressive impact of outside reform drives. But what are the political possibilities open to these societies when the constraints and opportunities created by oil are weighed? Can a sustainable political order be created out of oil wealth, its institutional patterns and political behaviour? Does it make sense to see it in foundational terms, as the cornerstone of a developing idea of the state?

It is clear that many of the problems analysed in this study stem from, and feed into, unaccountable and despotic politics. (The Gulf of Guinea's regional experience with democracy is very recent and highly flawed, and it includes some of the world's nastiest regimes.) It is also clear that transparency and real democracy are inherently desirable outcomes for the Gulf of Guinea. The problem arises when they become the panacea presented by policymakers for the predicament of the tyrannical, self-sufficient and opaque oil state. If previous experience of petro-states that are or have been democratic is anything to go by, there does not seem to be an irrefutable link between democracy and development, or at any rate, between democracy and the lessening of rent-seeking behaviour centring on the state.[176] The case of Nigeria's Second Republic

---

175 It may seem that the Angolan government's keenness for the language of sovereignty is absent from such an important dimension of internal sovereignty as the feeding of an estimated 10 per cent of the population. But it comes back with a vengeance: WFP flights were recently grounded because they were not paying user charges to the state airports agency. See "Red tape could ground humanitarian flights", *IRIN*, 21 October 2004.

176 This is indeed one of the most sobering lessons of Africa's wave of democratisation of the early 1990s. The rentier nature of politics was not fundamentally challenged even in states where democracy succeeded in a durable way. Karl makes this point in her study of post-1958 Venezuela, which has been called a "pacted democracy", that is, one in which the democratic outcome comes about as a consequence of an intra-

(1979-83) under Shehu Shagari is sobering, for this democratic interlude managed to heighten the degree of rent seeking, contract huge debts and exhaust the resources of the state to the extent that it ensured its own demise within four years.[177] Furthermore, the recent experience of semi-democratisation in Nigeria, Gabon and Cameroon shows that election years broadly coincide with the moments in which budgetary lines are utterly ignored, unfavourable loans are contracted, and spending skyrockets. The electorate (or, at any rate, the part of it that matters) is both fickle and persistent in its claims on state resources and the timely distribution of some of the latter around election time has come to form an important part of Biya's, Bongo's and Obasanjo's grip on power. A democracy that does not begin to address the very nature of the oil state and the quality of its domestic politics cannot be expected to succeed or even take root, and it will certainly not be the qualitative step forward in solving the realities of oil dependency and underdevelopment.

Though sound short- and medium-term economic management is central in addressing macro-economic instability and the towering debt burden,[178] a meaningful departure from the oil-dependent condition can only be achieved through a significant diversification of domestic economic activity. Yet working within the realm of present conditions and limitations, diversification of economic activity seems quite unfeasible in most cases, at least to the extent needed to liberate the economy from dependence on oil revenue. There are huge regional differences in economic potential. But nowhere can one find appreciable options, even

---

elite agreement in regard to the sharing of spoils. According to Karl, such arrangements also favour containment of political conflict outside the immediate elite circle through "preemptive inclusion" in the form of extension of benefits to key constituencies (say, oil trade unions). This amounts to furtherance rather than restraining of the petro-state. Karl (1997: 93).

177 See Joseph (1987).

178 The management of oil wealth is the main avenue of IFI policy prescription for sorting out the Gulf of Guinea economies; see Katz *et al.* (2004) for a typical statement. Other voices (including those of Jeffrey Sachs of the Earth Institute at Columbia University, and Arvind Subramanian, Division Chief at the IMF speaking in his personal capacity), suggest the direct distribution of oil proceeds to the people as the only way of addressing the oil curse. For an example of this as applied to post-Saddam Iraq see Birdsall and Subramanian (2004).

if the state was willing to expand its fiscal net and, generally speaking, play a "developmental" role. Firstly, the prices for alternative, mostly mineral or agricultural[179] commodities that may be available are often much less favourable now than they ever were, and tax receipts from these will likely be very modest in comparison with those of the oil sector. As other states that export highly valuable, much sought-after commodities such as oil or narcotics have discovered, the incentive structures to get rid of a profitable, if cursed, sector and embark on the uncertain buildup of less profitable alternatives are very weak indeed. Furthermore, social changes since the advent of oil may well preclude some of the better options, especially in the agricultural sector. In places such as Gabon or Congo, policies premised on a return to the land will have to cope with a population that is now mostly composed of urban dwellers.

Secondly, the risks of investing in the region are very high. Because of the peculiar nature of the oil sector, oil multinationals do not perceive risk in the same manner as other investors, but once one leaves its confines and those of associated businesses like security, the normal rules apply. Corruption, poor or non-existent regulatory frameworks and the tiny size of domestic markets suddenly matter again. The foreign private sector generally does not want to get involved in productive sectors of the economy, especially those that are labour-intensive and involve a high degree of technical and infrastructure prerequisites that are thought to be absent. In most there simply are not enough people around with the education and training needed to man a complex organisation such as the bureaucracy of a modern state or a firm. There will always be a steady trickle of businessmen interested in the region, but this is likely to favour consumer-oriented import-export activities. The small indigenous private sector, which can scarcely be said to exist outside Nigeria and Cameroon, is not an alternative to FDI because it tends to be unproductive, rent seeking and incapable of real investment. And finally, most other sectors are not internationally competitive to start with. The Gulf of Guinea states present no obvious comparative advantages outside the extractive industries. They lack skills, infrastructure, bureaucracies, communications, sizeable internal markets (except Nigeria), etc.—and

---

179 Exceptions include raw materials recently sought by Asia's booming economies: Chinese demand for copper, for instance, has led to a veritable boom in Zambia.

they lack this even in terms of what existed in the late colonial and early post-independence periods. Even those with a past of greater economic diversification and means such as Angola cannot count on an economy and infrastructure that have withered away.[180]

What will remain, when the petroleum is exhausted, as it must as the finite commodity that it is? Most possibly nothing, other than the bruising experience of dreams had and opportunities lost and the memory of mad spending. Some believe that, contrary to locations such as Liberia, Sierra Leone or the DRC, where no clear idea of the state seems to exist any more, oil is "constructing" state-centric projects, state-bound elites, state-driven international politics. Contrary to the average African state, it is argued, the Gulf of Guinea oil exporters have both the means and the willingness for "stateness", if not for the creation of admirable institutions: here, at least, there are ideas of the state, expectations of a political order of sorts. "State formation" is at play.

This superficially convincing interpretation, I think, is wrong. The analysis of the petro-state shows that the institutions it creates are the first casualty when resources dry up. The analysis of commodity booms shows that what follows is a slump. The frontier town is abandoned; attention moves elsewhere. The institutions and habits built on the strength of boom money, because they are coterminous with it, are rendered obsolete by its disappearance. In some analyses, almost everything participates in state formation,[181] but this notion is much too vague: it simply means the manner in which past experiences configure the field of possibility for the present. Taking a slightly more robust notion that presupposes a "constructive" role in creating, if not "development", then at least long term political sustainability, it is clear that, on the one hand, the experience of oil is constitutive of the present and future of these societies. On the other hand, it is unlikely to create the basis for political order or expansion of the field of human possibility and dignity. For there is, unfortunately, one type of state building experience where the future of the oil states of the Gulf of Guinea is likely to lie: the

---

180 Most of the infrastructure and whole sectors of Angola's pre-1974 economy such as coffee growing are now beyond repair and would have to be rebuilt from scratch.

181 See my critique of the abusive use of "state formation" in section one of Chapter 1.

"state building fiasco".[182] Institutions, decision-making frameworks and political imaginations are created in a swirl of oil windfalls, and they are gone when the wind blows the other way, in a pattern reminiscent of Ibn Khaldun's descriptions of the cyclical (and sterile) state creation and destruction in medieval North Africa.[183] The point is not that nothing will ever change, but rather that there is little in present-day tendencies that we can observe and assess with any objectivity as implying an auspicious turn in Gulf of Guinea politics in the medium term, all things being equal. Modestly positive trends such as the recent growth of Angola's postwar economy, I argue, are epiphenomenal, and do not contradict a long-term trend towards decline.

It is undeniable that, in these circumstances, the demise of the political economy of oil will have a tremendous de-structuring effect. One need not speculate on this, for something of the kind is already taking place in Gabon, where oil production has fallen by a third in only six years.[184] In contrast with the "hot" oil states of the region, exploration results there have been very disappointing, with the only growth areas in small, marginal fields run by independent companies.[185] This is so despite frantic attempts at increasing the fiscal base while promoting ever more appealing terms for oil investment. Unimpressed, the majors are slowly backtracking from Gabon: once the centre of Elf's regional strategy, it is now a comparatively minor outpost with little of the enthusiasm for the new frontiers.[186] President Bongo's elaborate and expensive patronage game that has preserved Gabon from the ethnic strife of most of its neighbours is now at stake.[187] Xenophobia (a recurrent theme in Gabon[188])

---

182 Martinez (2000).

183 Lacoste (1998).

184 According to the US Energy Information Administration, this is from 371,000 bpd (1997) to 243,000 bpd (2003), a fall of 34 per cent. Although oil production there is declining even faster than in Gabon, Cameroon will not face the same level of disruption as its economy was never dependent on oil to the same extent. See Hugon (1996).

185 N. Shaxson, "Waning Star", *Business in Africa*, May 2000.

186 Many have noted the rapid disengagement of Elf from Gabon, "which signals [the company's] overt interest in other regions in Africa, particularly Angola". See *Jeune Afrique Economie* 312, 3-16 July 2000.

187 "Un système à bout de souffle", *La Lettre du Continent* 441, 10 February 2004.

188 See Gray (1998).

has increased with the diminishment of economic opportunities and the dashing of hopes of "thousands of young men who dream of state jobs".[189] Gabon, which has thus far avoided "excessive dependence upon the IFIs", will be less insulated from their demands.[190]

The current high prices have meant that this substantial decline in production has had only a negligible impact on rents. Yet the decline is so steep that even this will not prevent a fiscal crisis in the near future. Technological innovation and plain luck have before come to the rescue of Gabon, and despite occasional efforts at preparing the "after oil",[191] locals still cling on to the possibility of an eleventh-hour deliverance. In the early 1980s, gloomy talk of the end of oil was put aside when Shell found the gigantic Rabi-Kounga field. In 1986, the crisis over the collapse of oil prices was very serious but the country and the regime survived it.[192] The problem this time is that diminishing exploration by companies means that spectacular findings are less rather than more likely by the day. Owing to its low population and consequent high GDP per capita, Gabon is considered a middle-income country and does not qualify for the IMF's Poverty Reduction and Growth Facility (PRGF) funding or for debt write-offs under the HIPC initiative. As other states in the region enter their petroleum age, Gabon may soon find itself the hapless pioneer of a post-oil economy devoid of viability.

Conclusion

Oil has afforded the Gulf of Guinea a presence in the world economy that virtually no other region of tropical Africa can match. While the rest of the continent is either forgotten or branching into illegal linkages to the outside or passively receiving its debris (as is the case with the

---

189 Atenga (2003: 122-3).

190 Atenga (2003: 123)

191 "L'après pétrole", in the French original. Attempts at diversification are mainly in other extractive industries, including iron ore and manganese. More recently, Gabon has received FAO funding for a "fish revolution": see "Government plans fish revolution with UN funds", *IRIN*, 29 July 2004 and also "Au bout de son pétrole, le Gabon cherche des idées", *Libération*, 8 December 2003.

192 See "Le Gabon face à la crise pétrolière", *Marchés Tropicaux et Méditerranéens*, 27 June 1986 and "Gabon's oil refuses to run out", *Financial Times*, 24 August 1989.

deadly availability of small arms), the Gulf of Guinea is in possession of today's most vital and sought-after commodity. For oil importers, the Gulf of Guinea remains a "zone of extraction",[193] a one-dimensional space of resource exploitation feeding technological change and ever-wealthier lifestyles elsewhere: an "Oil Coast", following in the footsteps of so many other West African commodity coasts of old.[194] For oil states and their leaderships, the international system remains a source of wealth and power, and one fantastically grander and more valuable than for many of their neighbours that do not have oil. Yet the oil sector has a very limited impact on the local economy and goes on excluding populations trapped in pre-capitalist bare subsistence and with no discernible link to global markets.

In the twenty-first century, oil endures as the underpinning of a global economic order that is in fact less "new" that often claimed, based as it is in the continuation and (in East Asia, for instance) expansion of the oil-fuelled industrial civilisation.[195] The pattern of political and economic interaction between the international system and the oil-producing states of the Gulf of Guinea is defined by the oil needs of the world economy, the region's plentiful reserves and the desperate search for economic viability by local elites. All actors involved are locked into this relationship, albeit in different manners: it is unlikely that it will be structurally upset in any meaningful way while these three conditions stay in place. This means that the next decades will most likely be characterised by the deepening of this political economy, increasing mutual reliance (especially from the viewpoint of oil states) and the furtherance of the calamitous governance record. One day—when oil is exhausted or is no longer craved—the

---

193 Cooper (2002: 130).

194 Awareness of the *longue durée* should not come at the cost of ignoring the comparative newness and peculiarity of the oil economy. Yet the Gulf of Guinea's dependent relationship with the world economy can only be fully grasped if understood historically, i.e. in terms of its integration within global capitalism over the past five centuries. As several authors have noted, the proliferation of coastal potentates in the pre-colonial period, where local elites prospered on the basis of facilitating the extraction of slaves and other resources from the hinterland, has important similarities with the present time.

195 See Harvey (1990) for the argument that the postmodern moment is but an instance of late industrial capitalism and therefore, strictly speaking, a part of late modernity rather than something qualitatively different and new.

Gulf of Guinea's relevance to the outside world will again decrease, and it may find itself hurled once more back into the tenuously connected "*Afrique inutile*". While it lasts, oil will set the terms of the Gulf of Guinea's insertion in the world economy.

CONCLUSION

This book has investigated the key paradox at the centre of present-day Gulf of Guinea politics. The multilevel crisis festering throughout every one of the region's states ought to cast doubts over their sustainability, the political survival of ruling cliques, and the willingness of outsiders to consider any sort of capital-intensive, long-term commercial involvement. However, the presence of petroleum radically changes this equation: the negative dynamics of state failure and widespread violence affect the general population but spare the oil nexus. The material and political resources made available by oil allow the state to survive regardless of bad policies, permit elite material success regardless of reckless management, earn international allies regardless of erratic domestic conduct, and make companies want to invest regardless of risk. The recent oil boom only strengthens this paradoxical viability: making possible what is arguably the largest inflow of resources into Africa in history, it is of a different order from the short-term viability afforded by the exploitation of other natural resources. Nonetheless, the venal character of the partnership between insiders and outsiders that permits the extraction of oil is not conducive to positive long-term outcomes in institution-building or broad-based economic growth. Highly dependent on uninterrupted money flows and beset by various destabilising trends, the political economy of oil in the Gulf of Guinea is poised in a state of "permanent crisis".[1] This conclusion outlines the most important features of this vast, unfolding process and explores its major implications.

---

1 Hont (1994).

The oil state

From an empirical viewpoint, Gulf of Guinea oil states are near or actual cases of failure, bringing together the worst pathologies of African postcolonial states with those of oil producers in the developing world. But because the overriding interests of multiple internal and external actors are premised on its skeletal existence, the state is constrained to go on enduring, despite its ebbing for most practical purposes. The unwavering international need to have sovereign partners underwrite the exploitation of oil means that its tenuous existence is buttressed by international support from all quarters. For their part, elites understand that the state is their vehicle for the external recognition that allows access to oil rents and, even while neglecting or redefining many of its dimensions in practice, continue to cultivate a state-centric discourse. The oil relationship thus conditions the way the Gulf of Guinea state is represented as well as the calculus of its survival.

The present-day oil state is a hollowed-out version of what it was three decades ago. Although mismanagement, patrimonial conduct and corruption were always at the heart of the oil state, the post-1973 boom saw attempts at shaping institutional structures and societies through the disbursement of oil revenues. In the intervening years, however, the state's mission, ways of wielding power, institutional goals and capabilities have changed considerably, whether discharged onto non-state actors or simply abandoned. An astonishing measure of neglect has set in, with many of the white elephant projects of old falling into irreparable decay. Large sections of nominal state territory have ceased to be administered from the centre; in fact, only a comparatively small portion of territory—"useful spaces" such as oil-producing enclaves, elite home towns and national and state capitals—is subjected to anything resembling an everyday state presence. The populations of the oil state's "useless spaces" have either migrated to the cities in search of prosperity or to flee war, or stayed behind, away from the gaze of the state. In addition to this territorial exclusion, there is a process of human separation between the vast majority that is cut out from even oblique benefits of the oil bonanza and the small number of those who benefit somewhat or considerably. Throughout the region, the provision of public goods by the state has withered, and other entities such as NGOs, the UN system, churches and even oil companies have imperfectly taken up that role. This contraction of the scope of the state is due both to capacity decline and to the fact that empowered political actors no longer believe that many of the functions hitherto exercised by the state should go on being ascribed to it.

However, there are three significant dimensions in which the downsized oil state goes on working fairly well: those essential for domestic participation in the oil partnership, maximisation of the opportunities it affords, and guaranteeing that incumbents are secure from armed challenges. The first dimension pertains to negotiations with all manners of foreign business interlocutors and is in most cases achieved by the NOC. NOCs deal on a day-to-day basis with oil firms, traders and bankers—with whom they strive for reliability—and are the key policy-makers, planners, tax collectors and technical supporters for the government. In the Gulf of Guinea context of state implosion, some aspects of NOC capacity are notable. Despite flaws, NOCs are essential to keep the downsized state and its end of the oil partnership running in a strange but seemingly successful marriage of financial expertise and oil markets savvy-ness with the narrow enrichment goals of failed state leaderships. The second dimension is the access to external finance or goods (e.g. weaponry) through the mortgaging of future oil production. Though the terms of such deals are onerous for the state, oil permits states in possession of it to bypass many of the credit limitations that other weak or failed states cannot. The third dimension pertains to the continuing vitality of the means of violence as well as their multiplication into privatised armed forces, militias, private security companies and other such entities.

Elites

The only domestic beneficiaries of the oil economy are elites who have found therein an international partnership that allows them to supersede internal decline and gain access to vast resources. This is achieved through legal transactions of the world's most strategic commodity rather than the unrecorded or illegal exchanges that increasingly predominate in Africa's external economic relations. In the oil states they run, clientelistic networks have shrunk to include only a minority of the population, as the state's fiscal intake and sources of legitimacy reside in the international system rather than at home. In analysing the role of elites, it is wrong to give exclusive attention to the institutional lock-in and external constraints that supposedly cast the oil nexus in stone, and neglect the role of actors: agency, as well as structure, matters greatly. Oil is almost always the bearer of governance patterns and external linkages that are negative for the majority, detrimental to sustainable institutions and limiting of policy leeway. This said, it is undeniable that reasonably sophisticated local elites behave in a manner that is damaging for the population but wealth-enhancing for them. They take consequential, if often constrained, decisions and are no one's puppets. Responsibility for the dire poverty of the oil state and for current human indicators—frequently lower than at independence—belongs as much to elite mismanagement and theft as to structural disadvantages such as oil price volatility or the so-called "resource curse".

Elites, for their part, are spared the hindrances their methods have created. By using the state as a tool of private accumulation but never taking full possession of it, elites can enjoy the prizes of the state without being burdened with its liabilities.[2] In this context, even the worst developments across the region such as indebtedness or high-intensity conflict can become profitable resources in private hands. There is, however, a fundamental weakness to Gulf of Guinea elites: their political project does not go much beyond resource control, short- or medium-term enrichment, and the survival of individual ruling cliques. There are no real mechanisms for power transfer, if one excludes the (fragile) ongoing democratic experiments. There are no unifying ideas for the state and no vision of national political communities. As they are shaky coalitions premised on illicit accumulation, few of the region's elites have an *esprit de corps* that is sturdy enough to stave off challenges from the many, inside and outside the state, who wish to access their share of the "national cake". Pressure is thus never-ending. Yet, on account of the resources they control, incumbents are unusually resilient.

---

2 See section I of Chapter 3 for this argument.

Oil companies

Multinational oil companies, along with numerous affiliated businesses such as service companies, security providers, and the major western banks that launder stolen revenues have likewise greatly benefited from their engagement at the regional level. In comparison with elsewhere in the developing world, state-firm relationships in the low-tax Gulf of Guinea are stable and even amicable and have provided the basis for oil exploitation for decades. Although very little FDI flows to sub-Saharan Africa, the biggest share of that goes to the Gulf of Guinea oil sector. Oil companies active there are well networked, long-established actors that are indispensable for the viability of Gulf of Guinea states to an unprecedented extent, providing up to 90 per cent of tax revenue in some cases and having absolute control over the technology, finance and know-how in all. Furthermore, they play a pivotal role in burnishing the international credibility of their sovereign partners by lobbying home states for favourable bilateral policies, toning-down of governance and human rights criticism and intervention on their behalf with the IFIs. Far from undermining the sovereignty of weak states, oil companies are their single most important source of internal "stateness" and external status.

The argument pursued in this study is not an anti-corporate one. It is undeniable that appropriately regulated FDI is a positive force, and that oil companies will be "only as useful or as dangerous as the host-state allows [them] to be".[3] Modish emphasis on CSR notwithstanding, multinational firms remain profit-seeking entities; they are neither good nor bad. But corporations can impact badly in the context of weak or failed states when partnering with predatory elites with no interest or capacity to pursue enlightened policies. The incidence of such *Realpolitik* partnerships is highest in the extractive industries and in non-labour-intensive economic sectors, where they hold scant chances of benign outcomes for the majority. The depth of the governance shortcomings in the Gulf of Guinea is primarily attributable to its internationally recognised states, not to an oil industry that is geologically bound to operate where oil happens to exist. But by unproblematically partnering with domestic actors essentially driven by the misappropriation of state resources,

---

3 Fielding (1999: 267).

companies provide the financing for the pursuit of particular policies with tragic consequences for the population. These policies, and the survival and prosperity of those enacting them for their own benefit, could not conceivably be pursued in the absence of the material and political resources made available by oil companies. In addition to the popular hostility in oil-producing areas and the burdensome accumulation by companies of obligations states fail to fulfil, there is a growing awareness in the West of oil firms' complicity in oil state governance. However, this has not changed the profitability of their engagement in the Gulf of Guinea, which will continue.

The oil nexus and the international system

This unfaltering interest attests to the way in which the Gulf of Guinea's present-day insertion in the international economy is different from that of the majority of African states. Bayart is right in noting that practically all, oil-rich or not, increasingly do so "through economies of extraction or predation in which many of the leading operators are foreigners".[4] But a leveling approach that does not differentiate diamond smugglers and warlord scavengers from multibillion-dollar, legally mediated engagements that are sustained over many decades fails to gauge the specificity of the oil rapport. Petroleum is a substance like no other by virtue of the revenues it originates, its state-centric character and its unique condition as the lifeline of industrial civilisation for over a century. Because of oil, foreign governments that import it have not relegated Gulf of Guinea states to Africa's post-Cold War marginality, and their ECAs remain generous supporters of investment there. Their fear of over-reliance on Middle East energy sources and the search for alternatives to it consolidates this interest. IFIs have been equally enthusiastic about oil sector investment in the Gulf of Guinea, with occasional concerns about transparency and good governance rarely playing a decisive role in their actions. More generally, many Northern actors have been torn between the progressive agenda that has appeared in response to the governance tragedies of the Gulf of Guinea and the *Realpolitik* agenda that sees security of energy supply and good relations with oil producers

---

4 Bayart, Ellis and Hibou (1999: 114).

as a priority. However, most attempts at reform have failed or gone unimplemented and the *Realpolitik* agenda remains the default option for policy-makers.

While not of the calibre of the Persian Gulf's, the oil wealth of the Gulf of Guinea is very significant. Investors in the region are far more sheltered than often assumed, and the incentives are of such magnitude that it is almost inconceivable that they will shy away from deepening involvement. This is why outside state and non-state actors, individually and collectively, across numberless venues and contexts nurture and invigorate the Gulf of Guinea oil nexus in general and the sick states at its centre in particular. This sovereignty-enhancing intervention occurs not because of conservative anxiety by the international community about state collapse, which is no longer a rarity, but for contextual, self-interested reasons. How else could one explain the centrality of the foreign private sector in this propping-up of the credibility of local actors and institutions?

The Gulf of Guinea political economy of oil and the partnerships it has elicited, while unique, are of great relevance for understanding the insertion of failed states in the international system more generally. Faced with dismay only a decade ago by an international community unsure of how to deal with it, state failure has become such a widespread occurrence that it no longer ruffles the establishment, even if sovereign de-certification remains a taboo.[5] The unexceptional character of failed states also means that diplomatic and material resources are not automatically made available to rescue them. Alternatively, individual failed or very weak states barter in the external arena the material and/ or political resources they possess in exchange for material and political support by the international system. The international system thus empowers failed states on an issue-by-issue, *à la carte* basis. Failed states that persevere on the basis of illegal or disruptive activities may do so in the interstices of the international system, benefiting from their nominal sovereign status but little else and permanently living in the expectation of international vilification. States such as those in the Gulf of Guinea,

---

5 The undeniable links between some failed states and a host of destabilising forces in world politics may have led to occasional international reoccupation but have not impelled a consistently interventionist role for outside powers.

which are in possession of an invaluable commodity and present very few failed state dynamics that are genuinely disruptive to the international system, will be sustained by sympathetic partners worldwide. The case is similar for states which can bring to the international system political resources—for instance, support in the war against terrorism—that will occasion their transformation from obscure backwaters into worthy allies.[6] In addition, the Gulf of Guinea experience shows that, far from being a species of endgame, state failure is an internal cataclysm that need not have any major impact on relations with the outside world. State failure in some contexts is the continuation of state politics by other means rather than its demise.

The study of the politics of oil in the Gulf of Guinea also shows how many of the dichotomies that structure widely held perceptions of international order turn out to be meaningless on closer inspection. These include the supposed contrast between legal activities, developed or optimistically developing economies and globally integrated spaces, on the one hand, and the illegal, the exploitative, the dysfunctional and the marginal, on the other. This study shows that a regional setting of unambiguous institutional decay, recurrent warfare and widespread impoverishment can also be the privileged ground for successful and sustained partnerships between disreputable elites and respectable members of the "proper" international community. The oil company executives, bankers, accountants and lawyers who haunt the lobbies of five-star hotels in the region are as constitutive of today's Gulf of Guinea as local actors. Resources that are criminally siphoned off by incumbents materialise in the West's sensible investment portfolios and tax havens.[7] And it is the Gulf of Guinea, and areas like it, that literally lubricate and feed the engine of world capitalism. A region long forgotten, or peripherally dismissed, by commentators with a transatlantic bias passing for global concern thus materialises as a key stage for the enactment of many themes that pervade world politics today. As Etienne Balibar notes, the lawless borders of the civilised world turn out to be "central" for its

---

6 This is the experience of a number of Central Asian states since the 11 September 2001 attacks.

7 See section two of Chapter 3.

constitution.[8] In this sense, the present study is part of a much wider rethinking of the insertion of peripheral locations in the international economy and of current debates on globalisation and marginality in global politics.

Future research

This book is the first to take the politics of oil in the Gulf of Guinea to comprise the region as a whole and not just its individual states. It is hoped that it makes a strong case for the urgency of the matters it describes and analyses. As noted in the Introduction, studying this subject has to take place amidst great paucity of authoritative secondary sources and with substantial access problems in regard to many of the relevant fieldwork locations and written primary sources. Furthermore, two of the central sets of actors in the oil rapport, companies and elites, are not only under-researched but also notoriously difficult to access on a basis that is not compromising of scholarly objectivity. It is therefore unsurprising that conspicuous lacunae abound.[9]

---

8 Balibar (2001: 16ff). I thank Ludmila du Bouchet for having brought this to my attention.

9 The following is not an exhaustive list of desiderata but merely lists the most obvious gaps I came across in my research. In the same way that this book draws on the insights of numerous disciplines, such a research agenda is, by definition, multidisciplinary. *From a policy perspective*, what lessons are to be drawn from an oil nexus where the rational pursuit of long-term interests by several sets of actors holds such tragic governance outcomes? Are the incentive structures in place for a reform of current methods? Is "soft" reform likely to bring about changes? What are the changes introduced to the Gulf of Guinea oil sector by the arrival of non-Western and/or non-brand-conscious companies for whom CSR does not even get rhetorical traction? What are the alternatives? *At the level of the state-firm partnership*, greater attention must be paid to, firstly, the study of contracts between companies and the state, and secondly, to the politics of oil trading and the constellation of actors that revolves around it. *At the level of the private sector*, the greatest need is for, firstly, monographs on oil company local subsidiaries since the beginning of their operations and their approaches to governance, especially in regard to local communities; secondly, studies of particular instances of "privatised" diplomacy whereby the embassies of Northern states are placed at the service of oil interests; and thirdly, a regional survey of oil firm security arrangements and the expanding relations with the privatized military industry. *At the level of petro-elites*, there is the need for ethnography as

An important set of questions that was not discussed at length in this study and deserves close scrutiny and dedicated fieldwork is the manner(s) in which petroleum has shaped the political imagination of the oil state and its inhabitants. It is improbable that these themes can be gauged regionally for they are country-specific or even locally embedded; this is the subject for detailed investigation, not broad, sweeping studies. And yet, even a cursory look will show the centrality of oil in the Gulf of Guinea to be awesome: in the oil state and the lives of its people, petroleum is to be found everywhere. It is at the heart of consumption patterns, electoral politics (where it exists), the makeup of society, regional and ethnic relations, self-images, political imaginations and readings of the past, dreams of the future, religious and occult discourses, rituals of the state. Oil is a mentality and an expectation, if not a certainty because of its inherently unstable nature. In view of the fact that oil has impacted on the region for decades and will go on doing so in a deepened manner for decades more, this is unquestionably the overriding political experience fashioning the lives, institutions, and political possibilities in the region. As happened to Andean countries haunted by their foundational silver booms centuries ago or Caribbean societies long shaped by the sugar needs of faraway consumers, the impact of the political economy of oil on the Gulf of Guinea will be acutely felt when oil profits are but a distant memory.

In many ways, the process described in this study is remarkably novel. The soaring international interest in the Gulf of Guinea's oil is responsible

---

on individual elites and studies of elite consumption and capital flight, including relations with Western banks and tax havens. Also, research on the links between elites and international criminality and the manner in which legal revenues circulate into criminal enterprise and vice versa should be pursued. *At the level of the state*, one needs more ethnographies of the oil state (in a similar mode to the works of Coronil and Apter) focusing particularly on the ongoing institutional mutations experienced by the newer oil states. Individual studies of the region's NOCs—whose centrality for the survival of the state and of ruling cliques cannot be overstated—are also required. Finally, there is *the politics of the subordinate social and political actors* that are barely discussed here. How does the vast majority cope with the wretchedness of their lives? Are there, in James Scott's memorable phrase, "weapons of the weak" to counter their disempowerment? Is there currently the space for progressive, emancipatory politics? Is the lack of legitimacy of power-holders likely to eventually upset their hold over the state?

for unprecedented revenue flows of the kind that only Nigeria has experienced over the years. It also allows a role for the region in today's rapidly globalising economy that is different from that of most other African states, with all this encompasses. In other ways, however, what is surprising is that there does not seem to be anything essentially new to the insertion of the Gulf of Guinea oil states in the world economy. In fact, it is distressingly familiar. A century ago, Joseph Conrad's impeccably modern *Nostromo* already warned of the perils of resource dependence and the lot reserved for the most distant corners of the global economy.[10] An apocryphal Latin American nation whose only wealth consists of a centuries-old silver mine in the hands of British entrepreneur Charles Gould, Costaguana[11] is the scene of perpetual, brutish power struggle for the available loot. For the outside world, the country does not exist as body politic, society, or territorial unit; it is simply coterminous with the "Gould Concession". Costaguana becomes, as it were, the silver mine with a pier attached to ease exports. Domestically, too, politics, personal relations and self-perceptions are inextricable from the overpowering mine hovering above all. Foreigners, locals, self-interested players of all kinds, in a rush for treasure, set against a backdrop of destitution, suffering and moral squalor: we have been here before.

---

10 Conrad (1994).

11 Literally, "the coast of bird excrement": the intertwining in the popular mind of manure and mineral extraction recurs across cultures and time.

BIBLIOGRAPHY

Academic Associates PeaceWorks, *Oil Prospecting and Communal Crisis: Case Studies of Private Corporations' Activities in the Niger Delta, Nigeria* (Port Harcourt: Peace and Security Secretariat, February 2006).

Achikbache, B. and F. Anglade, "Les villes prises d'assaut: les migrations internes", *Politique Africaine* 31 (1988), pp. 7-14.

Aerts, J.-J. *et al.*, *L'économie camerounaise: un spoir évanoui* (Paris: Karthala, 2000).

Africa Confidential, *Who's Who in Southern Africa* (Oxford: Blackwell, 1998).

Akinsanya, Adeoye A, "State Strategies Toward Nigerian and Foreign Business", in I.W. Zartman, ed., *The Political Economy of Nigeria* (New York: Praeger, 1983), pp. 145-84.

Amnesty International, "Equatorial Guinea: a Parody of a Trial in Order to Crush the Opposition", Amnesty International, July 2002 (AFR 24/014/2002).

Amuwo, Kunle, "General Babangida, Civil Society and the Military in Nigeria: Anatomy of a Personal Rulership Project", *Travaux et Documents* 48 (Bordeaux: CEAN, 1995).

Anderson, Benedict, *Imagined Communities: Reflections on the Origin and Spread of Nationalism* (London: Verso, 1983).

Anderson, Jon Lee, "Our New Best Friend: Who needs Saudi Arabia when you've got São Tomé?", *The New Yorker,* 7 October 2002, pp. 74-83.

Appadurai, Arjun, "Disjuncture and Difference in the Global Cultural Economy", *Public Culture* 2, no. 2 (1990), pp. 1-24.

Appiah, Anthony K., "Forward", in Saskia Sassen, *Globalization and its Discontents: Essays on the New Mobility of People and Money* (New York: The New Press, 1998), pp. xi-xv.

Apter, Andrew, "The Pan-African Nation: Oil Money and the Spectacle of Culture in Nigeria", *Public Culture* 8 no. 3 (1996), pp. 441-66.

——, "IBB=419: Nigerian Democracy and the Politics of Illusion", in John L. and Jean Comaroff, eds, *Civil Society and the Political Imagination in Africa: Critical Perspectives* (Chicago University Press, 1999), pp. 267-307.

——, *The Pan-African Nation* (Chicago University Press, forthcoming 2005).

Ardant, *Histoire de l'impôt* (Paris: Fayard, 1971-72).

Arnold, Guy, *Modern Nigeria* (London: Longman, 1977).

Aron, Raymond, "Classe sociale, classe politique, classe dirigeante", *Archives Européens de Sociologie* I (1960), pp. 265-81.

Ascherson, Neal, *The King Incorporated: Leopold the Second and the Congo* (London: Granta, 2001).

Assemblée Nationale, "Audition de M. Jean-François Bayart, directeur du CERI, et de M. Luis Martinez, chercheur au CERI", 20 January 1999, in *Rapport d'information n° 1859 sur le rôle des compagnies pétrolières dans la politique internationale et son impact social et environnemental - Sommaire des comptes rendus d'auditions* (Paris: Assemblée Nationale), pp. 129-37.

Atenga, Thomas, "Gabon: apprendre à vivre sans pétrole", *Politique Africaine* 92 (2003), pp. 117-28.

Auge, Axel, «Le recrutement social des élites politiques au Gabon: La place du lien ethnique et des autres liens» (unpublished doctoral thesis, University of Toulouse II, 2003).

Austen, Ralph, *African Economic History* (Oxford: James Currey, 1987).

Austen, R. and R. Headrick, "Equatorial Africa under Colonial Rule", in David Birmingham and Phyllis M. Martin, eds., *History of Central Africa* (London: Longman, 1983), Vol.2, pp. 27-94.

Auty, R., *Sustaining Development in Mineral Economies: The Resource Curse Thesis* (London: Routledge, 1993).

"Industrial Policy in Six Newly Industrializing countries: the Resource Curse Thesis", *World Development* 22, no. 1 (1994), pp. 11-26.

Badie, Bertrand, *La fin des territoires* (Paris: Fayard, 1995).

——, *Un monde sans souveraineté* (Paris: Fayard, 1999).

——, *The Imported State* (Stanford University Press, 2000).

Bain, William, "The Political Theory of Trusteeship and the Twilight of International Equality", *International Relations* 17, no. 1 (2003), pp. 59-77.

Bakary, T.D., *Les Elites africaines au pouvoir (problématique, méthodologie, états des travaux* (Bordeaux: CEAN, 1990).

Balibar, E., *Nous, citoyens d'Europe: Les frontières, l'Etat, le peuple* (Paris: La Découverte, 2001).

Ball, Nicole *et al.*, "Governance in the Security Sector", in N. Van de Walle and N. Ball, eds, *Beyond Structural Adjustment: The Institutional Context of African Development* (Basingstoke: Palgrave, 2003), pp. 263-304.

Bamberg, James, *The History of the British Petroleum Company: The Anglo-Iranian Years, 1928-1954* (Cambridge University Press, 1994).

_____, *British Petroleum and Global Oil: The Challenge of Nationalism* (Cambridge University Press, 2000).

Banfield, J., V. Haufler and D. Lilly, *Transnational Corporations in Conflict-Prone Zones* (London: International Alert, 2003).

Barber, Karin, "Popular Reactions to the Petro-naira", *Journal of Modern African Studies* 20, no. 3 (1982), pp. 431-50.

Barnes, Sandra T., "Global Flows: Terror, Oil, and Strategic Philanthropy", Presidential address to the African Studies Association, New Orleans, 12 November 2004.

Bartelson, Jens, *The Critique of the State* (Cambridge: Cambridge University Press, 2001)

Bates, Robert, *Markets and States in Tropical Africa: The Political Basis of Agricultural Politics* (Berkeley: University of California Press, 1981).

Bayart, J.-F., *L'Etat en Afrique: la politique du ventre* (Paris: Fayard, 1989).

_____, "L'Etat", in Christian Coulon and Denis-Constant Martin, eds, *Les Afriques politiques* (Paris: Editions la Découverte, 1991), pp. 213-28.

_____, "L'historicité de l'État importé", *Les Cahiers du CERI* 15 (1996).

_____, "Africa Within the World: a History of Extraversion", *African Affairs* 99, no. 395 (2000), pp. 217-267.

_____, ed., *La réinvention du capitalisme* (Paris: Karthala, 1994).

Bayart, J.-F., A. Mbembe and C. Toulabor, *Le politique par le bas* (Paris: Karthala, 1992).

Bayart, J.-F., Stephen Ellis and Béatrice Hibou, *The Criminalization of the State in Africa* (Oxford: James Currey, 1999).

Bazenguissa-Ganga, Rémy, *Les voies du politique au Congo: essai de sociologie historique* (Paris: Karthala, 1996a).

_____, "Milices politiques et bandes armées à Brazzaville: Enquête sur la violence politique et sociale des jeunes déclassés", *Les Études du CERI* 13 (1996b).

Beblawi, H. and G. Luciani, *Nation, State and Integration in the Arab World: The Rentier State* (New York: Croom Helm, 1987).

Behrend, H., "War in Northern Uganda: The Holy Spirit Movements of Alice Lakwena, Severino Lukoya and Joseph Kony (1986-1997)", in Christopher Clapham, ed., *African Guerrillas* (Oxford: James Currey, 1998), pp. 107-18.

Bernault, Florence, *Démocraties ambiguës en Afrique centrale. Gabon, Congo-Brazzaville, 1940-65* (Paris: Karthala, 1996).

_____, *No Condition is Permanent* (Madison: University of Wisconsin Press, 1993a).

Berry, Sara, "Coping with Confusion: African Farmers' Responses to Economic Instability in the 1970s and 1980s", in T. Callaghy and J. Ravenhill, eds, *Hemmed In: Responses to Africa's Economic Decline* (New York: Columbia University Press, 1993b), pp. 248-78.

Bevan, David L., Paul Collier and J.W. Gunning, *The Political Economy of Poverty, Equity and Growth: Nigeria and Indonesia* (Oxford University Press, 1999).

Bhagwati, Jagdish, *In Defense of Globalization* (Oxford University Press: 2004).

Bienen, Henry, "Nigeria: From Windfall Gains to Welfare Losses?", in Alan Gelb and Associates, *Oil Windfalls: Blessing or Curse?* (New York: World Bank and Oxford University Press, 1988), pp. 227-61.

Biersteker, T.J., *Multinationals, the State and the Control of the Nigerian Economy* (Princeton University Press, 1987).

Birdsall, Nancy and A. Subramanian, "Saving Iraq from its Oil", *Foreign Affairs* 83, no. 4 (2004), pp. 77-89.

Birnbaum, Pierre and Bertrand Badie, *Sociologie de l'Etat* (Paris: Grasset, 1978).

Booker, Salih and William Minter, "The US and Nigeria: thinking beyond oil" (2003), paper available at www.greatdecisions.org (accessed 5 January 2004).

Booth, Anne, ed., *The Oil Boom and After: Indonesian Economic Policy in the Suharto Era* (New York: Oxford University Press, 1992).

Boone, C., *Political Topographies of the African State: Territorial Authority and Institutional Choice* (Cambridge University Press, 2003).

Bottomore, Tom, *Elites and Society* (London: Routledge, 1993).

Bouquerel, J., "Le pétrole au Gabon", *Cahiers d'Outre-Mer* 20 (April-June) (1967), pp. 186-99.

Boulaga, F.E., *Les conférences nationales en Afrique noire: une affaire à suivre* (Paris: Karthala, 1993).

Boyce, J.K. and L. Ndikumana, "Is Africa a Net Creditor? New Estimates of Capital Flight from Severely Indebted Sub-Saharan African Countries, 1970-96", *Journal of Development Studies* 38, no. 2 (2001), pp. 27-56.

Bratton, Michael and N. van de Walle, *Democratic Experiences in Africa: Regime Transitions in Comparative Perspective* (Cambridge University Press, 1997).

British Petroleum, *BP's 2004 Statistical Review of World Energy* (London: British Petroleum, 2004).

Bruhns, Hinnerk, "Max Weber, l'économie et l'histoire", *Annales* 51, no. 6 (1997), pp. 1259-87.

Brumberg, D., "The Trap of Liberalized Autocracy", *Journal of Democracy* 13, no. 4 (2002), pp. 56-68.

Bruso, Joseph M. *et al.*, "Geology will Support Further Discoveries in the Gulf of Guinea's Golden Rectangle", *Oil and Gas Journal*, 16 February 2004, pp. 1-3.

Buijtenhuijs, Robert, "Chad in the Age of the Warlords", in D. Birmingham and P. M. Martin, eds, *History of Central Africa III: the Contemporary Years* (Harlow: Longman, 1998), pp. 21-40.

Bull, Hedley, *The Anarchical Society: A Study of Order in World Politics* (Basingtoke: Palgrave, 2002).

Burckhardt, Jacob, *The Civilization of the Renaissance in Italy* (New York: The Modern Library, 1995).

Byé, Maurice, "Self-Financed Multiterritorial Units and their Time Horizon", *International Economic Papers* 8, June (1957), pp. 147-78.

Caley, Cornélio, "Os Petróleos e a Problemática do Desenvolvimento em Angola: uma visão histórico-política" (Lisbon: ISCTE, unpublished MA dissertation, 1997).

Callaghy, Thomas M., "The State as Lame Leviathan: The Patrimonial Administrative State in Africa", in Z. Ergas, ed., *The African State in Transition* (New York: St Martin's Press, 1987), pp. 87-116.

————, "Africa and the World Political Economy: Still Caught Between a Rock and a Hard Place", in John W. Harbeson and Donald Rothchild, eds, *Africa in World Politics: Post-Cold War Challenges* (Boulder: Westview Press, 1995), pp. 41-68.

Callaghy, Thomas and John Ravenhill, eds, *Hemmed in: the Dilemmas of African Development* (New York: Columbia University Press, 1993).

Callaghy, Thomas, Ronald Kassimir and Robert Latham, eds, *Intervention and Transnationalism in Africa: Global-Local Networks of Power* (Cambridge University Press, 2001).

Campbell, Ian, "Army Reorganization and Military Withdrawal", in Keith Panter-Brick, ed., *Soldiers and Oil: the Political Transformation of Nigeria* (London: Frank Cass, 1978), pp. 58-100.

Canard Enchaîné, "Les dossiers du *Canard Enchaîné* 67: Elf, l'empire d'essence", April 1998.

Cardoso, Fernando Henrique, "The Consumption of Dependency Theory in the United States", *Latin American Research Review* 12, no. 3 (1977), pp. 7-24.

Cardoso, F.H. and E. Faletto, *Dependency and Development in Latin America* (Berkeley: University of California Press, 1979).

Carrier, James C., "Consumption", in A. Bernard and J. Spencer, eds, *Encyclopedia of Social and Cultural Anthropology* (London: Routledge, 1996), pp. 128-9.

Cassidy, John, "Pump Dreams: is energy independence an impossible goal?" *New Yorker*, 10 October 2004, pp. 42-7.

Castells, Manuel, *End of Millennium* (Oxford: Blackwell, 1998).

Cezar, E., J.S. Morrison and J. Cooke, "Alienation and Militancy in Nigeria's Niger Delta", *CSIS Africa Notes* 16 (2003).

Chabal, Patrick, "African Politics: the French Way", *International Affairs* 76, no. 4 (2000), pp. 825-32.

———, ed., *Political Domination in Africa* (Cambridge: Cambridge University Press, 1986).

Chabal, Patrick and Jean-Pascal Daloz, *Africa Works: Disorder as a Political Instrument* (London: James Currey, 1999).

Chafer, T., "Franco-African Relations: No Longer So Exceptional?", *African Affairs* 101, no. 404 (2002), pp. 343-63.

Charap, J. and C. Harm, "Institutionalized Corruption and the Kleptocratic State", IMF Working Paper, July 1999.

Christensen, John, "Hooray Hen-Wees", *London Review of Books*, 6 October 2005 (www.lrb.co.uk, accessed 13 May 2006).

Chaudhry, Kiren, *The Price of Wealth* (Ithaca: Cornell University Press, 1997).

Christian Aid, *Fuelling Poverty: Oil, War and Corruption* (London: Christian Aid, 2003).

—— *Behind the Mask: the Real Face of CSR* (London: Christian Aid, 2004).

Clapham, Christopher, *Third World Politics* (London: Routledge, 1985).

_____, *Revolutionary Change and Continuity in Ethiopia* (Cambridge University Press, 1988).

_____,"The Longue Durée of the African State", *African Affairs* 93, no. 372 (1994).

_____, *Africa and the International System* (Cambridge University Press: 1996a).

_____, "Governmentality and Economic Policy in Sub-Saharan Africa", *Third World Quarterly* 17, no. 4 (1996b), pp. 809-824.

_____,, ed., *Private Patronage and Public Power: Political Clientelism and the Modern State* (London: Frances Pinter, 1982).

Clarence-Smith, William G., "Les investissements belges en Angola, 1912-1961", in Catherine Coquery-Vidrovitch, ed., *Entreprises et entrepreneurs en Afrique, XIX-XX siècles* (Paris: L'Harmattan, 1983), pp. 79-91.

Clark, John, "Congo: Transition and the Struggle to Consolidate" in John Clark and David Gardinier, eds, *Political Reform in Francophone Africa* (Boulder: Westview Press, 1997), pp. 62-85.

_____, "The Neo-Colonial Context of the Democratic Experiment of Congo-Brazzaville", *African Affairs* 101, no. 403 (2002), pp. 171-92.

Coakley, George J., *et al.,* "The Mineral Industries of Africa", in *US Geological Survey Minerals Handbook* (Washington DC: Energy Information Administration, 2000), pp. 1-15.

Cohen, Abner, *The Politics of Elite Culture* (Berkeley: California University Press, 1979).

Collier, Paul and A. Hoeffler, *Greed and Grievance in Civil War* (Washington, DC: World Bank, Development Research Group, 2000).

Collier, Paul and Catherine Patillo, eds, *Reducing the Risk of Investment in Africa* (London: Macmillan, 2000).

Conrad, Joseph, *Nostromo* (London: Penguin, 1994 [1904]).

Cooper, F., "What is the Concept of Globalisation Good For? An African Historian's Perspective", *African Affairs* 100, no. 399 (2001), pp. 189-213.

_____, *Africa Since 1940: The Past of the Present* (Cambridge University Press, 2002).

Cooper, Frederick and Ann Laura Stoler, eds, *Tensions of Empire* (Berkeley: University of California Press, 1997), pp. 9-24.

Copinschi, Philippe, Pierre Noel and Ricardo Soares de Oliveira (contributing writer), *La politique africaine des compagnies pétrolières américaines* (Paris: IFRI for the Délégation aux Affaires Stratégiques, ministère de la Défense, 2004).

Coquery-Vidrovitch, C., *Le Congo au temps des grandes compagnies concessionaires 1898-1930* (The Hague: Mouton, 1972).

———, "The Process of Urbanization in Africa: From the Origins to the Beginning of Independence", *African Studies Review* 34, no. 1 (1991), pp. 1-98.

———, ed., *Entreprises et Entrepreneurs en Afrique, XIX-XX siècles* (Paris: L'Harmattan, 1983).

Coronil, Fernando, *The Magical State: Nature, Money, and Modernity in Venezuela* (Chicago: Chicago University Press, 1997).

Courade, Georges, *Le désarroi camerounais: l'épreuve de l'économie-monde* (Paris: Karthala, 2000).

Creveld, Martin Van, *The Rise and Decline of the State* (Cambridge: Cambridge University Press, 2002).

Daloz, Jean-Pascal, ed., *Le (non-) renouvellement des élites en Afrique subsaharienne* (Bordeaux: Centre d'Étude d'Afrique Noire, 1999).

Daojiong, Zha, "China's Energy Security: Domestic and International Issues", *Survival* 48, no. 1 (2006), pp. 179-90.

Daverat, G., "Un producteur africain du pétrole, le Gabon", *Cahiers d'Outre-Mer*, January-March (1977), pp. 31-56.

Davis, Stephen and Dimieari Von Kemedi, "Illegal Oil Bunkering in the Niger Delta" in *Niger Delta Peace and Security Secretariat: Background Papers for PaS Working Group* (Port Harcourt: Niger Delta International Centre for Reconciliation, Academic Associates PeaceWorks, February 2006), pp. 36-51.

———, "The Current Stability and Future Prospects for Peace and Security in the Niger Delta", in *Niger Delta Peace and Security Secretariat: Background Papers for PaS Working Group* (Port Harcourt: Niger Delta International Center for Reconciliation, Academic Associates PeaceWorks, February 2006), pp. 11-23.

Dean, Mitchell, *Governmentality: Power and Rule in Modern Society* (London: Sage, 1999).

Diamond Works, "DiamondWorks Ltd - Sao Tome oil deal formalised, expanded", Press Release, DiamondWorks Ltd 18 February 2004.

Dictionnaire Encyclopédique Quillet, "Guinée", volume 3.

Dietrich, Chris, "The Commercialization of Military Deployment in Africa", *African Security Review* 9, no. 1 (2000), http://www.iss.org.za/Pubs/ASR/9No1/Commerciallisation.html (accessed 18 January 2004).

Dos Santos, Daniel, "Cabinda: the Politics of Oil in Angola's Enclave" in Robin Cohen, ed., *African Islands and Enclaves* (London: Sage, 1983), pp. 101-17.

Douglas, Oronto, V. Kemedi, I. Okonta and M. Watts, "Alienation and Militancy in the Niger Delta: a Response to CSIS on Petroleum, Politics and Democracy in Nigeria", Foreign Policy in Focus Special Report (2003), www.fpif.org (accessed 18 November 2003).

Drennen, B., "Offshore Petroleum", *Geotimes*, July (2002) http://www.agiweb.org/geotimes (accessed 10 February 2003).

Dumoulin, Michel, *Petrofina: Un groupe pétrolier international et la gestion de l'incertitude Tome I (1920-1979)* (Louvain: Université de Louvain/Editions Peeters, 1997).

Dunn, John, "Conclusion", in D. Cruise O'Brien, J. Dunn and R. Rathbone, eds, *Contemporary West African States* (Cambridge University Press, 1989), pp. 181-92.

Dzurek, Daniel J. "Gulf of Guinea Boundary Disputes", *Boundary and Security Bulletin* 7 (1999), pp. 98-104.

Ebel, Robert E., *China's Energy Future: The Middle Kingdom Seeks a Place in the Sun* (Washington DC: CSIS, 2005).

Eboko, Fred, "Les élites politiques au Cameroun: renouvellement sans renouveau?", in Jean-Pascal Daloz, ed., *Le (non-) renouvellement des élites en Afrique subsaharienne* (Bordeaux: Centre d'Étude d'Afrique Noire, 1999), pp. 99-133.

ECON, *TotalFinaElf and CRS* (Oslo: ECON Analysis, 2004).

Ellis, Stephen, *The Mask of Anarchy* (New York University Press, 2001).

———, "Briefing: West Africa and its Oil", *African Affairs* 102, no. 406 (2003), pp. 135-38.

Ellis, Stephen and Janet MacGaffey, "Research on Sub-Saharan Africa's Unrecorded International Trade: Some Methodological and Conceptual Problems", *African Studies Review* 39, no. 2 (1996), pp. 19-41.

Emmott, Bill, *20:21 Vision* (London: Penguin, 2003).

Energy Information Administration, *Annual Energy Outlook 2007* (Washington DC: EIA, US Department of Energy, 2006).

Ennes Ferreira, Manuel, "La reconversion économique de la *nomenklatura* pétrolière", *Politique Africaine* 57 (1995), pp. 13-25.

Environmental Defense *et al.*, "The Chad-Cameroon Oil and Pipeline Project: A Call for Accountability, a joint multi-authored report by Environmental Defense, USA, the Centre pour l'Environnement et le Développement,

Cameroon, and the Association Tchadienne pour la Promotion et la Défense des Droits de L'Homme", Chad, June 2002.

European Commission, *Promoting a European Framework for Corporate Social Responsibility*, Green Paper, Directorate-General for Employment and Social Affairs, European Commission, Brussels, July 2001.

European Union, "Congo Election Présidentielle, 10 Mars 2002", Mission d'Observation Electorale de la Union Européenne, Rapport Final (Brussels, 2002).

Evans, Jonathan, Robert Hull and Stephen Davis, "Money Laundering and Nigeria" in *Niger Delta Peace and Security Secretariat: Background Papers for PaS Working Group* (Port Harcourt: Niger Delta International Centre for Reconciliation, Academic Associates PeaceWorks, February 2006), pp. 129-48.

Exxon-Mobil, *Exxon-Mobil Corporate Citizenship Report* (2003).

Fardmanesh, Mohsen, "Dutch Disease and the Oil Syndrome: an Empirical Study", *World Development* 19, no. 6 (1991), pp. 711-17.

Fearon, James D. and David D. Laitin, "Ethnicity, Insurgency and Civil War", *American Political Science Review* 97, no. 1 (2003), pp. 75-90.

Federal Republic of Nigeria, "Constitution of the Federal Republic of Nigeria, 1999" (www.nigeria-law.org/ConstitutionOfTheFederalRepublicOfNigeria. htm accessed 3 January 2006).

Federal Republic of Nigeria, "Niger-Delta Development Commission (Establishment etc) Act of 2000" (Act No 6, Laws of the Federation of Nigeria) (www.nigeria-law.org accessed 16 April 2006).

Ferguson, James, *The Anti-Politics Machine* (Minneapolis: Minnesota University Press, 1994).

Ferrier, R.W., *The History of the British Petroleum Company, Volume 1: The Developing Years, 1901-1932* (Cambridge University Press, 1990).

FIAS, *Guinée Equatoriale – Diagnostic sur le climat des investissements et la maximisation des investissements pétroliers* (Washington DC: Foreign Direct Investment Service/World Bank Group, 2002).

FIDH, "Elections en trompe-l'oeil au Congo Brazzaville" (Paris: Fédération Internationale des Ligues des Droits de l'Homme, 2002).

Fieldhouse, D.K., *The West and the Third World* (Oxford: Blackwell, 1999).

Foltz, William, "Reconstructing the State of Chad", in I.W. Zartman, ed., *Collapsed States: the Disintegration and Restoration of Legitimate Authority* (Boulder: Lynne Rienner, 1995), pp. 15-32.

Forbes, R.J., and D.R. O'Beirne, *The Technical Development of the Royal Dutch/ Shell, 1890-1940* (Leiden: E.J. Brill, 1957).

Foucault, Michel, "La gouvernementalité", in *Dits et écrits II* (Paris: Gallimard, 2001), pp. 635-57.

Frankel, Paul, *Mattei: Oil and Power Politics* (New York: Praeger, 1966).

Freire Antunes, J., *Os Americanos e Portugal 1962-1974: Nixon e Caetano, promessas e abandono* (Lisboa: Difusão Cultural, 1992).

Freund, Bill, *The Making of Contemporary Africa* (Basingstoke: Macmillan, 1998).

Friedrich, C.J., "A Critique of Pareto's Contribution to the Theory of Political Elites" *Cahiers Vilfredo Pareto* 5 (1965), pp. 259-67.

Friends of the Earth, "Broken Promises: The Chad Cameroon Oil and Pipeline Project: Profit at any Cost?" (New York: Friends of the Earth International, 2001).

Frynas, J.G., "Political Instability and Business: Focus on Shell in Nigeria", *Third World Quarterly* 19, no. 3 (1999), pp. 457-79.

_____, *Oil in Nigeria: Conflict and Litigation between Oil Companies and Village Communities* (Hamburg: LIT, 2000).

_____, "Corporate and State Responses to Anti-Oil Protests in the Niger Delta", *African Affairs* 100, no. 398 (2001), pp. 27-54.

_____, "Social and Environmental Litigation Against Transnational Firms", *Journal of Modern African Studies* 42, no. 3 (2004),pp. 363-88.

_____, "The Oil Boom in Equatorial Guinea", *African Affairs* 103, no. 413 (2004), pp. 527-46.

Frynas, J.G., G. Wood and R. Soares de Oliveira, "Business and Politics in São Tomé and Príncipe: from Cocoa Monoculture to Petro-State", *African Affairs* 102, no. 406 (2003), pp. 51-80.

Gaffney, Cline and Associates, "Operational & Function Audit of SNH Cameroon prepared for the Ministry of Economy and Finance, Cameroon" (2000).

Gambetta, Diego, "Mafia: the Price of Distrust", in D. Gambetta, ed., *Trust: The Making and Breaking of Cooperative Relations* (Oxford: Blackwell, 1988), pp. 158-75.

Gardinier, David, "The Development of a Petroleum-Dominated Economy in Gabon", *Essays in Economic and Business History* 17 (1999), pp. 1-24.

Gary, Ian and Terry Lynn Karl with R. Soares de Oliveira (contributing writer), *Bottom of the Barrel: Africa's Oil Boom and the Poor* (Baltimore: Catholic Relief Services, 2003).

Gary, Ian and Nikki Reisch, *Chad's Oil: Miracle or Mirage? Following the Money in Africa's Newest Petro-State* (Washington DC: Catholic Services and Bank Information Center, February 2005).

Gaulme, François, *Le Gabon et son ombre* (Paris: Karthala, 1987).

Geertz, Clifford, *The Interpretation of Cultures* (New York: Basic Books, 1973).

Geffray, Christian, *La Cause des Armes* (Paris: L'Harmattan, 1990).

Gelb, Alan and Associates, *Oil Windfalls: Blessing or Curse?* (New York: World Bank and Oxford University Press, 1988).

Gellner, Ernest and J. Waterbury, eds., *Patrons and Clients in Mediterranean Societies* (London: Duckworth, 1977).

Gereffi, Gary, R. Garcia-Johnson and Erika Sasser, "The NGO-Industrial Complex", *Foreign Policy* 125 (2001), pp. 56-65.

Gerretson, Frederik, *The History of the Royal Dutch Company* (four volumes, Leiden: E.J. Brill, 1953-57).

Geschiere, Peter and Piet Konings, eds, *Itinéraires d'accumulation au Cameroun* (Paris: Karthala, 1993).

Geuss, Raymond, *History and Illusion in Politics* (Cambridge University Press, 2001).

Gide, André, *Voyage au Congo* (Paris: Gallimard, 1927).

Gilpin, Robert, *The Challenge of Global Capitalism* (Princeton University Press, 2000).

Glaser, Antoine, and Stephen Smith, *Ces messieurs Afrique* (Paris: Calmann-Lévy, 1992).

_____, *Ces messieurs Afrique 2* (Paris: Calmann-Lévy, 1997).

Global Witness, *A Crude Awakening: The Role of the Oil and Banking Industries in Angola's Civil War and Plunder of State Assets* (London: Global Witness: 1999).

_____, *All the Presidents' Men: The Devastating Story of Oil and Banking in Angola's Privatized War* (London: Global Witness, 2002).

_____, "Does US bank harbour Equatorial Guinea's oil millions in secret accounts? US Department of Justice must investigate" (London: Global Witness press release, 20 January 2003).

_____, *Time for Transparency* (London: Global Witness, 2004).

Goldwyn, David L., paper presented at the conference on "Oil and Gas in Africa", Royal Institute of International Affairs, London, 24-25 May 2004.

Goldwyn, David and Robert Ebel, "Africa Policy Panel: Crafting a US Energy Policy for Africa" (Washington DC: Center of Strategic and International Studies, 2004).

Goody, Jack, *Technology, Tradition and the State in Africa* (Oxford University Press, 1971).

Gordon, David F., "Debt, Conditionality and Reform: The International Relations of Economic Policy Restructuring in Sub-Saharan Africa" in T. Callaghy and J. Ravenhill, eds, *Hemmed in: the Dilemmas of African Development* (New York: Columbia University Press: 1993), pp. 90-129.

Gore, Charles and David Pratten, "The Politics of Plunder: the Rhetorics of Order and Disorder in Southern Nigeria", *African Affairs* 102, no. 407 (2003), pp. 211-40.

Gorozpe, Iñaki, "Reinvidicacíon política y particularismo en Annobón", *Lusotopie* (1995), pp. 251-9.

Government of São Tomé and Príncipe, "Memorandum of Understanding with Chrome Oil (15 March 2003).

Graf, W.D., *The Nigerian State: Political Economy, State Class and Political System in the Postcolonial Era* (London: James Currey, 1988).

Gray, Christopher J., "Cultivating citizenship through xenophobia in Gabon, 1960-1995", *Africa Today* 45, no. 3/4 (1998), pp. 389-409.

Grayson, L., *National Oil Companies* (Chichester: John Wiley and Sons, 1981).

Guest, Robert, "How to Make Africa Smile: A Survey of Sub-Saharan Africa", *The Economist,* 17 January 2004.

Gulf Oil, *Gulf Oil Corporation Annual Report* (1976).

Gustafson, Thane, *Capitalism, Russian-style* (Cambridge University Press, 2002).

Guyer, Jane I., *Feeding African Cities: Studies in Regional Social History* (Manchester University Press, 1987).

———, "Briefing: the Chad-Cameroon Petroleum and Pipeline Project", *African Affairs* 101, no. 402 (2002), pp. 109-15.

Halliday, Fred, *Arabia without Sultans* (London: Saqi Books, 2002).

Hannerz, Ulf, "The World of Creolization", *Africa* 57, no 4 (1987), pp. 546-59.

Harbinson, D., R. Knight and J. Westwood, "West African Deep Water Developments in a Global Context", *The Hydrographic Journal* 98 (2000), http://www.hydrographicsociety.org (accessed 20 October 2003).

Harding, Jeremy, "The Mercenary Business", *London Review of Books,* 1 August 1996, pp. 3-9.

Harnischfeger, Johannes, "The Bakassi Boys: Fighting Crime in Nigeria", *Journal of Modern African Studies* 41, no. 1 (2003), pp. 23-49.

Harvey, David, *The Condition of Postmodernity* (Oxford: Blackwell, 1990).

Hast, Adele, ed., *International Directory of Company Histories* (Chicago: St James Press, 1991), volume 4.

Hawthorn, Geoffrey, *Enlightenment and Despair: A History of Social Theory* (Cambridge University Press, 1986).

Herbst, Jeffrey, "Responding to State Failure in Africa", *International Security* 21, No. 3 (1996), pp. 120-144.

_____, *States and Power in Africa* (Princeton University Press, 2000).

Hertner, P. and G. Jones, *Multinationals: Theory and History* (Gower: Aldershot, 1986).

Hibou, Béatrice, ed., *La privatisation des Etats* (Paris: Karthala, 1998a).

_____, "The Political Economy of the World Bank's Discourse: From Economic Catechism to Missionary Deeds (and Misdeeds)", *Les Études du CERI* 39 (1998b).

_____, "The 'Social Capital' of the State as an Agent of Deception", in Jean-François Bayart, Stephen Ellis and Béatrice Hibou, *The Criminalization of the State in Africa* (Oxford, James Currey, 1999), pp. 69-113.

_____, "L'historicité de la construction européenne : le secteur bancaire en Grèce et au Portugal", *Les Etudes du CERI* 85-86 (2002).

_____, ed., Privatising the State (London : C.Hurst & Co., 2004)

Hidy, R., and M. Hidy, *Pioneering in Big Business* (New York: Harper, 1955).

Hirschman, Albert O., *A Bias for Hope: Essays on Development and Latin America* (New Haven: Yale University Press, 1971).

_____, "Beyond Asymmetry: Critical Notes on myself as a Young Man and on Some Other Old Friends", *International Organization* 32, no. 1 (1978), pp. 45-50.

_____, "On Hegel, Imperialism and Stagnation" in *Essays in Trespassing: Economics to Politics and Beyond* (Cambridge University Press, 1981), pp. 167-76.

Hochschild, A., *King Leopold's Ghost* (New York: Houghton Mifflin Company, 1999).

Hodges, Tony, *Angola from Afro-Stalinism to Petro-Diamond Capitalism* (Oxford: James Currey, 2001).

Hoeven, R. van der, and F. van der Kraaj, eds., *Structural Adjustment and Beyond in Sub-Saharan Africa* (London: Heinemann, 1994).

Hoffman, David, *The Oligarchs: Wealth and Power in the New Russia* (New York: Public Affairs, 2002).

Hont, Istvan, "The Permanent Crisis of a Divided Mankind: 'Contemporary Crisis of the Nation State' in Historical Perspective", *Political Studies* XLII (1994), pp. 166-231.

Hopkins, A.G., "Imperial Business in Africa. Part I: Sources", *Journal of African History* 17, no. 1 (1976), pp. 29-48.

_____, "Imperial Business in Africa. Part II: Interpretations", *Journal of African History* 17, no. 2 (1976), pp. 267-90.

Horsnell, Paul, "Oil Company Histories", *Journal of Energy Literature* V, no. 2 (1999), pp. 3-31.

Howarth, Stephen, *A Century in Oil: The "Shell" Transport and Trading Company 1897-1997* (London: Shell, 1997).

Hugon, Philippe, "Sortir de la Récession et préparer l'après-pétrole: le préalable politique", *Politique Africaine* 62 (1996), pp. 35-44.

Human Rights Watch, *Nigeria, the Ogoni Crisis: a Case-study of Military Repression in Southeastern Nigeria* (New York: Human Rights Watch, 1995).

_____, *The Price of Oil: Corporate Responsibility and Human Rights Violations in Nigeria's Oil Producing Communities* (New York: Human Rights Watch, 1999).

_____, *Military Revenge in Benue: a Population Under Attack* (New York: Human Rights Watch, April 2002).

_____, Angola's Oil Diagnostic - a Backgrounder (Washington DC: Human Rights Watch, May 2002).

_____, *The Niger Delta: No Democratic Dividend* (New York: Human Rights Watch, October 2002).

_____, *The Warri Crisis: Fueling Violence* (New York: Human Rights Watch, 2003).

_____, *Sudan, Oil and Human Rights* (New York: Human Rights Watch, 2003).

_____, *Some Transparency, No Accountability: The Use of Oil Revenue in Angola and Its Impact on Human Rights* (New York: Human Rights Watch, 2004).

_____, "Angola: Between War and Peace in Cabinda" (New York: Human Rights Watch, December 2004).

IEA, *International Energy Agency's World Energy Outlook 2000* (Paris: IEA, 2000).

IEA, *International Energy Agency's World Energy Outlook 2004* (Paris: IEA, 2004).

IMF, "São Tomé and Príncipe: Staff Report for the 2001 Article IV Consultation and Staff-Monitored Program" (Washington DC: IMF, February 2002).

_____, "Republic of Congo: 2003 Article IV Consultation and a New Staff-Monitored Program Staff Report; Staff Supplement; and Public Information Notice on the Executive Board Discussion", IMF Country Report No. 03/193 (June 2003).

_____, "Equatorial Guinea: 2003 Article IV Consultation - Staff Report" (Washington DC: IMF, 9 December 2003).

_____, "IMF Executive Board Approves US$84.4 million PRGF Arrangement for the Republic of Congo", IMF Press Release No. 04/262 (Washington DC: IMF, 7 December 2004).

Imle, John, "Multinationals and the New World of Energy Development: A Corporate Perspective", *Journal of International Affairs* 53, no. 1 (1999), pp. 263-80.

Institute of Advanced Strategic and Political Studies, *African Oil: A Priority for US National Security and African Development* (Washington D.C.: IASPS, 2002).

International Bar Association, "Equatorial Guinea at the Crossroads" (Report of a Mission to Equatorial Guinea by the IBA Human Rights Institute, 2003).

International Consortium of Investigative Journalists, "The Curious Bonds of Oil Diplomacy" (2002), http://www.public-i.org/ (accessed 5 January 2003).

International Institute for Strategic Studies, *The Military Balance 2003-2004* (Oxford: IISS and Oxford University Press, 2003).

Jackson, Robert H., *Quasi-states: Sovereignty, International Relations and the Third World* (Cambridge University Press, 1990).

Jackson, Robert H. and Carl Rosberg, "Sovereignty and Underdevelopment: Juridical Statehood and the African Crisis", *Journal of Modern African Studies* 24, no. 1 (1986), pp. 1-31.

Jackson, R. and C. Rosberg, "The Political Economy of African Personal Rule", in D. Apter and C. Rosberg, eds, *Political Development and the New Realism in Sub-Saharan Africa* (Charlottesville: University Press of Virginia, 1994), pp. 291-322.

Joly, E., *Est-ce dans ce monde là que nous voulons vivre* (Paris: Les Arènes, 2003).

Jones, Geoffrey, *The State and the Emergence of the British Oil Industry* (London: Macmillan, 1981).

_____, *The Evolution of International Business: an Introduction* (London: Routledge, 1996).

Joseph, Richard, *Democracy and Prebendal Politics in Nigeria: the Rise and Fall of the Second Republic* (Cambridge University Press, 1987).

_____, ed., *State, Conflict and Democracy in Africa* (Boulder: Lynne Rienner, 1999).

Jua, Nantang, "State, Oil and Accumulation", in P. Geschiere and P. Konings, eds, *Itinéraires d'accumulation au Cameroun* (Paris: Karthala, 1993), pp. 131-59.

Kahler, Miles, "Political Regime and Economic Actors: the Response of Firms to the End of Colonial Rule", *World Politics* 33, no. 3 (1981), pp. 383-412.

Kansteiner, Walter and J. Stephen Morrison, *Rising US Stakes in Africa: Seven Proposals to Strengthen US-Africa Policy* (Washington DC: CSIS, 2004).

Kaplan, Robert, "The Coming Anarchy: How Scarcity, Crime, Overpopulation and Disease are Rapidly Destroying the Social Fabric of Our Planet", *Atlantic Monthly* (February).

Karl, Terry Lynn, *The Paradox of Wealth* (Berkeley: University of California Press, 1997).

———, "The Perils of the Petro-State: Reflections on the Paradox of Plenty", *Journal of International Affairs* 53 no.1 (1999), pp. 31-48.

Katz, Menachem, *et al. Lifting the Oil Curse: Improving Petroleum Revenue Management in Sub-Saharan Africa* (Washington DC: International Monetary Fund, 2004).

Keay, M., *Oil and Governance: Focus on West Africa* (Briefing Paper, Sustainable Development Programme, Royal Institute of International Affairs, 2002).

Kennedy, C.R., "Relations between Transnational Corporations and Governments of Host Countries: a Look at the Future", *Transnational Corporations* 1, no. 1 (1992), pp. 67-91.

Kensington Ltd vs Republic of Congo (2005) EWHC 2684.

Khan, Khameel I.F., "National Oil Companies: Form, Structure, Accountability and Control", in K.I.F. Khan, ed., *Petroleum Resources and Development* (London: Belhaven Press, 1987), pp. 185-96.

Khan, Sarah Ahmed, *Nigeria: the Political Economy of Oil* (Oxford University Press, 1994).

Kirk, Robin, *More Terrible than Death: Massacres, Drugs and America's War in Colombia* (New York: Public Affairs, 2003).

Kirk-Greene, A. and D. Rimmer, *Nigeria since 1970: a Political and Economic Outline* (London: Hodder and Stoughton, 1981).

Klieman, Kairn, "Oil, Politics and Development in the Formation of a State: The Congolese Petroleum Wars, 1963-68", unpublished paper, 2006.

Kobrin, S.J., "Expropriation as an Attempt to Control Foreign Firms in LDCs: Trends from 1960 to 1979", *International Studies Quarterly* 28 (1984), pp. 329-48.

Kohli, Atul, *State-Directed Development: Political Power and Industrialization in the Global Periphery* (Cambridge University Press, 2004).

Kopytoff, I., ed., *The African Frontier: the Reproduction of Traditional Societies* (Bloomington: Indiana University Press, 1987).

Krasner, Stephen, *Sovereignty: Organized Hypocrisy* (Princeton University Press, 1999).

Lacoste, Yves, *Ibn Khaldoun: Naissance de l'Histoire, passé du tiers monde* (Paris: La Découverte, 1998).

Larson, H., *et al., New Horizons* (New York: Harper & Row, 1971).

Lebigre, J.M., "Production vivrière et approvisionnement urbain au Gabon", *Les Cahiers d'Outre-Mer* 130 (1980), pp. 167-85.

Leite, Carlos and Jens Weidmann, "Does Mother Nature Corrupt? Natural Resources, Corruption and Economic Growth", IMF Working Paper WP/99/85 (1999).

Lenin, V.I., *Imperialism, the Highest Stage of Capitalism* (New York: International Publishers, 1993 [original publication 1916]).

Lewis, Peter M., "Economic Statism, Private Capital, and the Dilemmas of Accumulation in Nigeria", *World Development* 22, no. 3 (1994), pp. 437-51.

———, "From Prebendalism to Predation: the Political Economy of Decline in Nigeria", *Journal of Modern African Studies* 34, no. 1 (1996), pp. 79-103.

Liniger-Goumaz, *Small is Not Always Beautiful: The Story of Equatorial Guinea* (London: C. Hurst & Company, 1988).

———, *Historical Dictionary of Equatorial Guinea* (Lanham, MD, and London: The Scarecrow Press, 2000).

Litvin, Daniel, *Empires of Profit* (New York: Texere, 2003).

Lonsdale, John, "States and Social Processes in Africa: a Historiographical Survey", *African Studies Review* 24, no. 2 (1981), pp. 139-225.\

Lubeck, Paul and Michael J. Watts, "An Alliance of Oil and Maize? The Response of Indigenous and State Capital to Structural Adjustment in Nigeria", in Bruce Berman and Colin Leys, eds, *African Capitalists in African Development* (Boulder: Lynne Rienner, 1994), pp. 204-34.

Luckham, Robin, *The Nigerian Military* (Cambridge University Press, 1971).

Luxemburg, Rosa, *The Accumulation of Capital* (London: Penguin, 1972).

Mabeko-Tali, Jean-Michel, "La question de Cabinda: Séparatismes éclatés, habilités luandaises et conflits en Afrique centrale", *Lusotopie* (2001), pp. 49-62.

Mabro, Robert, "OPEC Behaviour 1960-98: A Review of the Literature", *Journal of Energy Literature* IV, no. 1 (1998), pp. 3-27.

Maier, Karl, *Angola: Promises and Lies* (London: Serif, 1996).

Malkki, Liisa, *Purity and Exile: Transformations in Historical-National Consciousness among Hutu Refugees in Tanzania* (Chicago University Press, 1994).

Mamdani, M., *Citizen and Subject: Contemporary Africa and the Legacy of Late Colonialism* (Princeton University Press, 1996).

Mañe, Damian Ondo, "Emergence of the Gulf of Guinea in the Global Economy: Prospects and Challenges", IMF Working Paper WP/05/235 (Washington, DC: IMF, December 2005).

Mann, Michael, "The Autonomous Power of the State: its Origins, Mechanisms and Results", *Archives Européens de Sociologie* XXV (1984), pp. 185-213.

_____, "Has Globalization Ended the Rise and Rise of the Nation-State?", *Review of International Political Economy* 4, no. 3 (1997), pp. 472-96.

Manning, Robert A., *The Asian Security Factor: Myths and Dilemmas of Energy, Security and the Pacific Future* (New York: Palgrave, 2000).

Marquez, Gabriel Garcia, *Vivir para contarla* (Madrid: Planeta, 2002).

Martin, Phyllis, "The Cabinda Connection", *African Affairs* 76, no. 302 (1976), pp. 47-59.

Martinez, Luis, *The Algerian Civil War 1990-1998* (London: C. Hurst & Co., 2000).

Mauss, Marcel, *Sociologie et Anthropologie* (Paris: Presses Universitaires de France, 1950).

Mayall, James, "Oil and Nigerian Foreign Policy", *African Affairs* 75, no. 300 (1976), pp. 317-30.

_____, *Nationalism and International Society* (Cambridge University Press, 1990).

Mbembe, Achille, "At the Edge of the World: Boundaries, Territoriality and Sovereignty in Africa", *Public Culture* 12, no. 1 (2000), pp. 259-84.

_____, *On the Postcolony* (Berkeley: University of California Press, 2001).

McCloy, J.J., *The Great Oil Spill: The Inside Story of Gulf Oil's Bribery and Political Chicanery* (New York: Chelsea House Publishers, 1976).

McKinsey & Company, "Assessing the Global Compact's Impact", Report, 11 May 2004.

McPhail, Kathryn, "How Oil, Gas and Mining Projects Can Contribute to Development", *Finance and Development* 37, 4 (2000), http://www.imf.org/external/pubs/ft/fandd/2000/12/mcphail.htm (accessed 5 November 2001).

McPherson, Charles, "National Oil Companies: Evolution, Issues, Outlook", Conference on Fiscal Policy and Implementation in Oil Producing Countries, IMF, Washington DC, 5-6 June 2002.

Médard, Jean-François, "La corruption internationale et l'Afrique sub-saharienne: un essai d'approche comparative", *Revue Internationale de Politique Comparée* 4, no. 2 (1997), pp. 413-40.

_____, "Oil and War: Elf and "Françafrique" in the Gulf of Guinea", paper presented at the Transparency International Annual Conference, Prague, May 2001.

Messiant, Christine, "Angola, les voies de l'ethnisation et de la décomposition- I- De la guerre à la paix (1975-1991): le conflit armé, les interventions internationales e le peuple angolais", *Lusotopie* (1994), pp. 155-210.

_____, "La Fondation Eduardo dos Santos: à propos de l'investissement de la société civile par le pouvoir politique", *Politique Africaine* 73 (1999), pp. 82-101.

_____, "L'Angola? Circulez, il n'y a rien à voir!" *Lusotopie* (2000), pp. 9-26.

_____, "Why did Bicesse and Lusaka Fail? A Critical Analysis", *Accord* 15 (2004), pp. 16-23.

_____, "Une capitale dans et à l'abri de la guerre: guerre, ordre politique et violences à Luanda", unpublished paper.

Miller, J.C., *Way of Death: Merchant Capitalism and the Angolan Slave Trade 1730-1830* (Madison: University of Wisconsin Press, 1988).

Mintz, S.W., *Sweetness and Power: The Place of Sugar in Modern History* (London: Penguin Books, 1985).

Misser, François and Olivier Vallée, *Les gemmocraties* (Paris: Desclée de Brouwer, 1997).

Mitchell, John *et al.*, *The New Economy of Oil: Impacts on Business, Geopolitics and Society* (London: Earthscan and Royal Institute of International Affairs, 2001).

Mitchell, T., "The Limits of the State: Beyond Statist Approaches and their Critics", *American Political Science Review* 85, no. 1 (1991), pp. 77-96.

Moore, Barrington, *Social Origins of Dictatorship and Democracy: Lord and Peasant in the Making of the Modern World* (Boston: Beacon Press, 1993).

Moran, Theodore, *Governments and Transnational Corporations* (New York: Routledge, 1992).

Morgan, W.B., "Food Supply and Staple Food Imports of Tropical Africa", *African Affairs* 76, no. 303 (1977), pp. 167-76.

Morrison, J. S. and J. Cooke, eds., *Africa Policy in the Clinton Years: Critical Choices for the Bush Administration* (Washington DC: Africa-America Institute, 2001).

Morrison, J. Stephen and Princeton N. Lyman, "Countering the Terrorist Threat in Africa", in W. Kansteiner and J.S. Morrison, *Rising US Stakes in Africa: Seven Proposals to Strengthen US-Africa Policy* (Washington DC: CSIS, 2004), pp. 104-18.

Mosley, Paul *et al.*, *Aid and Power: the World Bank and Policy-based Lending* (London: Routledge, 1991).

Naipaul, V.S., *In a Free State* (London: Picador, 1971).

Navaro-Yashin, Yael, *Faces of the State: Secularism and Public Life in Turkey* (Princeton University Press, 2002).

Ndzana, Vianney Ombe, *Agriculture, pétrole et politique au Cameroun* (Paris: L'Harmattan, 1987).

Neff, Richard and George Williford, eds, *Oil History: a Selected and Annotated Bibliography* (Houston: International Association of Drilling Contractors, 1995).

Newitt, Malyn, *History of Mozambique* (London: Hurst, 1995).

Ngu, Joseph, "The Political Economy of Oil in Cameroon", in P. Geschiere and P. Konings, eds, *Conference on the Political Economy of Cameroon: Historical Perspectives* (Leiden, June 1988), pp. 109-46.

Nies, Susanne, "Une géographie planétaire discontinue: Les enclaves: "volcans" éteints ou en activité", *La Revue Internationale et Stratégique* 49 (2003), pp. 111-20.

Noel, Pierre, "La stratégie americaine de sécurité et le pétrole du Moyen-Orient", Working Paper 10-03-01 (Paris: Institut Français de Relations Internationales, October 2003).

Norland, Donald R., "Innovations of the Chad/Cameroon Pipeline Project: Thinking Outside the Box", *Mediterranean Quarterly* 14, no. 3 (2003), pp. 46-59.

O'Brien, Kevin A., "Private Military Companies and African Security, 1990-98", in Abdel Fatau Musah and J. 'Kayode Fayemi, eds, *Mercenaries: an African Dilemma* (London: Pluto Press, 2000), pp. 43-75.

O'Brien, P.J., "A Critique of Latin American Theories of Dependence", in I. Oxall *et al.*, eds, *Beyond the Sociology of Development* (London: Routledge & Keegan Paul, 1975), pp. 50-85.

O'Keefe, P., "Toxic Terrorism", *Review of African Political Economy* 15, no. 42 (1988), pp. 84-90.

O'Leary, B., I. S. Lustick and Thomas Callaghy, *Rightsizing the State: The Politics of Moving Borders* (Oxford University Press, 2002).

OCLD, *Dossier Noir du pétrole camerounais* (Paris, L'Harmattan/Organisation Camerounaise de Lutte pour la Démocratie, 1982).

OECD, *African Economic Outlook of 2003/2004* (Paris: OECD, 2003).

Okruhlik, Gwen, "Rentier Wealth, Unruly Law and the Rise of Opposition: the Political Economy of Oil States", *Comparative Politics* 31, no. 3 (1999), pp. 295-315.

Olukoju, Ayodeji, "Never Expect Power Always: Electricity Consumers' Response to Monopoly, Corruption and Inefficient Services in Nigeria", *African Affairs* 103, no. 410 (2004), pp. 51-71.

Olukoshi, Adebayo and Tajudeen Abdulraheem, "Nigeria Crisis Management under the Buhari Regime", *Review of African Political Economy* 34 (1985), pp. 95-101.

Omitoogun, Wuyi, *Military Expenditure Data in Africa*, SIPRI Research Paper 17, (Stockholm, International Peace Research Institute, 2003).

Onishi, Norimitsu, "Deep in the Republic of Chevron", *New York Times Magazine*, 4 July 1999.

Osaghae, Eghosa, "The Ogoni Uprising: Oil Politics, Minority Agitation and the Future of the Nigerian State", *African Affairs* 94, no. 376 (1995), pp. 325-44.

_____, *Crippled Giant* (London: Hurst, 1998).

Ottaway, Marina, "Reluctant Missionaries", *Foreign Policy*, July-August (2001), pp. 45-54.

Oxford Institute of Energy Studies, "National Oil Companies and International Oil Companies", *Oxford Energy Forum* 57 (2004), pp. 12-19.

Oyerdiran, O., ed., *Nigerian Government and Politics under Military Rule, 1966-79* (Lagos: Macmillan, 1979).

Panter-Brick, Keith, ed., *Soldiers and Oil: the Political Transformation of Nigeria* (London: Frank Cass, 1978).

Péan, Pierre, *affaires africaines* (Paris: Fayard, 1983).

Pearson, Scott R., *Petroleum and the Nigerian Economy* (Stanford University Press, 1970).

Pécaut, Daniel, "Les configurations de l'espace, du temps et de la subjectivité dans le contexte de terreur: l'exemple colombien", *Cultures et Conflits*, 24-5 (2002), http://www.conflits.org (accessed 5 June 2003).

Peel, Michael, "Crisis in the Niger Delta: How Failures of Transparency and Accountability are Destroying the Region", Chatham House, Africa Programme Briefing Paper, AFP BP 05/02, July 2005.

Pegg, Scott, "The Cost of doing Business: Transnational Corporations and Violence in Nigeria", *Security Dialogue* 30, no.4 (1999), pp. 473-84.

Penrose, Edith T., *The Large International Firm in Developing Countries: The International Petroleum Industry* (London: Allen & Unwin, 1968).

_____, "Africa and the Oil Revolution", *African Affairs* 75, no. 300 (1976), 277-83.

Perez Alfonzo, J.P., *Hundiendos en el Excremento del Diablo* (Caracas: Colleción Venezuela Contemporánea, 1976).

PFC Energy, "Appendix A: West Africa Petroleum Sector: Oil Value Forecast and Distribution", in Walter H. Kansteiner III and J. Stephen Morrison, eds, *Rising US Stakes in Africa: Seven Proposals to Strengthen US-Africa Policy* (Washington DC: CSIS, 2004), pp. 151-61.

Polanyi, Karl, *The Great Transformation* (New York: Beacon Press, 1957).

Politique Africaine, special issue on "Liberia, Sierra Leone, Guinée: la régionalisation de la guerre" (2002).

Pourtier, Roland, *Le Gabon* (Paris: L'Harmattan, 1989, two volumes).

Proceedings of "Resolving International Border Disputes for Commercial Success in the Oil and Gas Industry" (Two-Day International Conference, London, 10-11 July 2000).

Prunier, G., *The Rwandan Crisis* (London: Hurst, 1995).

Prunier, Gérard and Rachel Gisselquist, "The Sudan: a Successfully Failed State", in R. I. Rotberg, ed., *State Failure and State Weakness in a Time of Terror* (Washington DC: Brookings Institution, 2003), pp. 101-27.

Reed, Michael C., "Gabon: a Neo-Colonial Enclave of Enduring French Interests", *Journal of Modern African Studies* 25, no. 2 (1987), pp. 283-320.

Reno, William, *Corruption and Politics in Sierra Leone* (Cambridge University Press, 1995).

_____, *Warlord Politics and African States* (Boulder: Lynne Rienner, 1998).

_____, "How Sovereignty Matters: International Markets and the Political Economy of Local Politics in Weak States" in T. Callaghy *et al.*, *Intervention and Transnationalism in Africa: Global-Local Networks of Power* (Cambridge University Press, 2001), pp. 197-215.

República Democrática de São Tomé e Príncipe, "Investigation and Review, Second Bid Round, Joint Development Zone, Nigeria and São Tomé and Príncipe", Office of the Attorney General, São Tomé and Príncipe, 2 December 2005.

Ricardo, David, *The Principles of Political Economy and Taxation* (London: Dent, 1973).

Rich, Bruce, Korinna Horta and Aaron Goldzimer, *Export Credit Agencies in Sub-Saharan Africa: Indebtedness for Extractive Industries, Corruption and Conflict* (Washington DC: Environmental Defense, 2001).

Richards, Paul, *Fighting for the Rain Forest* (Oxford: James Currey, 1995).

Risse-Kappen, Thomas, ed., *Bringing Transnational Relations Back In* (Cambridge University Press, 1995)

Roberts, Paul, *The End of Oil: The Decline of the Petroleum Economy and the Rise of a New Energy Order* (London: Bloomsbury, 2004).

Rodrik, Dani, *The New Global Economy and Developing Countries: Making Openness Work* (Washington, DC: Overseas Development Council, 1999).

Roitman, Janet "The Garrison-Entrepot", *Cahiers d'Etudes Africaines* XXVIII, no. 4 (1998), pp. 297-329.

Roitman, Janet, and Gerard Roso, "Guinée-Equatoriale: être offshore pour rester nationale", *Politique Africaine* 81 (2001), pp. 121-42.

Roque, Fatima *et al.*, *Economia de Angola* (Lisbon: Bertrand Editora, 1991).

Rose-Ackerman, Susan, *Corruption and Government* (Cambridge University Press, 1999).

Ross, Michael, *Extractive Sectors and the Poor* (Washington DC: Oxfam America, 2001).

"Does Oil Hinder Democracy?", *World Politics* 53, no. 3 (2001), pp. 325-61.

Rotberg, Robert I., ed., *When States Fail: Causes and Consequences* (Princeton University Press, 2004).

Ruggie, J.C., "Territoriality and Beyond: Problematizing Modernity in International Relations", *International Organization* 47, no. 1 (1993), pp. 139-74.

_____, "The Theory and Practice of Learning Networks", *Journal of Corporate Citizenship* 5 (2002), pp. 27-37.

Sachs, Jeffrey D. and Andrew M. Warner, "Natural Resource Abundance and Economic Growth", National Bureau of Economic Research Working Paper (Cambridge, MA, 1995).

Sampson, Anthony, *The Seven Sisters: The Great Oil Companies and the World They Made* (New York: Viking Press, 1975).

Santiso, Javier, *Amérique Latine: Révolutionnaire, libérale, pragmatique* (Paris: Autrement, 2005).

Sassen, Saskia, *Globalization and its Discontents: Essays on the New Mobility of People and Money* (New York: The New Press, 1998).

Schmitz, C., ed., *Big Business in Mining and Petroleum* (Aldershot: Gower, 1995).

Schumpeter, J., "The Crisis of the Tax State", in J. Schumpeter, *The Economics and Sociology of Imperialism* (Princeton University Press, 1991), pp. 99-140.

Schmidt, Steffen W. *et al.*, *Friends, Followers and Factions: a Reader in Political Clientelism* (Berkeley: University of California Press, 1977).

Schraeder, Peter J., *United States Foreign Policy towards Africa: Incrementalism, Crisis and Change* (Cambridge University Press, 1995).

———, "Cold War to Cold Peace: Explaining US-French Competition in Francophone Africa", *Political Science Quarterly* 115, no. 3 (2000), pp. 395-419.

Scott, James C., *Weapons of the Weak: Everyday Forms of Peasant Resistance* (New Haven: Yale University Press, 1985).

———, *Seeing Like a State* (Yale University Press, 1998).

Seibert, Gerhard, "São Tomé e Príncipe: Military Coup as Lesson?", *Lusotopie* (1996), pp.71-80.

———, *Comrades, Clients and Cousins: Colonialism, Socialism and Democratization in São Tomé and Príncipe* (Leiden: CNWS Publications, 1999).

———, "Coup d'état in São Tomé and Príncipe: Domestic Causes, the Role of Oil and Former "Buffalo" Battalion Soldiers" (Johannesburg: Africa Security Analysis Programme Occasional Paper, Institute for Security Studies, 2003).

———, "São Tomé and Príncipe Update" (Johannesburg: Africa Security Analysis Programme Occasional Paper, Institute for Security Studies, 26 April 2004).

———, "São Tomé and Príncipe: the Difficult Transition from International Aid Recipient to Oil-Producer", in Matthias Basedau and Andreas Mehler, eds, *Resource Politics in Sub-Saharan Africa* (Hamburg: Institute of International Affairs, 2005), pp. 223-50.

Shaxson, Nicholas, "The Elf trial: Political Corruption and the Oil Industry", in *Global Corruption Report 2004* (Berlin: Transparency International, 2004), pp. 67-71.

———, "New Approaches to Volatility: Dealing with the Resource Curse in Sub-Saharan Africa", *International Affairs* 81, no. 2 (2005), pp. 311-24.

———, *Poisoned Wells: The Dirty Politics of African Oil* (London: Palgrave Macmillan, 2007).

Shell-BP, *The Shell-BP Story* (Port Harcourt: Shell-BP, 1965).

Shore, Cris and Stephen Nugent, eds, *Elite Cultures: Anthropological Perspectives* (London: Routledge, 2002).

Silverstein, Ken, "US Oil Politics in the 'Kuwait of Africa'", *The Nation*, 22 April 2002, pp. 11-20.

———, "Chad Sees First Trickle of Cash from Pipeline", *The Nation*, 26 December 2003.

Sindzingre, Alice, "Etat et intégration internationale des Etats d'Afrique subsaharienne: l'exemple de la fiscalité", *Afrique Contemporaine*, 3rd quarter (2001), pp. 63-77.

Singer, H.W., "The Distribution of Trade between Investing and Borrowing Countries", *American Economic Review* 40, no. 2 (1950), pp. 473-85.

Singer, Peter W., *Corporate Warriors: the Rise of the Privatized Military Industry* (Ithaca: Cornell University Press, 2003).

Skeet, I., *OPEC: Twenty-five Years of Prices and Politics* (Cambridge University Press, 1988).

Skidelsky, Robert, "The Mystery of Growth", *New York Review of Books* L 4, 13 March 2003, pp. 28-31.

Skinner, Quentin, "The State", in Terence Ball *et al.*, eds, *Political Innovation and Conceptual Change* (Cambridge University Press, 1988), pp. 90-131.

Sklar, Richard, "The Nature of Class Domination in Africa", *Journal of Modern African Studies* 17, no. 4 (1979), pp. 531-52.

Skocpol, Theda, *States and Social Revolutions* (Cambridge University Press, 1979).

———, "Bringing the State Back In?", in Peter Evans *et al.*, eds, *Bringing the State Back In* (Cambridge University Press, 1985), pp. 3-37.

Skocpol, T., ed., *Vision and Method in Historical Sociology* (Cambridge University Press, 1984).

Soares de Oliveira, Ricardo M.S., "The Creole Elite and Political Power in Late Colonial and Early Post Independence Angola" (unpublished BA dissertation, Department of Politics, University of York, 1998).

———, "Portuguese Relations with Angola and Mozambique during the Peace Processes, 1990-94" (unpublished MPhil dissertation, Centre of International Studies, Cambridge University, 1999).

———, "Aussenpolitik Angolas", in Jurgen Bellers and Thorsten Benner, eds, *Handbuch der Aussenpolitik von Afghanistan bis Zypern* (Munich: R. Oldenbourg Verlag, 2001), pp. 699-703.

_____, "The Astonishing Survival of the Failed State", paper delivered at the African Studies Association Annual Conference, Houston, 18 November 2001.

_____, "Nigeria Field Report, September 2002", Catholic Relief Services (2002a).

_____, "Chad Field Report, October 2002", Catholic Relief Services (2002b).

_____, "National Oil Companies and the Privatized State in the Gulf of Guinea", unpublished paper (2003a).

_____, "Relações do Fundo Monetário Internacional e do Banco Mundial com os Países Ricos em Petróleo do Golfo da Guiné", paper presented at IMVF Annual Conference, Lisbon, 22 November (2003b).

_____, "Context, Path Dependency and Oil-Based Development in the Gulf of Guinea", in Michael Dauderstadt and Arne Schildberg, eds, *Dead Ends of Transition: Rentier Economies and Protectorates* (Frankfurt/Main: Campus, 2006).

Sorensen, Georg, "Sovereignty: Change and Continuity in a Fundamental Institution", *Political Studies* 47, no. 3 (1999), pp. 590-604.

Soyinka, Wole, *Open Sore of a Continent* (Oxford University Press, 1996).

Spruyt, Hendrik, *The Sovereign State and its Competitors: an Analysis of Systems Change* (Princeton University Press, 1994).

Stevens, Paul, "Resource Impact: Curse or Blessing? A Literature Survey", *Journal of Energy Literature* IX, no. 1 (2003), pp. 3-42.

Strange, Susan, "States, Firms and Diplomacy", *International Affairs* 68 (1992), pp. 1-15.

_____, *States and Markets* (London: Pinter, 1994).

_____, *The Retreat of the State* (Cambridge University Press, 1996).

Suberu, Rotimi T., "The Struggle for New States in Nigeria, 1976-1990", *African Affairs* 90, no. 361 (1991), pp. 499-522.

Svevo, Italo, *As a Man Grows Older* (New York: New York Review of Books, 2001).

Sweet, G. and E. H. Knowlton, *The Resurgent Years* (New York: Harper, 1956).

Taussig, Michael, *Shamanism, Colonialism and the Wild Man: a Study in Terror and Healing* (Chicago University Press: 1987).

Thompson, V. and T. Adloff, *Historical Dictionary of the Congo* (Metuchen, NJ: Scarecrow Press, 1994).

Thomson, Janice E., *Mercenaries, Pirates and Sovereigns* (Princeton University Press, 1994).

Tilly, Charles, *Big Structures, Large Processes, Huge Comparisons* (New York: Russell Sage, 1984).

_____, "War Making and State Making as Organized Crime", in P. Evans *et al.*, *Bringing the State Back In* (Cambridge University Press, 1985), pp. 169-91.

_____, *Coercion, Capital and European States, 990-1992* (Oxford: Blackwell, 1992).

Traub-Merz, R. and Douglas Yates, eds, *Oil Policy in the Gulf of Guinea: Security & Conflict, Economic Growth, Social Development* (Bonn: Friedrich Ebert Stiftung, 2004), pp. 51-9.

Turner, Bryan S., "Max Weber's Historical Sociology: a Bibliographical Essay", *Journal of Historical Sociology* 3, no. 2 (1990), pp. 192-208.

Turner, Terisa, "The Working of the Nigerian National Oil Corporation", in Paul Collins, ed., *Administration for Development in Nigeria* (Lagos: African Education Press, 1980), pp. 98-134.

Turner, Thomas, "Angola's Role in the Congo War", in John C. Clark, ed., *The African Stakes in the Congo War* (Basingstoke: Mcmillan, 2002), pp. 75-92.

UNCTAD, *World Investment Report 2005: Transnational Corporations and the Internationalization of R & D* (New York and Geneva: United Nations, 2005).

United States Department of Justice, "Nigeria: Drug Intelligence Brief", Drug Enforcement Administration, Intelligence Division (Washington DC: US Department of Justice, August 2001).

United States House of Representatives, hearings on "The Chad Cameroon Pipeline Project: a New Model for Natural Resource Development" (Washington DC: US House of Representatives International Relations subcommittee on Africa, 18 April 2002).

United States Senate, "Money Laundering and Foreign Corruption: Enforcement and Effectiveness of the Patriot Act. Case Study Involving Riggs Bank", Report Prepared by the Minority Staff of the Permanent Subcommittee on Investigations (Washington DC: US Senate, 15 July 2004).

Uriz, Genovena Hernandez, "To Lend or Not To Lend: Oil, Human Rights, and the World Bank's Internal Contradictions", *Harvard Human Rights Journal* 14 (2001), pp. 197-392.

Vail, Leroy and Landeg White, *Capitalism and Colonialism in Mozambique* (London: Heinemann, 1980).

Valle, Vicente, "US Policy towards the Gulf of Guinea", in R. Traub-Merz and Douglas Yates, eds., *Oil Policy in the Gulf of Guinea: Security & Conflict, Economic Growth, Social Development* (Bonn: Friedrich Ebert Stiftung, 2004), pp. 51-9.

Vallée, Olivier, "Les cycles de la dette", *Politique Africaine* 31 (1988), pp. 15-21.

———, *Pouvoirs et politiques en Afrique* (Paris: Desclée de Brouwer, 1999a).

———, "La dette publique est-elle prive?", *Politique Africaine* 73 (1999b), pp. 50-67.

Van de Walle, Nicholas, "Rice Politics in Cameroon: State Commitment, Capability, and Urban Bias", *Journal of Modern African Studies* 27, no. 4 (1989), pp. 579-99.

———, "Decline of the Franc Zone: Monetary Politics in Francophone Africa", *African Affairs* 90, no. 360 (1991), pp. 383-405.

———, *African Economies and the Politics of Permanent Crisis 1979-1999* (Cambridge University Press, 2001).

Vansina, Jan, *Paths in the Rainforest* (London: James Currey, 1990).

Vellut, Jean-Luc, "Mining in the Belgian Congo", in David Birmingham and Phyllis M. Martin, eds, *History of Central Africa* (London: Longman, 1983), pp. 126-62.

Vennetier, P., "L'urbanisation et ses conséquences au Congo-Brazzaville", *Cahiers d'Outre-Mer* 63 (1963), pp. 263-380.

Vernon, Raymond, ed., *The Oil Crisis* (New York: W.W. Norton, 1976).

Verschave, François-Xavier, *La Françafrique* (Paris: Stock, 1999).

———, *Noir Silence* (Paris: Les Arènes, 2000).

———, *L'envers de la dette: Criminalité politique et économique au Congo-Brazza et en Angola* (Marseilles: Agone, 2001).

Veyne, Paul, *Le Pain et le cirque: sociologie historique d'un pluralisme politique* (Paris : Editions du Seuil, 1976).

———, "Foucault révolutionne l'histoire", in *Comment on écrit l'histoire* (Paris: Editions du Seuil, 1978), pp. 383-429.

Vieille, Paul, "Le pétrole comme rapport sociale", *Peuples Méditerranéens* 26 (1984), pp. 3-29.

Villalon, L.A. and P.A. Huxtable, *The African State at a Critical Juncture* (Boulder: Lynne Rienner, 1998).

Vines, Alex, *Renamo: From Terrorism to Democracy in Mozambique?* (Oxford: James Currey, 1996).

_____, "Mercenaries, Human Rights and Legality", in Abdel Fatau Musah and J. 'Kayode Fayemi, eds, *Mercenaries: an African Dilemma* (London: Pluto Press, 2000), pp. 169-97.

Volman, Daniel, "The Bush Administration and African Oil: The Security Implications of US Energy Policy", *Review of African Political Economy* 30, no. 98 (2003), pp. 573-84.

WAC Global Services, *Peace and Security in the Niger Delta: Conflict Expert Group Baseline Report*, December (London: WAC Global Services, 2003).

Wade, Robert Hunter, "US Hegemony and the World Bank: the Fight over People and Ideas", *Review of International Political Economy* 9, no. 2 (2002), pp. 215-43.

Wall, Bennett, *Growth in a Changing Environment: A History of Standard Oil (New Jersey) 1950-1972 and Exxon Corporation 1972-1975* (New York: McGraw-Hill, 1988).

Watts, Michael, "Oil as Money: the Devil's Excrement and the Spectacle of Black Gold", in S. Corbridge, R. Martin and N. Thrift, eds, *Money, Power and Space* (Oxford: Blackwell, 1994), pp. 406-45.

_____, ed., *State, Oil and Agriculture in Nigeria* (Berkeley: University of California Press, 1987).

Watts, Michael, and Paul Lubeck, "The Popular Classes and the Oil Boom: a political economy of rural and urban poverty", in I. W. Zartman, ed., *The Political Economy of Nigeria* (New York: Praeger, 1983), pp. 107-35.

Wauthier, Claude, *Quatre présidents et l'Afrique - De Gaulle, Pompidou, Giscard d'Estaing, Mitterrand* (Paris: Seuil, 1995).

Weber, Max, "Politics as Vocation", in H.H. Gerth and C. Wright Mills, eds, *From Max Weber. Essays in Sociology* (London: Routledge, 1948), pp. 77-128.

_____, *Economy and Society* (Berkeley: University of California Press, 1978).

_____, *Histoire économique - esquisse d'une histoire universelle de l'économie et de la societé* (Paris: Gallimard, 1991).

White House, *Security of Energy Supply: Report of the National Energy Policy Development Group* (Washington DC: Government Printing Office, 2001), http://www.whitehouse.gov/energy/ (accessed 6 July 2002).

Wilkins, Mira, "Modern European Economic History and the Multinationals", *The Journal of European Economic History* 6 (1977), pp. 577-95.

Williams, Gavin, *State and Society in Nigeria* (Idanre: Afrografika, 1980).

Williams, M.L, "The Extent and Significance of the Nationalization of Foreign-owned Assets in Developing Countries, 1956-1972", *Oxford Economic Papers* 27, no. 2 (1975), pp. 260-73.

World Bank, *Accelerated Development in Sub-Saharan Africa: an Agenda for Action* (Washington DC: World Bank, 1981).

_____, "Gabon: Issues and Options in the Energy Sector", Report No.6915-GA (Washington DC: ESMAP/World Bank Energy Sector Assessment Program, July 1988).

_____, "Nigeria: Issues and Options in the Energy Sector", Report No. 11672-UNI (Washington DC: ESMAP, World Bank Western Africa Department and Industry and Energy Division, 1993).

_____, "Implementation Completion Report, Republic of Equatorial Guinea – Second Technical Assistance Project for the Petroleum Sector", 5 June 1998.

_____, "Report on the Angola-Luanda Water Supply and Sanitation Project" (Washington DC: World Bank, 1999).

_____, "Project Appraisal Document on Proposed International Bank for Reconstruction and Development Loans in Amounts US$ 39.5 Million to the Republic of Chad and US$53.4 Million to the Republic of Cameroon and on Proposed International Financial Corporation Loans in Amounts of US$100 Million in A-Loans and up to US$300 Million in B-Loans to the Tchad Oil Transportation Company, S.A., and the Cameroon Oil Transportation Company, S.A., for a Petroleum and Pipeline Project", 13 April 2000.

_____, *World Development Indicators 2000* (Washington DC: World Bank, 2000).

_____, "Project Performance Assessment Report: Equatorial Guinea – Second Petroleum Technical Assistance Project, Report No: 24430" (Washington DC: World Bank, 1 July 2002).

_____, "Chad: Petroleum Development and Pipeline Report: Investigation Report", The Inspection Panel, Report No. 23999 (Washington DC: World Bank and International Development Association, 17 July 2002).

_____, "Management Report and Recommendation in Response to the Inspection Panel Investigation Report No. 23999" (Washington DC: World Bank, 21 August 2002).

————, "Angola: Transition Strategy, Technical Assistance, and Postconflict Support", Press Release 2003/274/AFR (Washington DC: World Bank, 27 March 2003).

————, "Republic of Congo Reaches Decision Point Under the Enhanced HIPC Debt Relief Initiative", News Release No: 2006/301/AFR, World Bank, Washington DC, 9 March 2006.

Wright, George, *US Policy towards Angola* (London: Zed Books, 1998).

Xu, Xiaojie, *Petro-Dragon's Rise: What it Means for China and the World* (Ficecchio: European Press Academic Publishing, 2002).

Yates, Douglas, *The Rentier State in Africa: Oil Rent Dependency and Neocolonialism in the Republic of Gabon* (Trenton, NJ: Africa World Press, 1996).

Yengo, Patrice, "'Chacun aura sa part': les fondements historiques de la (re)production de la "guerre" à Brazzaville", *Cahiers d'Études Africaines* 150-152 (1998), pp. 471-503.

Yergin, Daniel, *The Prize: the Epic Quest for Oil, Money and Power* (London: Pocket Books, 1991).

Yergin, Daniel and Michael Stoppard, "The Next Prize", *Foreign Affairs* 82, no. 6 (2003), pp. 103-14.

Yolla, Eustache Mandjouhou, *La politique étrangère du Gabon* (Paris: L'Harmattan, 2003).

Young, Crawford, *The Colonial State in Comparative Perspective* (New Haven: Yale University Press, 1994).

————, "The End of the Postcolonial State in Africa? Reflections on Changing African Political Dynamics", *African Affairs* 103, no. 410 (2004), pp. 23-49.

Zartman, I.W., ed., *Collapsed States: the Disintegration and Restoration of Legitimate Authority* (Boulder: Lynne Rienner, 1995).

————, "L'administration Clinton et l'Afrique: une appréciation d'ensemble", *Afrique Contemporaine* 197 (2001), pp. 3-11.

VIDEO

Meurice, Jean-Michel and Fabrizio Calvi, dirs., *Une Afrique sous influence* (Paris: ARTE France and MK2 TV, 2000), 2h16m.

NEWS AND DATA SOURCES

Africa Analysis (Gauteng, South Africa)
Africa Confidential (London)
Africa Contemporary Record
Africa Report (Washington DC)
African Business (London)
African Energy (London)
Africa Energy Intelligence (Paris)
Agence France Press (Paris)
Angola News (Luanda)
Angolense (Luanda)
Argus (London)
Atlantic Monthly (Washington DC)
Bloomberg News (New York)
Boston Globe
Business in Africa (Rivonia, South Africa)
Business Week (New York)
Canard Enchaîné (Paris)
Christian Science Monitor (Boston)
Daily Champion (Lagos)
Economist (London)
Economist Intelligence Unit *Country Profiles* and *Country Reports* (London)
El País (Madrid)
Energy Information Administration Country Briefs (Energy Information Administration,
Department of Energy, US Government, Washington DC)
EurAfrica (Geneva)
Expresso (Lisbon)
Figaro (Paris)
Financial Times (London)
Forbes (New York)
Fortune (New York)
Guardian (Lagos)
Guardian (London)
Houston Chronicle
Independent (London)
Insider Weekly (Lagos)
International Herald Tribune (Paris)

IRIN (UN Office for the Coordination of Humanitarian Affairs)
Jeune Afrique / L'Intelligent (Paris)
Jornal de Angola (Luanda)
Les Echos (Paris)
Lettre du Continent (Paris)
Libération (Paris)
Los Angeles Times
Mail & Guardian (Johannesburg)
Marabout (Ouagadougou)
Marchés Tropicaux et Méditerrannéens (Paris)
Miami Herald
Nation (Washington DC)
New Republic (Washington DC)
New York Times
N'djaména Hebdo
Ottawa Citizen
Oxford Analytica
Petroleum Economist (London)
Público (Lisbon)
Reuters News (London)
Semanário Angolense (Luanda)
Tell (Lagos)
This Day (Lagos)
Time (New York)
Times (London)
Upstream (Oslo)
Vanguardia (Barcelona)
Washington Post
West Africa (London)

INDEX